Risk Management for Food Allergy

Food Science and Technology International Series

A complete list of books in this series appears at the end of this volume.

Risk Management for Food Allergy

Edited by

Charlotte Bernhard Madsen
DVM Research Leader Division of Toxicology
and Risk Assessment National Food Institute
Technical University of Denmark

René W. R. Crevel
Safety and Environmental Assurance Center,
Unilever, Sharnbrook, Bedfordshire, UK

Clare Mills
Manchester Institute of Biotechnology,
University of Manchester, UK

Steven L. Taylor
Food Allergy Research & Resource Program,
University of Nebraska, Lincoln, NE, US

AMSTERDAM • BOSTON • HEIDELBERG • LONDON
NEW YORK • OXFORD • PARIS • SAN DIEGO
SAN FRANCISCO • SINGAPORE • SYDNEY • TOKYO
Academic Press is an imprint of Elsevier

Academic Press is an imprint of Elsevier
The Boulevard, Langford Lane, Kidlington, Oxford, OX5 1GB, UK
225 Wyman Street, Waltham, MA 02451, USA

First published 2014

British Library Cataloguing in Publication Data
A catalogue record for this book is available from the British Library

Library of Congress Cataloguing in Publication Data
A catalogue record for this book is available from the Library of Congress

ISBN: 978-0-12-381988-8

For information on all Academic Press publications
visit our website at **store.elsevier.com**

Printed and bound in the United States

14 15 16 17   10 9 8 7 6 5 4 3 2 1

Working together
to grow libraries in
developing countries

www.elsevier.com • www.bookaid.org

Contents

SECTION 2 ALLERGEN THRESHOLDS AND RISK ASSESSMENT

SECTION 3 RISK MANAGEMENT OF GLUTEN

SECTION 4 PRACTICAL FOOD ALLERGEN RISK
MANAGEMENT

List of Contributors

Anton Alldrick Campden BRI, Chipping Campden, UK

Ricardo Asero Ambulatorio di Allergologia, Clinica San Carlo, Paderno-Dugnano, Milano, Italy

Barbara K. Ballmer-Weber Allergy Unit, Department of Dermatology, University Hospital Zürich, Switzerland

Susan A. Baranowsky Consumer Affairs, Campbell Soup Company, Camden, NJ, US

Joseph L. Baumert Department of Food Science & Technology and Food Allergy Research & Resource Program, University of Nebraska, Lincoln, NE, US

Erna Botjes Stichting Voedselallergie (Dutch Food Allergy), Nijkerk, The Netherlands

Peter Burney National Heart Lung Institute, Imperial College, London, UK

Stella Cochrane Unilever Safety and Environmental Assurance Center, Sharnbrook, Bedfordshire, UK

René W.R. Crevel Safety and Environmental Assurance Center, Unilever, Sharnbrook, Bedfordshire, UK

Ross Critenden Valio Ltd, Helsinki, Finland

Anthony E.J. Dubois Department of Paediatric Pulmonology and Paediatric Allergy, University Medical Center Groningen, University of Groningen, Groningen, The Netherlands

George E. Dunaif Food Safety & Technical Services, Grocery Manufacturers Association, Washington, DC, US

Audrey DunnGalvin Department of Paediatrics and Child Health, University College Cork, Cork, Ireland

Montserrat Fernández-Rivas Allergy Department, Hospital Clínico San Carlos, IdISSC, Madrid, Spain

B.M.J. Flokstra-de Blok Department of General Practice, University Medical Center Groningen, University of Groningen, Groningen, The Netherlands

Steven M. Gendel Food and Drug Administration, Center for Food Safety and Applied Nutrition, US

Linus Grabenhenrich Institute for Social Medicine, Epidemiology and Health Economics, Charité University Medical Center, Berlin, Germany

Kirsten Grinter Nestle Oceania, Rhodes, NSW, Australia

Sue Hattersley Food Standards Agency, London, UK

Geert Houben Food & Nutrition, TNO, Zeist, The Netherlands

Steffen Husby Hans Christian Andersen Children's Hospital at Odense University Hospital, Denmark

Anneli Ivarsson Department of Public Health and Clinical Medicine, Epidemiology and Global Health, Umeå University, Sweden

Phil E. Johnson Institute of Inflammation and Repair, Manchester Academic Health Science Center, Manchester Institute of Biotechnology, University of Manchester, Manchester, UK

Thomas Keil Institute for Social Medicine, Epidemiology and Health Economics, Charité University Medical Center, Berlin, Germany

Rita King British Beer and Pub Association, London, UK

André C. Knulst Department of Dermatology/Allergology, University Medical Center, Utrecht, The Netherlands

I.S. Leitch Environmental Health Department, Omagh District Council, Omagh, Co. Tyrone, Northern Ireland

Charlotte Bernhard Madsen DVM Research Leader Division of Toxicology and Risk Assessment National Food Institute Technical University of Denmark

J. McIntosh Safefood, Eastgate, Little Island, Co. Cork, Ireland

Clare Mills Institute of Inflammation and Repair, Manchester Academic Health Science Centre, Manchester Institute of Biotechnology, University of Manchester, UK

Pia Nørhede Division of Toxicology and Risk Assessment, National Food Institute, Technical University of Denmark, Lyngby, Denmark

Jonathan O'B. Hourihane Department of Paediatrics and Child Health, Clinical Investigations Unit, Cork University Hospital, Wilton, Cork, Ireland

Cecilia Olsson Department of Food and Nutrition, Umeå University, Sweden

Sylvia Pfaff Food Information Service (FIS) Europe, Bad Bentheim, Germany

David Reading Food Allergy Support Ltd, Aldershot, UK

Ben C. Remington Department of Food Science & Technology and Food Allergy Research & Resource Program, University of Nebraska, Lincoln, NE, US

Robin Sherlock FACTA, Tennyson, Queensland, Australia

Dan Skrypec Kraft Foods, Glenview, IL, US

Steven L. Taylor Food Allergy Research & Resource Program, University of Nebraska, Lincoln, NE, US

Marjan van Ravenhorst Allergenen Consultancy, Scherpenzeel, The Netherlands

Jean-Michel Wal INRA, Unité d'Immuno-Allergie Alimentaire, Jouy-en-Josas, France

Rachel Ward r.ward Consultancy Limited, Nottingham, UK

Gary Wong Department of Paediatrics, Chinese University of Hong Kong, New Territories, Hong Kong, China

Laurian Zuidmeer-Jongejan Department of Experimental Immunology, Laboratory of Allergy Research, Academic Medical Center, University of Amsterdam, Amsterdam, The Netherlands

Foreword

I first became aware of food allergen issues in the late 1990s when I was Director of the Center for Food Safety and Applied Nutrition (CFSAN) at the US Food and Drug Administration (FDA). The agency had contracted with two states to survey local food manufacturing plants that made food products (such as cookies), some of which were intended to contain, and some of which were intended not to contain, common food allergens such as milk and eggs. The states collected samples of the products *not* intended to contain food allergens and tested them to verify this. The results were astounding, as an alarming percentage of products actually contained milk or eggs when they were not supposed to.

This survey caught everyone's attention. It helped explain two things: (1) why undeclared food allergens were the number one cause of Class 1 recalls (those recalls presenting the most serious risk to health) and (2) why the government and the food industry needed to do much more to reduce the likelihood of this from happening. After all, a clear and dependable food label is the only means that food allergic individuals (and parents of food allergic children) have to prevent illness and injury. Accurate food labels are truly their lifeline.

In the ensuing 10+ years, much has been done to advance the awareness of food allergen issues and to put in place systems to better protect food allergic individuals.

- US Congress passed the Food Allergen Labeling and Consumer Protection Act (FALCPA) of 2004, which required the prominent listing − in plain English − of the eight most common food allergens: peanuts, tree nuts, milk, eggs, fish, shellfish, soy, and wheat. Today, food allergic consumers (and their parents) have a much easier time determining which foods are safe for them to eat.
- The US National Institutes for Health (NIH) have significantly increased the amount of funding devoted to studying potential cures for food allergies, thereby recognizing that food allergies are a significant health issue that needs to be addressed.
- There is much greater public awareness that a food allergic individual can be placed in a life-threatening situation if the wrong food allergen is consumed. This awareness has led to greater vigilance in public schools and the passage of state laws directing that ambulances carry the drug epinephrine − which is needed almost immediately for emergency treatment.
- Most recently, in late 2010, US Congress passed sweeping food safety legislation − called the US FDA Food Safety Modernization Act (FSMA) − which included a clear mandate for greater control of food allergens during the food manufacturing process.

Such progress does not happen by accident. The hard work of many, many people and organizations has been brought to bear. Three stand out for special mention.

- First was the creation of the Food Allergy and Anaphylaxis Network (FAAN) over 20 years ago by two parents of a food allergic child, Anne Munoz-Furlong and Terry Furlong. Having nowhere to turn for reliable information, they decided to research the issue themselves and become a clearing-house for objective, scientific information to share with other parents like themselves. The organization grew into a membership of over 30,000, and FAAN became the world leader in food allergy education. I had the honor to serve on the FAAN Board of Directors for six years (serving as Chair for one year) and can attest to the values and dedication this group has brought to bear. More recently, FAAN merged with the Food Allergy Institute (FAI) group, with great success in raising money for clinical research to form the consolidated group called Food Allergy Research and Education (FARE).

- Second was the establishment of a clinical research program at Mt. Sinai hospital in New York. Headed by Hugh Sampson, M.D., this facility has become the national leader in food allergy clinical research. It has close ties to FARE and both of its predecessor organizations. It is hoped that more research programs will arise around the country to add to our base of expertise, and the increased number of US National Institute of Health grants referenced above should facilitate reaching this goal.

- Third is a group at the University of Nebraska, headed by Stephen L. Taylor, Ph.D., called the Food Allergy Research and Resource Program (FARRP). This group is the national leader in understanding how to detect and measure food allergens in food and in assessing the risk to health, if any, of tiny amounts of food allergens. It is hoped that research of this type will help establish 'thresholds' for food allergens — meaning, safe levels that food companies can test against.

Despite such progress, many challenges remain. Awareness of the food allergy issues needs to be maintained and even enhanced, education of new parents with food allergic children is a continuing necessity, advocacy for stronger laws continues, and research needs to be continued until a cure is found.

What I have learned most over the past decade is that with food allergies, there are no villains — only victims — but also many champions trying to protect them. Thankfully, the number of champions is constantly growing. Those contributing to this book are high among them and deserve our collective admiration and gratitude.

Joseph A. Levitt[1]

---

[1] Mr. Levitt is a partner in the Washington, D.C., office of Hogan Lovells US LLP, where he counsels food companies on, among other matters, food allergen-related issues. He is the former Director of FDA's Center for Food Safety and Applied Nutrition (CFSAN). He is also a former Board Member and Chair of the Food Allergy and Anaphylaxis Network (FAAN).

Introduction

You can't have no risk at all you know, even if the child never leaves the house, so you have to deal with risk ... we just want a better way

(Mother of Carla, aged 10, US).

like sometimes you can't find the cause [of a reaction] ... it just happens, you know ... not knowing makes you worried and unsure of yourself ... when I have a first bite like, if I'm not at home, I think is this it?.... will I die? .... what can you do?

(Fran, aged 12, Ireland)

These statements from chapter 1 describe the reality of living with severe food allergy.

Managing food allergy on an individual level is the responsibility of the individual and those looking after that individual. In the modern world this is very, very difficult without help from society.

Food is an essential part of our lives. We eat approximately five times a day, very often away from home. Food allergic individuals have to rely on the information about the food they eat from the persons producing their food. They have to trust that food producers, both in industry and catering, know how crucial it is that the information they provide is correct. This means that, although the individual has to manage his or her food allergy, the food producers need to manage allergenic foods.

Food producers need to manage many different risks. This is costly, and an important driving force in prioritizing has been legal and regulatory requirements.

HISTORY OF 'ALLERGY' LABELING

In the eighties, international food labeling was extremely focused on food additives. This resulted in labeling rules where ingredients such as milk or wheat did not have to be declared on the label if they were constituents of compound ingredients (the so-called 25% rule), whereas food additives always had to be labeled. This rule could result in ingredient lists dominated by additives and made it almost impossible for food allergic individuals to get appropriate information from ingredient lists.

In order to change international food labeling rules to make them more helpful to food allergic individuals, a Nordic initiative led by Norway in 1993 presented a document (Consideration of Potential Allergens in Food) to the **xix**

Codex Committee on Food Labeling. The documents suggested changing the 25% rule on compound ingredients to a 5% rule and suggested a list of allergenic foods that should always be declared. The matter was discussed again in 1994 and 1995, and in November 1995 a FAO Technical Consultation on Food Allergens was held. The recommendations from this consultation were to change the 25% rule as suggested. The suggested list of allergenic foods was slightly modified. After several years of further discussion the Codex Alimentarius Commission adopted the proposal in June 1999.

Several countries changed their food labeling rules in accordance with the Codex. In the EU the 25% rule was totally abandoned in 2004.

The change in labeling rules and the increased focus on allergenic foods has been an advantage for the allergic consumer, but it also created unforeseen problems. These arise from insufficient scientific knowledge on safe levels of food allergens. The European Food Safety Authority concluded in 2004 that:

> *The doses of allergens capable of triggering food allergic reactions are variable and can be very small, i.e., in the milligram or microgram range. The information currently available is insufficient to draw firm conclusions regarding the lowest dose that could cause an adverse effect (threshold).*

For this reason, the authorities were not able to advise industry on what amounts of food allergen could be considered effectively harmless and help them to develop operational standards.

This again led industry to develop various labels such as 'may contain nuts' or 'manufactured in a facility that also handles nuts', because they were not able to guarantee the total absence of allergenic food in their products and no one could tell them when their products were safe enough. A further consequence of the lack of guidance was that criteria for using such labeling varied across the food industry.

WHERE ARE WE NOW — THE BOOK

Unfortunately we have not solved all the questions that arise when dealing with allergenic food in food production, but much has happened. This book presents the newest knowledge on food allergy and food allergen management and includes suggestions for practical management of food allergens.

The book is organized in four sections. Section 1, *Food Allergy: Causes, Prevalence, and Impacts*, provides a background for understanding the context and rationale for food allergy as a problem in society. It gives an overview of how patients experience daily life with food allergy and how it impacts their lives. It describes food allergy as a disease and lists which foods cause allergy as well as the epidemiology of food allergy. Section 2, *Allergen Thresholds and Risk Assessment*, describes how clinicians determine the amount of allergenic food causing a reaction. It suggests quantities of different allergenic foods that can be considered to present minimal risk and describes how they are derived and how these data are used in risk assessment, both theoretical and in practice. Section 3, *Risk Management of Gluten*, gives an overview of the gluten-induced disease celiac disease with emphasis on diagnosis, prevalence, prevention, and management. Section 4, *Practical Food Allergen Risk Management*, focuses on the practical aspects, including how allergenic food is managed in a factory and in catering businesses. It has stories illustrating how concrete problems with food allergens in production were handled and explains the role of health service professionals. It describes the analytical detection methods for food allergens and the ways that processing can alter the allergenicity of foods. It covers effective communication with consumers including the use of 'may contain' labeling. Lastly it gives a short overview of legislation and useful places to keep updated.

THE AUTHORS

The authors are clinicians, researchers, and public and industrial risk assessors and risk managers. Many of the authors have been partners of the EU-funded research project *The Prevalence, Cost and Basis of Food Allergy in Europe* (EuroPrevall) and present data from the project.

Charlotte Bernhard Madsen, René W. R. Crevel, Clare Mills, Steven L. Taylor

Food Allergy: Causes, Prevalence, and Impacts

Chapter one

Living with Food Allergy: Cause for Concern

Audrey DunnGalvin[1], Anthony E.J. Dubois[2],
B.M.J. Flokstra-de Blok[3], Jonathan O'B. Hourihane[4]

[1]*Department of Paediatrics and Child Health, University College Cork, Cork, Ireland*
[2]*Department of Paediatric Pulmonology and Paediatric Allergy, University Medical Center Groningen, University of Groningen, Groningen, The Netherlands*
[3]*Department of General Practice, University Medical Center Groningen, University of Groningen, Groningen, The Netherlands*
[4]*Department of Paediatrics and Child Health, Clinical Investigations Unit, Cork University Hospital, Wilton, Cork, Ireland*

CHAPTER OUTLINE

INTRODUCTION

Since, at present, there is no 'cure' for food allergy, avoidance of the responsible allergenic food and emergency management in the form of injectable epinephrine (EpiPen or Anapen), in case a food allergen is accidentally ingested, is the only reliable therapy offered to those living with such conditions. However, 'avoidance' is not as straightforward as it might first appear. Firstly, it is complicated by the fact that foods like peanuts, nuts, or soy can be found in many foods (e.g., breads, muffins, pastries, biscuits, cereals,

3

Risk Management for Food Allergy. http://dx.doi.org/10.1016/B978-0-12-381988-8.00001-4

soups, ice creams, seasoning, sauces) and in different forms. Living with a food allergy also means constantly reading food ingredient labels, concern for cross-contamination, vigilance in a variety of social activities, and immediate access to an auto-injector [1]. Secondly, symptoms may occur within minutes of ingesting a food allergen, include itching and swelling of the lips, tongue, and soft palate as well as nausea, abdominal pain, vomiting, and diarrhea. Anaphylaxis refers to a sudden, severe, potentially fatal, systemic allergic reaction that can involve the skin, respiratory tract, gastrointestinal tract, and cardiovascular system. The most dangerous symptoms include breathing difficulties and a drop in blood pressure, or shock, which are potentially fatal. Therefore, although the life-threatening nature of anaphylaxis makes prevention the cornerstone of therapy, it also has implications for the health-related quality of life of the children, teens, and adults living with the allergy.

In the past, the medical community defined health as an absence of disease. It is now recognized that health consists of physical, psychological, and social aspects. Because the concept of health has changed, the way we measure health or the impact of any disease has also changed. Health professionals now know that it is essential to use outcome measures that reflect the patient's perspective in order to gain a truly meaningful picture of the impact of a disease on a patient's everyday life [2]. We call this health-related quality of life (HRQL). The perception of HRQL is influenced by the individuality and subjectivity of experience and response, and may depend on many factors, such as age, gender, context, and culture [3−7]. Therefore, physiological measures often relate poorly with *perceived* physical well-being [8], and patients with the same clinical criteria often have dramatically different responses. To give an example, two patients with the same prognosis following an operation for a heart bypass can have two very different perspectives on how their lives have changed. For one it may be an opportunity, for others it may be perceived as a catastrophic event that changes how they see themselves, how they interact with others, and how they perceive the overall quality of their everyday lives. In turn, this can impact on how well they follow medical advice for their future health. It has become increasingly important, therefore, for researchers and healthcare professionals to understand how the perceptions, experience, and impact of a chronic disease might influence a patient's interpretation and response to it, so that we in turn can respond more appropriately. Furthermore, involving children as well as adults and parents in research is important, because children are now acknowledged to have rights in the determination of medical decisions that affect them [9]. This has encouraged research to be undertaken with children themselves to understand their views on the impact of a disease on their experiences and relationships.

Although a growing number of families must live and cope with food allergy on a day-to-day basis, it is only in recent years that the socio-emotional impact of food allergies on children, teens, adults, and parents has been researched in depth. The EuroPrevall project (europrevall@bbsrc.ac.uk) gave great impetus to research in the area of HRQL. In addition to clinical research on the prevalence, mechanisms, and causes of food allergy, research output in the area of psychosocial impact included HRQL measures for all age groups, and an examination of its socio-economic impact.

HRQL is measured by two major types of instruments; generic and disease-specific. Generic HRQL instruments are not specific to any particular disease and are therefore useful for comparing HRQL across different conditions, whereas disease-specific questionnaires focus on issues pertinent to one disease. However, generic instruments are necessarily more 'general' and therefore less sensitive to the particular problems associated with a particular condition [9]. For example, asking parents of children with food allergy if their disease impacts on their children's ability to run up and down stairs will provide meaningless results. In contrast, asking the same parents if children feel left out at birthday parties because of their food allergy provides a picture of its impact on the children's ability to take part fully in everyday social events — activities that children without food allergy enjoy without much thought or restriction. Disease-specific HRQL questionnaires provide an in-depth picture of the day-to-day concerns of patients and are also able to capture small changes in HRQL that may occur as a result of clinical or therapeutic treatment.

Several disease-specific measures have been developed under the aegis of EuroPrevall to assess quality of life in children and teens. These include the Food Allergy Quality of Life Questionnaire — Parent Form (parent-administered for children aged 0—12 years); the Food Allergy Quality of Life Questionnaire — Child/Teen Form (self-administered for children and teens aged 8—17 years) and the Adult Form (for those aged 18+). These questionnaires were developed according to gold standard methodologies [10—15].

Health-related quality of life instruments capture the impact of food allergy; however, the manner in which it is experienced and managed every day (coping) must also be evaluated [1,3,7]. Coping has not only been shown to be related to patient HRQL, but is also strongly linked to health behavior [16], having both the short- and long-term impacts. To illustrate this, we return to our earlier example of the two patients with the same prognosis following heart bypass surgery. They may, for example, be unable to return to their previous employment because of their changed health circumstances. Whereas one may cope with this experience by viewing the surgery as an opportunity to change their lives, to become healthier, to experience new challenges, others could 'give up,' become depressed, reject the company of friends, and feel that they no longer have a meaningful contribution to make to society. Here we have the same prognosis, but very different coping strategies. Research tells us that children with any chronic condition have twice the risk of developing mental health disorders as do healthy children, even without an accompanying physical disability [17]. Therefore, efforts have increasingly been made to assess how well children and adolescents cope with chronic conditions. In the context of research in children, qualitative research also provides an opportunity to tap into the richness of children's thoughts and feelings about themselves, their environments, and the world in which they live.

In this chapter, we will look at the impact of food allergy on HRQL and subsequent risk management. Firstly, we will let the children, teens, and families describe, in their own words, what it is like to live and cope with a food allergy every day. Qualitative studies were also carried out under EuroPrevall, both in the initial focus groups put in place to generate items for the questionnaires, and independently thereafter. The findings will be presented in the context of a developmental model that captures the pathway from childhood to adolescence and explains why some children are 'anxious' while others are 'risky.'

We will then discuss some scientific research on HRQL, the majority generated over the life of the EuroPrevall project. Research on factors (such as risk perception) that are related to, and impact on, HRQL are also examined. There is a strong emphasis throughout on developmental considerations in food allergies, from infants to adults. We conclude by offering some recommendations for future research and practice in food allergy risk management, based on the findings in this review.

QUALITATIVE STUDIES ON THE IMPACT OF FOOD ALLERGY ON HRQL

Experience and coping in any chronic disease is an intricate pattern of 'facts' and 'feelings' interwoven into a child's developmental pathway from birth to adulthood. Lay perceptions of risk may seem irrational to some clinicians, but have their own logic and validity from the perspective of those living and coping with food allergy. Here we use the patients' and parents' own voices to explore what it is like to live with food allergy, in order to better understand the decisions they make about managing their condition.

The findings will be framed within an integrated developmental framework [3] to explain the onset, development, and maintenance of food allergy—related cognitions, emotions, and behavior. In order to develop this framework we interviewed 120 children/teenagers aged 6—18 years in 15 age-appropriate focus groups. Fifty-two percent of the children were female. Parents were also interviewed. All children were physician diagnosed with IgE-mediated food allergy and had been issued an Anapen/EpiPen. Developmentally appropriate techniques such as vignettes or stories (where children could comment on characters in the third person) and activity books were designed to stimulate discussion, maintain interest, and minimize threats to the child's self-esteem.

Analyses of the data encompassed precipitating events (stressful events in the children's lives caused by food allergy—related factors); psychological impact (cognitive appraisal and emotional effects); and behavioral consequences or coping strategies. Our findings indicated that experience and coping in food allergy situations is complex and dynamic, comprising a series of interactive processes (both age-, gender-, and disease-specific) that are embedded in a child's developmental path.

Subsequently, we also analyzed data from focus groups and interviews held in Australia (N=60), the UK (N=72), Italy (N=45), Singapore (N=20), and the US (N=45). The themes that emerged from other countries were strikingly similar to our previous research, including the impact of *living with uncertainty*, with *difference*, with *rules* and the coping strategies used (Figure 1.1). The findings [3−4,18−21] are discussed below using direct quotes from parents, children, teens, and young adults.

CHILDREN AND TEENS: THE EVOLUTION OF UNCERTAINTY

Because children are rapidly changing in response to physiological, social, and psychological influences, the developmental process plays an important role in shaping and determining their health and HQRL. Children's social and emotional experiences are essential in shaping how the child will manage and live with their illness, both in the short and long term. Sensitive transition points occur along the developmental pathway when physiological

The developmental model:
Maximization/Balanced Adaptation/Minimization

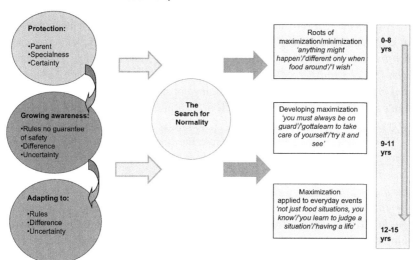

FIGURE 1.1 An illustration of the developmental model. Because they were diagnosed when infants, young children feel that they are 'the same' as other children and parents help them to feel normal and protected in their everyday lives. They therefore have an illusory perception of control and certainty. As children become more aware of the rules as restrictive, together with a growing awareness of difference and uncertainty, the search for normality becomes stronger and children evolve strategies in order to cope. Although their roots may be discerned in children in the youngest age group, by adolescence, children's coping strategies become more defined, and in some cases more rigid, and an expanding social world gives further impetus to the search for normality. Dunn-Galvin and Hourihane JACI, 2009.

(e.g., puberty) or environmental (school change) variables may have a significantly higher impact.

Middle childhood is when children begin to gain autonomy and self-belief in their ability to control events in their lives. An important transition point during this developmental period occurs when children learn or feel that parents (and therefore children themselves) cannot conclusively prevent an allergic reaction, after which we see a change in thinking, emotions, and behaviors *'Mum did read the ingredients but I still got sick after'* (Lucy, age 8, Ireland). The impact of attempting to cope every day with uncertainty above and beyond that faced by most children at this age without food allergies may result in increased levels of anxiety or risk-taking behavior.

Living with uncertainty is an important concept that affects children's sense of control, beliefs about risk, level of vigilance, and confidence in safety. Young children have an illusory perception of control because of parental protection, but we see the roots of uncertainty even in very young children. These children are very aware of parental anxiety, and speak about the possibility of a reaction occurring at any time; *'because you never know what might happen.'* Similar responses were found in UK, US, and Australian data; *'I need to kind of live my life on the risk that something is going to happen or something might never happen'* (Kathy, age 16, UK). Even when following the rules carefully, children often cannot pinpoint why a reaction occurred.

Always being aware and alert to the possibility of danger is a heavy burden for children in their everyday lives. Becky (age 10, Ireland) explains *'because food is always around it is hard to forget about it'*; and Matt (age 11, Ireland) says, *'... I can't just eat something like my friends ... or be with people without thinking about what they are eating.'* Being constantly vigilant also affects children's enjoyment of social events: *'well ... it means you can never relax at a party and just enjoy it'* (Kevin, age 11, Ireland).

Older children and teens emphasize the uncertainty of living with food allergy and the consequent feeling of a loss of control: *'like sometimes you can't find the cause [of a reaction] ... it just happens, you know ... not knowing makes you worried and unsure of yourself ... when I have a first bite like, if I'm not at home, I think is this it? ... will I die? ... what can you do?'* (Fran, age 12, Ireland). Adolescence is another important transition point with increased stresses related to age-specific challenges (peer pressure, the need to 'fit in,' issues relating to identity, the physiological changes of puberty), in addition to the burden of food allergy. Grace (age 13) captured the feelings of many teens when she describes why she feels anxious: *'when I get up in the morning I can't be sure I won't have a reaction that day.'*

Perception of risk and sense of threat are heightened along the developmental pathway. For example, uncertainty is compounded by a general lack of awareness and understanding in society. In many countries, children and teens described a low level of awareness and understanding in schools, restaurants, coffee shops, and other social arenas.

'The restaurants, hundred percent, even if they say it's nut free I mean you can't tell because they'll be cooking things. If they, say they shove some cashew nuts in a wok, shove the dish out and then my dish goes in there's going to be some bits of the cashew nuts still left' (Sally, age 16, UK).

'McDonald's we know that doesn't have any nuts in it but Pizza Hut'... we asked them ...'and they said that there's like in the ingredients of what they use for the dough there's some nut oils and stuff like that' (Jimmy, age 13, UK).

'I don't usually go to friends' houses, I only go to those that I've known for a long time ... in case anything happens' (Kim, age 9, Singapore).

'We don't tend to eat out and if we do me and [brother] will have like chips and garlic bread kind of thing because we just don't trust restaurants' (Cara, age 19, US).

A lack of awareness can also be found among peers, teachers, and other adults in children's lives: *'other kids don't get it ... they think it's a bit of a rash ... if anything bad happens, I worry they won't know to help me'* (Jilly, age 8, Australia). The experiences described by participants in Ireland, UK, US, Singapore, and Australia were very similar: *'sometimes they just joke around and they say 'ohh, there's nuts in this .... makes me [sad] ... I ask them to stop. ... and sometimes they don't stop'* (Jack, age 10, Australia). Among schoolchildren without food allergies, we found only a vague awareness of what food allergy means in terms of symptoms and lifestyle, and how to help in an emergency. This finding applied to all age groups: *'I don't know what would happen if he got a reaction ... maybe they start having breathing problems, like they start gasping or something ... there's nothing about it in the school ... the teacher hasn't said anything ... yeah we can share food'* (Calum, age 12, Australia).

Uncertainty also impinges on developing children's beliefs and subsequent coping strategies. Children respond by beginning a search for normality. For some 'normality' may mean assurance that they are safe at all times and are accepted and understood by particular friends, for others it means being able to interact freely and being accepted as normal *'in the real world,'* and for the remainder, it means finding a balance between the two. The roots of these strategies may be discerned in the youngest children. For example, in Italy over 75% of children (age 5−11 years) claim to have a monotonous diet, and school-aged children are significantly less interested in tasting new foods than younger children. Eighteen percent of children never attend parties. Other children, by *'eating just a little bit'* and seeing how they react, appear to be trying to determine their own risk thresholds: *'you'd have a small bit now and then and see what happens'* (Johnny, age 11). It may also be a way for children to exert control over uncertain conditions. Coping strategies were found to lie on a maximization/avoidance to minimization/risk continuum. They may be emotion focused or problem focused; often they are both. Some are actions, interactions, or beliefs. Their defining quality is that they are used in clusters by particular children, as demonstrated by the responses above.

By adolescence, children's coping strategies become more defined, and in some cases more rigid: *'I don't really want to be in a situation where I'm worried every day about what I eat and have to take all my own food, it's just not worth it, I'm not going to go ... So she's making decisions and limiting her own life'* (Frances, Mother of Pattie, age 15, UK). Other teens and young people appear to take risks as a means of coping in a social situation and to counteract feelings of difference in a search for normality. *'I have been having reactions since I was six, and I am really tired of trying to live normally with the labeling'* (Janes, age 20, UK); *'She just wants to be seen as normal, she disnae want tae, like create a fuss and ask in restaurants, you know, is there nuts in that'* (Mother of Gillian, age 19, UK). Frustration with labeling is also clear: *'When it says 'may have traces of nuts' I sometimes still eat them, because it's on everything, that is on absolutely everything and it's like if I can't eat that then what can I eat? ... if I actually went by that I wouldn't be able to eat anything'* (Gerry, UK, age 15).

Not telling others you have food allergy also forms part of the risk cluster of emotions and behaviors, for example, in terms of new relationships, *'why would I tell anyone ... it isn't like a cool talent or something'* (Jamie, age 14, Ireland).

'Because I remember I went back to this one girl's house and she ate Nutella and I was like, 'ach, no.' I didn't want to say anything so I just didn't say anything but I was fine, I mean nothing happened ... I'm just worried in case if I did kiss her I would have a reaction and 'oh no,' but nothing happened' (David, age 19, UK).

Research shows that adolescents with severe allergies are at particularly high risk of severe and fatal anaphylactic reactions [6]. The factors contributing to this are unclear as there has been no systematic research into the attitudes and experiences of this group. The observed high rates of morbidity and mortality may be due to a combination of limited allergen avoidance and poor emergency management among adolescents. There is clear consensus in the research literature that auto-injectors are under-used by patients of all ages [22−24]. However, the reasons for this must be fully explored in order to inform improvements in clinical practice. Because auto-injectors are central to emergency management, ensuring their correct use is a priority for clinicians. To do this, clinicians need to understand how and why adolescents respond in the way they do, taking account of the social context and the developmental transitions of adolescence.

In the developmental model, taking risks with medication is termed 'minimization': *'I try to 'forget' the stupid pen, but Mum makes me get it'* (Danny, age 10, Ireland);

'I forget it all the time. I'm quite bad. I keep it in my car, my car's normally quite close to me all the time but I'm quite bad at carrying it about with me. It's just the sheer hassle of having to take it, you know, it's like, I mean they're quite big and they don't fit in any of my pockets, so I just leave them in the car, although I know I should, but ... I just put them in my glove box' (Martin, age 19, UK).

'You would remember it in the usual situations ... it's just that if it's some-thing out of the ordinary, you know or like if you are going somewhere with friends that isn't like a restaurant, that's ok then, isn't it?' (Julie, age 15, UK). *'I hate bringing [the pen] because you can't hide it and it reminds me of being allergic'* (Tom, age 16, US).

These responses clearly demonstrate that, although children are aware that they should bring the auto-injector with them at all times, there are many bar-riers to full compliance. In many cases, 'at all times' does not generalize to non-'usual' occasions or activities.

Confusion concerning how much allergen it takes to induce a reaction rein-forces a sense of uncertainty and interacts with the search for normality to also play a part in risky behavior: *'I don't want to take risks ... but sometimes, like, I do if I'm out. I know it's hard for restaurants and your friends ... to get their head around it. When I was younger, I wouldn't touch a thing, but I'm old enough now to know that I can have so much and it's ok. If I could know how much exactly, I could look after myself so much better ... without that scared feeling you get in your stomach sometimes'* (Chris, age 17, UK).

As we will show next, parents share many of the same experiences, con-cerns, anxieties, as their children and teens.

THE PARENTAL PERSPECTIVE: LIVING WITH UNCERTAINTY

The impact of a chronic illness on members varies greatly among families, but it is clear that the family plays a pivotal role in determining how children with chronic conditions adapt to their condition and how it impacts upon their HRQL [25]. It is generally accepted that parents of children with chronic conditions potentially have lower HRQL because of the additional demands and stressors placed on them by their children's condition. In the case of food allergy, like many other chronic conditions, parental HRQL may be affected by the child's diagnosis, as families have to deal with the day-to-day management and emotional strain of the illness, as well as additional costs in terms of time and money and disruption to everyday household routines [26]. Further-more, parents have a valuable insight into the impact of food allergy on children in their everyday lives. Therefore, in order to improve care and support for both parents and children, it is necessary to identify and understand parents' con-cerns for themselves, their families, and their children.

The potential impact of food allergy on children's social and emotional impact is a concern for parents in terms of identity: *'he asked me 'when will I be normal?' and I was shocked; I didn't realize he felt like that'* (Mother of Jack, age 7, Ireland); confidence: *'you know, she has to put so much more thought into every social occasion that she's aware of in advance, which I think takes away from her confidence'* (Mother of Donna, age 15, Ireland) and social integration: *'it's harder for them the older they get, they just want to fit in with their friends … Danny used to get bullied and now he's very conscious of his allergy ... I worry about him and how he's going to cope when he goes to secondary school'* (Mother of Danny, age 10, Ireland).

Children's <u>growing autonomy</u> presents parents with particular challenges: *'he has a lot of new friends now and I worry ... but he gives out to me if I mention it when they come over ... he's embarrassed about it'* (Father of Peter, age 12, Ireland).

'He is relatively sensible, I mean he could be a lot less sensible than he is, but he does take risks and sometimes you find a Cadbury's Dairy Milk wrapper in his pocket or something like that when he knows full well that the risks of that are there. I think also he feels acutely aware that if there's sweeties being handed out at swimming or something like that often they're things that he can't have and sometimes he'll just chance it I suspect rather than be different from other folk' (Mother of John, age 10, UK).

'If he's gone out for a drink and they've been eating and, you know, like curry, there could be curries, there could be satay and, you know, he's drunk and doesn't know what he's eating, that's my concern — would he be able to look after himself then and give himself an injection' (Mother of Shane, age 19, UK).

<u>A lack of awareness</u>, across a wide range of public settings, impacts negatively on both parents' and children's enjoyment of social occasions.

'I've had pretty negative experiences, some very negative practice, you know, as soon as they hear they'll say "well it's best if she doesn't eat anything" or "we don't really want to serve her anything," or "the kitchen's too small and there's stuff everywhere"' (Mother of Jenny, age 8, UK).

'... and they brought a piece of nice fish that had been fried in olive oil and she assured us the fish were fried in olive oil and nothing else in the oil. And when the chips came she started eating them, there was a peanut sitting right in the middle of the chips, so it was just that, God I give up, we were both really depressed by that, you know' (Mother of Emily, age 12, UK).

<u>Living with risk and coping everyday</u> needs to be negotiated carefully, and engenders emotions such as confusion, anxiety, uncertainty, frustration, and some anger, as the following quotes illustrate;

'I am absolutely terrified that I would buy something with nuts in it by mistake ... if anything happened, I would never get over it' (Mother of Jimmy, age 6, Ireland).

'I try to introduce as much variety as possible ... I don't want them to develop problems with food ... but it's difficult when you are trying to be so careful at the same time. I just have to think about labels ... the blanket labeling is terrible ... and I get anxious and frustrated' (Mother of Sandra, age 5, US).

'I get confused and anxious trying to get him not to worry too much about it ... and then I worry that he's not worried enough' (Mother of Matt, age 10, Ireland).

'I would have loved to have built up even a relationship with a single restaurant who knew her and were careful and that you didn't feel that every mouthful

she took was a risk. Because eating out it's not pleasurable because it's such a Russian Roulette' (Mother of Deirdre, age 8, UK).

Transition points along the developmental pathway, such as the move from junior school to high school or secondary school, can be particularly stressful. Parents struggle with ways to support children's independence while controlling their own anxiety and genuine fear of risk.

'I made up reasons for him not to be out and I was very very protective of him and wouldn't let him have the freedom, I was so paranoid that something was going to happen to him' (Mother of Peter, age 13, UK).

'I am so scared for him ... when he leaves the house at all for anything ... it's always there ... sometimes in the background ... sometimes strong' (Mother of Jen, age 14, Singapore).

This can sometimes cause family conflict: *'he was furious when I brought [food allergy] up when he had his friends over ... I don't think he tells people any more ... it worries me'* (Mother of Zack, age 13, US).

'She tells me to stop nagging her ... but she often tries to leave the house without the pen' (Mother of Christy, age 15, US).

Taken together, these findings suggest that children, teens, and families living with food allergy need to cope with normal developmental changes and as well as their condition, placing them under increased psychosocial stress and leading to possible maladaptive coping strategies and consequent risk. For both families and individuals, their food allergy has direct and indirect effects on emotional adjustment, social interaction and social life, confidence in coping with risk, stress, and overall quality of life. Particular concerns include 'labeling,' dietary restrictions, confusion over how much allergen can cause a reaction, general lack of awareness, and balancing children's growing independence while ensuring their safety. Two of these issues in particular impact on confidence in coping with everyday risk. Firstly, the confusion about 'threshold' (the amount of allergen required to induce a reaction in a patient), is a significant source of uncertainty and stress: *'when she was diagnosed .... you can't take it in ... just don't let her eat this or this or this ... I mean, what can she eat ... how is she going to grow properly ... how much will kill her ... why don't they tell you?'* (Mother of Jane, age 6, UK). No information is typically given on individual threshold dose for a patient; thus a typical response concerning a child's allergy is often very vague: *'my child is very allergic.'* In addition, anaphylaxis is poorly described and subject to variable interpretation, with emphasis on one extreme of a continuum of severity. Secondly, 'labeling' on food products is perceived as untrustworthy, too inclusive, and not personally relevant; *'what's it based on anyway ... the labels ... everything says 'may contain' ... how do we know that's it's right, it's really confusing ...'* (Mother of Sam, age 8, Ireland). Although our findings show that it is understood and accepted by clinicians, parents, children, and teens that zero risk for food allergic persons is not a realistic or attainable option, those living with a food allergy do seek a way to translate emerging new scientific findings, on 'thresholds,' for example, into meaningful strategies to improve their quality of life: '*You can't have no risk at all you know, even if the child never leaves the house, so you have to deal with risk ... we just want a better way*' (Mother of Carla, age 10, US).

In the next section we will review scientific research on HRQL, the majority of which was generated over the life of the EuroPrevall project. This supports the qualitative findings we have discussed above.

THE PARENTAL PERSPECTIVE: QUANTITATIVE

The first validated HRQL food allergy—specific measure, the Food Allergy Quality of Life — Parental Burden (FAQL-PB) questionnaire [25], measures the parental burden associated with having a child with a food allergy. Scores in the food allergic cohort were significantly lower for general health perception, parental distress and worry, and interruptions and limitations in usual family activities, than in healthy controls.

The FAQLQ-PF (parent form) is completed by parents on behalf of their children [10]. To ensure that the measure is developmentally appropriate, it caters to three age groups; 0—3 years (14 items); 4—6 years (26 items); 7—12 years (30 items). The core questionnaire has three subscales, calculated as the mean of each scale (Figure 1.2).

The subscales measure *Food Anxiety, Social and Dietary Limitations,* and *General Emotional Impact.* The total score is calculated as the mean of the three subscales. Supplementary sections contain questions on clinical and demographic variables; parental concern for their child's emotional and physical health; stress levels experienced by parents and family; impact on family activities; and expectation of outcome following accidental ingestion of allergen. The FAQLQ-PF has demonstrated very high reliability and validity (cross-sectional, cross-cultural, longitudinal) [10—11,15,27—28].

In the course of the development and validation of the FAQLQ-PF [10], we found a strong impact of food allergy on HRQL, in relation to many

---

Food Anxiety: EG
- My child is afraid to try unfamiliar foods
- Concerned by poor labelling on food products

Social & Dietary Limitations: EG
- My child has little variety is his/her diet because of food allergy
- Because of food allergy, my child's social environment is restricted because of limitations on restaurants we can safely go to as a family

General Emotional Impact: EG
- Is more worried in general than other children of his/her age
- Is not as confident as other children of his/her age in social situations
- My child feels different from other children

FIGURE 1.2 Examples of items and content in the three subscales of the Food Allergy Quality of Life Questionnaire; Parent Form (FAQLQ-PF). Three factors (emotional impact; food anxiety; social and dietary limitations) emerged following exploratory and confirmatory factor analysis in the development and validation of the Food Allergy Quality of Life Questionnaire; Parent Form (FAQLQ-PF). Reprinted from Dunn-Galvin et al. Clinical and Experimental Allergy, 2008: 38; 977—986.

psychosocial aspects of children's everyday lives (Figure 1.2). For example, in the initial focus groups put in place to generate items for the FAQLQ-PF, parents suggested that the anxiety associated with the risk of a potential reaction has more profound effects on emotional and social aspects of a child's everyday life than the clinical reactivity induced by food intake. The importance of a sub-scale assessing this aspect of anxiety was subsequently confirmed using clinical impact and factor analytic methodologies. Children were also found to be 'generally anxious' according to parents, that is, the anxiety associated with food often 'generalized' to non-food situations.

During the longitudinal validation of the FAQLQ-PF [11], we discovered that a food challenge (which is performed in a hospital setting in order to diagnose food allergy) may alleviate anxiety. In our design, we administered the FAQLQ-PF to parents of children 0–12 years before the child underwent a clinically indicated food challenge, and at 2 and 6 months post food challenge. Eighty-two children underwent a challenge in total (42 positive; 40 negative). Although significant differences were found between positive and negative groups on all subscales and total score at 6 months [$F_{(2,59)} = 6.221$, $p < 0.003$], we found HRQL improved significantly post challenge time points (all $p < 0.05$) for both positive and negative groups. A possible explanation for improvement in the 'positive' groups (long suspected but never documented) concerns the impact of uncertainty on perception of HRQL. 'Living with uncertainty' appears as a central theme for all age groups with a food allergy, including parents, and will be discussed further throughout this chapter.

Our findings suggest that a food challenge may be valuable, not only as an essential diagnostic tool, but as a therapeutic one. In effect, by providing a sense of certainty, a food challenge may have a positive impact on HRQL, irrespective of outcome. This positive impact may also have been reinforced by specialist consultation, personalized information, and interaction with other children with food allergies.

Tracking the impact of food allergy entails beginning at the earliest possible time on the developmental pathway from childhood to adulthood. Findings in the EuroPrevall birth cohort study demonstrate that the impact of a diagnosis of food allergy begins early and can be detected over the course of one year [28]. Iceland (N=60), UK (N=45), Germany (N=40), Spain (N=36), Netherlands (N=95), and Italy (N=25) administered the FAQLQ-PF before the infant was diagnosed with food allergy by food challenge and 12 months later. On average, 60% of the infants tested positive for at least one food type. The impact of food allergy on HRQL increased significantly for the positive group only (Figure 1.3). We found a similar pattern of responses across countries.

Overall, there were significant differences ($p < 0.05$) between positive/negative groups over 12 months. In this age group (3–6 years), the subscale measuring *food anxiety* showed the biggest increase in burden, from baseline to 12 months for the positive group (Figure 1.4). Even at this very young age, it appears that children are reluctant or afraid to try new foods, and have a

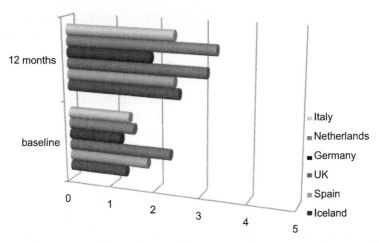

FIGURE 1.3 Impact of food allergy rises for children from baseline to diagnosis to 12 months in six European countries. Findings in the EuroPrevall birth cohort study demonstrate that the impact of a diagnosis of food allergy begins early and can be detected over the course of one year. The impact of food allergy on HRQL increased significantly for the positive group only. A similar pattern of responses was found across countries.

lack of variety in their diets. Children's ability to take part fully in social events is also adversely impacted compared to the negative group. Such responses may be due to a projection of parental anxiety, although this in itself is likely to have a profound impact on the children's own perception. We found similar results using the FAQLQ-PF in the US, Singapore, and Japan [4].

It is important also to take into account other factors, related to HRQL, which may provide a deeper understanding of the impact and outcomes of a diagnosis of food allergy. To this end, the FAQLQ-PF was used to examine

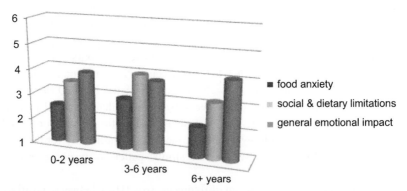

FIGURE 1.4 Age patterns in FAQLQ-PF subscale scores across three age groups in Ireland, the US, and Singapore. We found age-specific differences in the three subscales. The subscale measuring 'food anxiety' showed the biggest increase in burden, from baseline to 12 months, for the positive group aged 3–6 years. General emotional impact was highest for the 6+ age group.

specific psychological factors which may influence parents' decisions to take part in clinical studies [27]. Parents of food allergic children in the US were offered investigational oral immunotherapy (which attempts to desensitize children to a specific allergen) in the regular outpatient clinic. Forty parents (Group A) declined, and 25 parents (Group B) agreed to take part. Both groups completed the FAQLQ-PF.

Our results showed that parents who perceive that their child is at high risk of dying from a food allergy are more likely to enroll their child in an investigational trial in which the child will be given peanut immunotherapy (OR 6.75; CI 3.45−9.73). This is in spite of the fact that the experimental therapy is intensive and has attendant adverse risks including induction of anaphylaxis, compared to the routine clinical practice. The association was independent of the severity of symptoms, experience of anaphylaxis, and the perception of the impact of food allergy on HRQL. Socio-economic status was not a significant factor.

These findings may be explained by parental concern to avoid potentially life-threatening consequences of accidental ingestion in the often 'uncontrolled' environment of their child's everyday life. Research using the FAQLQ-PF has documented [11] parental perceptions of possible adverse outcomes if an allergen is accidentally ingested by children. Of the 100 parents participating in the study, none felt that there was 'no risk' of their children accidentally ingesting an allergen and/or dying from food allergy. Ten of the parents (Figure 1.5) reported 'a certain chance' that their child would die following such an event, underlining the uncertainty, sense of responsibility, and feelings of anxiety with which some parents live every day. This perceived level of threat may be an important factor in motivating parents to consent to their children taking part in investigational therapies in a 'controlled'

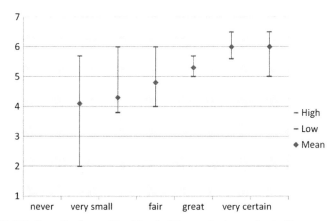

FIGURE 1.5 Results from the Food Allergy Independent Measure (FAIM): Parent estimation of their child's chance of dying if they accidentally eat a food to which they are allergic. The FAIM is part of the Food Allergy Quality of Life Questionnaire-Parent Form (FAQL-PF1). The questionnaire items are scored on a 7-point Likert scale ranging from 1 (no impact on HRQL) to 7 (extreme impact on HRQL).

environment, even though this involves a protocol in which reactions are more likely than if not in the trial.

Age and gender also influence the perception of HQRL. In the course of development and validation of the FAQLQ-PF, multivariate analysis showed an interaction between sex and age group for general emotional impact on HRQL scores [10—11]. In effect, parents of boys reported higher mean total scores up to the age of 6 years; parents of girls reported higher mean scores in the 6—12 years age group, particularly in the subscales 'general emotional impact' and 'food anxiety'; whereas boys had higher scores in the 'social and dietary limitations' subscale at all ages.

THE CHILD, ADOLESCENT, AND ADULT PERSPECTIVE: QUANTITATIVE

In 2009, the first disease-specific HRQL questionnaire for children with food allergy became available. This Food Allergy Quality of Life Questionnaire — Child Form (FAQLQ-CF) was demonstrated to be reliable, valid, and easy to use for children aged 8 to 12 years [12]. As quality of life focuses on the perception of the patients, questionnaires are usually completed by the patients themselves. This also holds for children. Although the understanding of HRQL is determined by the age, maturity, and cognitive development of a child, it has been reported that children aged 8 years and older are able to understand questions about their HRQL and to give reliable and valid answers [29]. It is, of course, important to take the level of development of the child into account when developing an HRQL questionnaire for children. Therefore, food allergic children were included in the development phase of FAQLQ-CF. Response categories of the FAQLQ-CF are illustrated by faces (smileys), which are more appropriate for the cognitive development of the child.

The FAQLQ-CF is complementary to the FAQLQ-PF (parent form), because the FAQLQ-PF is completed by the parents of the food allergic child. Thus, the FAQLQ-PF measures the quality of life of the child according to the parent (proxy-reported). It is known from the literature that children and their parents may differ in their views on and judgments of quality of life and on perception of risk [30]. On the other hand, it is obvious that in young children one can only make use of proxy-reported HRQL measures. Recently, the FAQLQ-CF and FAQLQ-PF were simultaneously completed by a sample of 74 food allergic children (aged 8—12 years) and their parents [31]. It was found that parents reported significantly less impact of food allergy on the quality of life of their child than the children themselves. However, perceived disease severity as measured with the Food Allergy Independent Measure (FAIM) [32] was comparable for the children and their parents. This may indicate that the parents underestimate the negative impact food allergy has on the quality of life of their child. It may also suggest that parents perceive that they take on much of the burden in terms of risk management, with consequent adverse impact on parent HRQL, particularly with regard to emotional issues

[4,27]. The challenge of risk management by parents was discussed earlier in this chapter, using qualitative research findings.

In addition to the FAQLQ-CF, two other disease-specific HRQL questionnaires have been developed for adolescents (13—17 years) and adults (18 years and older) with food allergies. These questionnaires are also self-completed and have been shown to be reliable and valid instruments for measuring HRQL in food allergic adolescents and adults, respectively [12—14,33—34,36—37].

With regard to food allergic adolescents, a moderate agreement was found in the impact reported by the adolescents themselves and their parents. Disagreement was mainly associated with the perceptions and characteristics of the adolescent rather than the perceptions and characteristics of the parent [35]. Therefore, in order to gain as complete a picture as possible of the impact of food allergy on the HQRL of children and teens, and to identify areas of disagreement, the proxy forms and the self-completed forms should be used together.

The FAQLQ-AF has been translated and validated in the US and a number of European countries. In the US, an online version of the FAQLQ-AF was used instead of a conventional paper version. The online version was found to be feasible, consistent and valid in the US. Moreover, when comparing the HRQL scores of the American food allergic patients with a comparable sample of Dutch food allergic patients, a greater impairment in HRQL was found for the American food allergic patients. As epinephrine auto-injector prescription rates differed remarkably between the two patients groups (but severity of reported symptoms were quite similar), the prescription of epinephrine auto-injectors and effect of this on HRQL are important targets for further research [33]. Another study compared the HRQL of food allergic adults as measured with the FAQLQ-AF in seven European countries: Iceland, the Netherlands, Poland, France, Spain, Italy, and Greece. The food allergic adults were recruited through the EuroPrevall project (community survey and outpatient clinic survey). The FAQLQ-AF was shown to be valid and consistent in these European countries and HRQL scores were comparable among the seven European countries studied [34].

DISCUSSION

Children's social and emotional experiences are essential in shaping how the child will manage and live with their illness, both in the short- and long-term. Health-related quality of life instruments are a powerful method of capturing the impact of a diagnosis of food allergies from the patient perspective. It is clear from the quantitative findings on HQRL in Europe, the US, and elsewhere that food allergy has a significant adverse impact on perceptions of everyday life and risk management. Qualitative methods can be used as a complementary method to explore areas about which little is known or to gain a novel understanding of a particular area. In addition, qualitative methods can be used to obtain the intricate details about phenomena such as feelings, thought processes, and emotions that are difficult to extract or learn about through conventional research methods such as questionnaires.

'Living with uncertainty' is a central theme when living with food allergy. As Katie (age 7) tells us, *'you never know what might happen.'* Allergic reactions are perceived to be unpredictable, sometimes they are mild, sometimes severe, and sometimes they happen when least expected, even if the individual is vigilant. This is not only because of the uncertain nature of reactions in food allergy, but also because of confusion and lack of transparency and specificity in food labeling, inconsistency around guidelines for use and prescription of auto-injectors, and lack of awareness and understanding among some schools, shops, restaurants, coffee shops, cinemas, and the general public. Children and teens can respond to these conditions by experiencing a loss of control over their condition and therefore, in some cases, becoming very anxious and avoidant in their emotions and behaviors or, in contrast, becoming frustrated or angry and taking risks with their safety.

The developmental process is intricately enmeshed in the evolution of uncertainty. Middle childhood is an important transition point when children begin to gain autonomy and self-belief in their ability to control events in their lives. We find increased levels of anxiety or risk-taking behavior follows this point, resulting from the negative impact of attempting to cope every day with challenges that are above and beyond those faced by most children in this age group who do not have a food allergy.

Adolescence is another important transition point, with increased stresses related to age-specific challenges in addition to the burden of food allergy. Parents also suffer high levels of stress and anxiety due to constant vigilance and feelings of guilt. Some of this worry is maladaptive (e.g., overprotection), thus inhibiting normal social development, and therefore may have a long-term impact on HQRL and positive coping ability.

A reduced public trust in the safety of food labeling and confusion about 'thresholds' (how much allergen is required to cause a reaction, and how severe this reaction might be) is a significant source of uncertainty (and stress) for children, teens, and parents. For example, in many cases, teens and young adults felt it was pointless and frustrating reading ingredients on labels and therefore take deliberate risks. This attitude was often formed during the middle childhood years. Although our research shows that it is understood and accepted by clinicians, parents, children, and teens that zero risk for food allergic persons is not a realistic or attainable option, they do want a better way.

IMPROVING RISK MANAGEMENT AND HRQL: A BETTER WAY?

To address and attempt to alleviate food allergy—related stressors, both quantitative and qualitative research suggests that alleviating uncertainty should be a major goal for health professionals working with children, teens, and families. Remarkable similarities in our findings on food allergy across countries suggest that policies and programs that address quality of life and risk management issues may be relevant to many different populations.

Transition points are a source of stress and uncertainty, particularly for parents of food allergic children. Whether or not the child has already been at nursery, entry to infant/primary school is a new trigger of anxiety, because the parent is required to hand over responsibility and control to a third party, and they need to put strategies in place to cope with this. As children grow and become more aware of difference and their social world expands, parents worry that their children may take risks in order to fit in with other children. Keeping children safe while helping them to be independent and self-reliant can be a challenging balancing act. Going to secondary school is a transition point that tests resources, increases uncertainty, and intensifies anxiety in parents. Parents can transmit their anxiety to their child, and just as children can pick up on parental anxiety, they can also respond to a parent's ability to stay calm in stressful situations [38]. Therefore, when treating children, it is important to address parental anxiety and to improve their understanding of their child's condition. Allaying parental anxiety reduces the child's anxiety and creates a positive feedback loop, which ultimately affects both the child and parent. Greater support and clear information are important at the time of diagnosis and at the different transition points along the developmental pathway. Specifically, parents have suggested that greater emphasis is needed on the social and emotional aspects of food allergy, on knowing what to expect, and on enhancing self-management skills that both children and their families can draw on and that generalize to both everyday and non-typical situations [3–4,7]. The development and validation of a psycho-educational intervention for parents, children, teens, and schools are currently under way by the Cork Research Group, in collaboration with the UK, the Netherlands, and the US.

Great strides are currently being made in research on population thresholds [39,40]. However, can the science be useful? Can it be translated into meaningful strategies to improve quality of life? To answer these questions, it is vital that perspectives from clinicians, food industry, support groups, and consumers are used to identify issues (regarding thresholds and labeling) which must be addressed in order to develop harmonized approaches and strategies that actually work, will be accepted, and can be communicated clearly. In an ongoing worldwide study under the aegis of FARRP (www.farrp.org/) on the effects of labeling on the concerns of the allergic consumer, we are investigating the acceptability by stakeholders (parents, young people, clinicians, producers) of alternative approaches, with the ultimate aim of improving a sense of control and hence risk management and quality of life in those living with food allergy. Risk communications that are not specific are more likely to increase anxiety without increasing awareness or confidence.

'You would have to know ... how they developed it, what's behind it. We're not stupid, we will get it ... if it's communicated properly' (Gwen, Mother of Sandra, age 6, and Tim, age 8).

'It would be good if there were some degree of risk implied by labels — also if possible, maximum allergen levels. I understand that this is complex but many parents view the labeling as a legal ploy to protect companies, so this must be addressed' (US allergist).

Trying to agree an *acceptable level of risk* is a 'red herring' because perception of risk is inherently very subjective and because it has negative associations that impede constructive discussion and meaningful progress. We must ask instead what is the best way to communicate what we know (and don't know) about thresholds, with the dual aims of reducing uncertainty as much as possible and translating new scientific discoveries into meaningful strategies that improve quality of life and risk management in food allergy.

REFERENCES

[1] DunnGalvin A, Hourihane JO'B. Developmental aspects of HRQL in food related chronic disease. The International Handbook of Behavior. US: Springer; 2011. Diet and Nutrition.

[2] Blok De, Vlieg-Boestra B, Oude-Elberink J, DunnGalvin A. A framework for measuring the social impact of food allergy across Europe. Allergy 2007:83.

[3] DunnGalvin A, Gaffney A, Hourihane JO'B. Developmental pathways in food allergy: a new theoretical model. Allergy 2009;64:560—8.

[4] DunnGalvin A. Food allergy: a challenge for patients and families. Symp presentation EAACI 2011. Istanbul.

[5] DunnGalvin A, Hourihane J, Frewer L, Knibb RC, Oude Elberink JNG, Klinge I. Incorporating a gender dimension in food allergy research: a review. Allergy 2006;61: 1336—43.

[6] Sampson M, Muñoz-Furlong A, Sicherer SH. Risk-taking and coping strategies of adolescents and young adults with food allergy. J Allergy Clin Immunol 2006; 117(6):312—8.

[7] DunnGalvin A, Hourihane JOB. Developmental trajectories in food allergy: a review. In: Taylor S, editor. Advances in Food and Nutrition Research 2009.

[8] Guyatt GH, Feeny DH, Patrick DL. Measuring health-related quality of life. Ann Intern Med 1993 04/15;118(8):622—9.

[9] Flokstra-de Blok BM, van der Velde JL, Vlieg-Boerstra BJ, Oude Elberink JN, DunnGalvin A, Hourihane JO, et al. Health-related quality of life of food allergic patients measured with generic and disease-specific questionnaires. Allergy 2010 Feb 1.

[10] DunnGalvin A, de Blok BMJ, Dubois A, Hourihane JO'B. Development and Validation of the Food Allergy Quality of Life — Parent Administered Questionnaire (FAQLQ-PF) for food allergic children aged 0—12 years. Clin Exp Allergy 2008;38:977—86.

[11] DunnGalvin A, Cullinane C, Daly D, Flokstra-de Blok BMJ, Dubois AEJ, Hourihane JO'B. Longitudinal validity and responsiveness of the Food Allergy Quality of Life Questionnaire — Parent Form (FAQLQ-PF) in children 0—12 years following positive and negative food challenges. Clin Exp Allergy 2010 Mar;40(3): 476—85.

[12] Flokstra-de Blok BMJ, DunnGalvin A, Vlieg-Boersta BJ, Oude Elberink JNG, Duiverman EJ, Hourihane JO, et al. Development and validation of a self-administered Food Allergy Quality of Life Questionnaire for children. Clin Exp Allergy 2009;39:127—37.

[13] Flokstra-de Blok BMJ, DunnGalvin A, Vlieg-Boerstra BJ, Oude Elberink JNG, Duiverman EJ, Hourihane JO, et al. Development and validation of the self-administered Food Allergy Quality of Life Questionnaire for adolescents. J Allergy Clin Immunol 2008 Jul;122(1):139−44, 144.e1−2.

[14] Flokstra-de Blok BMJ, van der Meulen GN, DunnGalvin A, Vlieg-Boerstra BJ, Oude Elberink JNG, Duiverman EJ, et al. Development and validation of the first disease-specific quality of life questionnaire for adults; The Food Allergy Quality of Life Questionnaire-Adult Form (FAQLQ-AF). Allergy 2009 Aug; 64(8):1209−17.

[15] Hourihane JO'B, Chiang WC, Laubach SS, DunnGalvin, Burks AWA. Psychometric validation of the FAQLQ-PF in a US sample of children with food allergy. JACI 2008;121−2(1):S106−7.

[16] Schmidt S. Coping with chronic disease from the perspective of children and adolescents − a conceptual framework and its implications for participation Child: Care. Health Dev 2003;29(1):63−75.

[17] Barlow JH, Wright C, Sheasby J, Turner A, Hainsworth J. Self-management approaches for people with chronic conditions: a review. Patient Educ Couns 2002;48:177−87.

[18] DunnGalvin A. Effects of labeling on the concerns of the allergic consumer. The 35th Annual Winter Meeting of The Toxicology Forum; February, 2010.

[19] DunnGalvin A, Du Bois B, De Blok J, Hourihane. Child vs. maternal perception of HRQL in food allergy: developmental trajectories and evolution of risk behavior. Allergy 2007;62(83). 70−16.

[20] DunnGalvin A, Hourihane JO'B. Self-assessment of reaction thresholds in food allergy: a new theory of risk taking which changes over time. J Allergy Clin Immunol 2009;123: S142.

[21] Bertalli N, Allen K, Hourihane JO'B, DunnGalvin A. Cross cultural comparisons of Irish and Australian children and teens living with food allergy: The SchoolNuts Study. Symp presentation. AAAAI 2011. San Francisco.

[22] Sicherer SH, Forman JA, Noone SA. Use assessment of self-administered epinephrine among food allergic children and pediatricians. Pediatrics 2000;105:259−362.

[23] Simons. FER first aid treatment of anaphylaxis to food: focus on epinephrine. JACI 2004;1134:837−44.

[24] Kim JS, Sinacore JM, Pongracic JA. Parental use of epipen for children with food allergies. JACI 2005;116:164−8.

[25] Cohen BL, Noonc NS, Munoz-Furlong A, et al. Parental burden in food allergy. J Allergy Clin Immunol 2004;114(5):1159−63.

[26] Primeau MN, Kagan R, Joseph L, Lim H, Dufresne C, Duffy C, et al. The psychological burden of peanut allergy as perceived by adults with peanut allergy and the parents of peanut-allergic children. Clin Exp Allergy 2000;30:1135−43.

[27] DunnGalvin A, Burks WJ, Dubois AEJ, Chang WC, Hourihane JO'B. Profiling families enrolled in food allergy immunotherapy studies. Pediatrics 2009;124:e503−9.

[28] DunnGalvin A, Hourihane JO'B, Dubois AE, Flokstra-DeBlok BM. Impact of food challenge tests on children's and parent's health related quality of life: A Time Series Case-Control Study. Submitted.

[29] Riley AW. Evidence that school-age children can self-report on their health. Ambul Pediatr 2004 07;4(4):371−6.

[30] Davis E, Nicolas C, Waters E, Cook K, Gibbs L, Gosch A, et al. Parent-proxy and child self-reported health-related quality of life: using qualitative methods to explain the discordance. Qual Life Res 2007 Jun;16(5):863−71.

[31] van der Velde JL, Flokstra-de Blok BM, DunnGalvin A, Hourihane JO, Duiverman EJ, Dubois AE. Parents report better health-related quality of life for their food allergic children than children themselves. Clin Exp Allergy 2011 May 16.

[32] van der Velde JL, Flokstra-de Blok BM, Vlieg-Boerstra BJ, Oude Elberink JN, DunnGalvin A, Hourihane JO, et al. Development, validity and reliability of the food allergy independent measure (FAIM). Allergy 2010 May;65(5):630−5.

[33] Goossens NJ, Flokstra-de Blok BM, Vlieg-Boerstra BJ, Duiverman EJ, Weiss CC, Furlong TJ, et al. Online version of the food allergy quality of life questionnaire-adult form: validity, feasibility and cross-cultural comparison. Clin Exp Allergy 2011 Apr;41(4):574–81.

[34] Goossens NJ, Flokstra-de Blok BM, van der Meulen GN, Asero R, Barreales L, Burney P, et al. Impact of food allergy on health-related quality of life in adult patients is comparable in seven European countries. Allergy 2010;65(s92):747.

[35] van der Velde JL, Flokstra-de Blok BM, Hamp A, Knibb RC, Duiverman EJ, Dubois AE. Adolescent-parent disagreement on quality of life of food allergic adolescents: Who makes the difference? 2011, Submitted.

[36] Flokstra-de Blok BM, Dubois AE, Vlieg-Boerstra BJ, Oude Elberink JN, Raat H, DunnGalvin A, et al. Health-related quality of life of food allergic patients: comparison with the general population and other diseases. Allergy 2010 Feb;65(2):238–44.

[37] van der Velde JL, Flokstra-de Blok BMJ, Duiverman EJ, Dubois AEJ. Longitudinal validity and responsiveness of the food allergy quality of life questionnaire-adult form; Quality of life improves significantly after double blind placebo controlled food challenges. Allergy 2010;65(s92):746.

[38] Clinch J, Dale S. Managing childhood fever and pain – the comfort loop. Child Adolesc Psychiatry 2007:1–7.

[39] Taylor S, et al. Food Chem. Toxicol 2009;47:1198–204.

[40] DunnGalvin A, Daly D, Cullinance C, Stenke E, Keeton D, Erlewyn-Lajeunesse M, et al. Highly accurate prediction of food allergen outcome using routinely available data. JACI 2011;127:633–9, e631–633.

Chapter two

Which Foods Cause Food Allergy and How Is Food Allergy Treated?

Montserrat Fernández-Rivas[1], Ricardo Asero[2]

[1]*Allergy Department, Hospital Clínico San Carlos, IdISSC, Madrid, Spain*
[2]*Ambulatorio di Allergologia, Clinica San Carlo, Paderno-Dugnano, Milano, Italy*

CHAPTER OUTLINE

Risk Management for Food Allergy. http://dx.doi.org/10.1016/B978-0-12-381988-8.00002-6

INTRODUCTION

Food allergy is the adverse reaction to foods where an immune mechanism is involved. Depending on the type of altered immunological response, food allergies can be divided into IgE mediated and non-IgE mediated [1]. IgE mediated food allergies are the most frequent and best known. They are characterized by quick onset reactions after food intake, generally appearing in the first hour, and called "immediate" in the medical literature. Symptoms can range from mild to severe, and in some cases lead to anaphylaxis, a severe and potentially life-threatening reaction. Non-IgE mediated reactions to food ingestion mostly induce gastrointestinal symptoms, generally appearing hours or days after the ingestion, and sometimes only after regular intake of the food. Quick onset life-threatening reactions is not seen. Although strict food avoidance is necessary in both IgE and non-IgE mediated reactions, the inherent risk of an accidental exposure is higher in the former, because of the potential risk of anaphylaxis. In this chapter we will address the foods and management of IgE mediated reactions to foods, and we will refer to them hereafter by the term "food allergy." In chapter 7 of this book, celiac disease, the most common and best known non-IgE mediated food allergy, is reviewed.

FOODS INVOLVED IN ALLERGIC REACTIONS

Any food has the potential to induce an allergic reaction, and as a matter of fact, more than 150 different foods have been implicated. However, the majority of reactions are induced by a small number of food items [2]. In 1995, a FAO technical consultation [3] identified the following eight food groups as the most common causes of allergy worldwide: milk, egg, peanut, tree nuts, wheat, soy, fish, and shellfish. These foods have been known since then as the 'big eight allergens' and are recognized as allergenic foods of public health importance and therefore included in regulatory allergen lists worldwide (a topic further developed in chapter 16). However, epidemiological studies have shown important age and geographical differences in the prevalence of allergy to individual foods, and thus different regions have different 'top ten' lists of allergenic foods.

When considering the most important foods involved in allergic reactions, it is important to bear in mind the limitations of the studies available. Firstly, more than 80% of the epidemiological studies on food allergy have been performed in Europe, the United States (US), Canada, Australia, and New Zealand. A few surveys have been carried out in Asia, and there is very little information about food allergy in Africa and Latin America. Secondly, recent meta-analyses and large scale reviews of food allergy epidemiology [4,5] have raised major issues in the comparability of studies due to marked heterogeneity in design, instruments applied, and outcome measurements. Depending on the definition of food allergy the overall prevalence changes, as well as the prevalence of individual food allergies. The highest estimates are found in those studies of self-reported reactions; they are lower when IgE sensitization is evaluated, further reduced when IgE sensitization is combined with consistent reported symptoms, and lowest in those studies where food allergy is confirmed by oral food challenges [4,5]. Thirdly, when IgE testing, by means of either skin prick tests or serum IgE determinations, is included in the definition of food allergy, the sensitivity and specificity of these tests have an impact on the estimation of the prevalence. It is well established that the sensitivity of skin and serology tests used to assess the presence of IgE varies with the food allergens involved, the age of the patients, the presence of associated atopic diseases, and between geographical areas [6—9]. When stable allergens are involved, the sensitivity of IgE tests is higher, compared to, for instance, labile allergens in fresh plant foods which are frequently altered with the extraction processes and may account for a great number of false negative results and hence and underestimation of the prevalence of that food allergy. Additionally, cross-reactivity is a common phenomenon among foods, and between foods and aeroallergens, and it can produce false positive results, resulting in an overestimation of the prevalence.

Milk

Milk allergy starts in the first year of life, and it is one of the most prevalent food allergies in children below 4 years of age in Western developed countries and Japan, where cow's milk (formula) is an essential food in the infants and children's diet [4,10—14]. The frequency of true milk allergy, confirmed by oral challenges in five European birth cohorts, ranges from 1.9% to 4.9% [11]. However, as aforementioned the perception is far more frequent and reaches 17% in some studies [4]. Milk allergy is frequently out-grown, although the rate of tolerance development varies among studies: in some European cohorts more than 80% of children develop tolerance by the age of 8 years [15—17], while in an American series tolerance is achieved by 79% at 16 years of age [18]. But even in the worst scenario, IgE mediated milk allergy is very uncommon in adults. Despite this, adults commonly report adverse reactions to milk. In a large European survey of more than 40,000 telephone contacts, 5 million European respondents claimed to be milk-allergic, with adult women forming the group making most of these claims [19]. Bearing in mind the low frequency of milk-specific IgE found in adult subjects [20], this may probably reflect lactose intolerance, which is far more prevalent in adults.

Egg

Together with milk, hen's egg allergy is one of the most common food allergies in infants and young children. It usually presents around 1 year of age, reflecting the typical age of introduction of egg into the child's diet. As for milk, tolerance to egg is achieved spontaneously, with resolution in 50% by age 3 years and in 66% by age 5 years [21]. However, a more recent study found a slower resolution: 12% by age 6 years, 37% by age 10 years, and 68% by age 16 years [22]. Heated egg is tolerated earlier than raw egg

[23, 24], and the introduction of heated egg into the diet seems to favor the development of tolerance to raw egg [25].

The prevalence of egg allergy confirmed by oral challenges with cooked egg in children 1 to 3 years of age varies between 1.3% and 2.5% [26–28]. In a recent study performed in 1-year-old infants in Australia [14], the prevalence of egg allergy confirmed by a challenge with raw egg was 8.9%. Of these patients, 80.3% tolerated baked egg, resulting in a prevalence of heated egg allergy of 1.7%, similar to the aforementioned studies. In the US, a prevalence of 1.8% in children 1–5 years of age was estimated using the 95% predictive value of serum-specific IgE to egg (7 kU/L) [13].

Peanut

Allergy to peanuts is common, and likely increasing in prevalence. It starts early in life, being most commonly diagnosed between 6 to 24 months of age, and it is more persistent than milk or egg allergies, with only 20% of patients developing tolerance [9,29,30].

Worldwide, the prevalence of peanut allergy is quite variable, with the highest rates (1–3%) in the US [13], Canada [31], Australia [14], and the United Kingdom (UK) [32–34]. The rates are lower (0.2–0.7%) in other westernized countries such as France [35,36] or Denmark [27], and even lower (<0.2%) in Israel [34]. Interestingly, Jewish children in the UK have a 10-fold higher prevalence (1.85%) than Jewish children in Israel [34]. Peanut allergy is much less common in Asia [12,28,37]. In Japan, peanut induces only 2% of all food allergic reactions [12]. In a survey carried out in schoolchildren in Singapore and the Philippines, the prevalence of peanut allergy among children of Asian origin was 0.43% to 0.67%, whereas in (Western born) expatriate children it was 1.2% [37].

Tree Nuts

The prevalence of tree nut allergy in the US and Canada using a random telephone survey was, respectively, 0.5% and 1% in adults and 1.1% and 1.6% in children [9,10,31,38]. In a study performed in the US population, only around 10% develop tolerance [39].

In a school survey in France, the point prevalence of tree nut allergy was 0.7% [36], and in UK children and teenagers it ranged from 1.2% to 2.2% [9,32,33]. In two studies performed in Germany that included oral challenges, the prevalence of allergy to walnut was 0.8% and to hazelnut varied from 0.7% to 4.3% [40,41]. Tree nut allergy is frequently observed in Europe in pollen allergic patients, especially in those sensitized to birch pollen, and is frequently associated with allergy to fresh fruits and vegetables [39–42]. The overall prevalence of hazelnut and Brazil nut allergy confirmed by oral challenges in an unselected population of Danish adults was 4.6% and 1.7%, respectively. Among those with a pollen allergy the prevalence increased to 19.2% for hazelnut and 7% for Brazil nut [27].

Similar to that mentioned for peanut, the prevalence of tree nut allergy in Asian children from Singapore and the Philippines was 0.3%, whereas in expatriate children it was 1.2% [37].

Wheat

The prevalence of allergy to wheat in children and adolescents according to European studies that included oral challenges was 0—0.5% [5,27,40]. In the US it has been found to be 0.5% in 1-year-old children [10]. In adults, specific IgE to wheat was found more frequently (3.6%) than reported reactions (<1%) [5], probably reflecting cross-reactive IgE antibodies with grass pollen without clinically relevant food allergy.

In Japan, wheat is the food inducing the third highest number of allergic reactions after egg and milk, in children and young adults [12]. The prevalence of wheat allergy in Japanese adults was found to be 0.21% [43]. It is also of note that buckwheat (a non-cereal grain used in soba noodles) accounted for 6% of all food allergic reactions in Japan, although in patients older than 7 years the frequency was double [12].

Wheat allergy is frequently out-grown. In a US study of children with IgE mediated wheat allergy, rates of resolution were 29% by 4 years, 56% by 8 years, and 65% by 12 years [44].

Soy

The prevalence of allergy to soy in two European studies that included oral challenges was 0% and 0.7% [27,40]. In the US it has been found to be 1.4% in 1-year-old children [10]. Sensitization is uncommon in the European population (≤0.2%), with the exception of two Swedish studies, in which serum IgE antibodies to soy were found in up to 2.9% of subjects. This finding might reflect cross-reactive IgE with birch pollen without clinically relevant soybean allergy, since reported reactivity to soy in Sweden was 0.3—1.3% [5].

Soy allergy is frequently out-grown. In a US study, the resolution rates were 25% by age 4 years, 45% by age 6 years, and 69% by age 10 years [45].

Sesame

The prevalence of sesame allergy has been investigated in Israel, the UK, Canada, Australia and the US [14,31,34,38,46,47]. The estimates range from 0% to 0.8%, with the highest found in 1-year-old infants from Australia (0.80%) [14], and in Jewish children in the UK (0.79%) [34]. In Israel, sesame allergy is the third most frequent food allergy in children (after egg and milk), with prevalence estimates of 0.1—0.2%, lower than in westernized countries [46,47]. It starts early in life, around 1 year of age, and the spontaneous loss is comparatively lower than those of milk and egg allergies, although there is only one longitudinal study with a short follow-up [47].

Fish

The prevalence of fish allergy in studies carried out in Europe, the US, and Canada varied between 0.2% to 0.6% [4,10,14,27,40]. In children from Spain, fish allergy starts in the second year of life and is the third most prevalent food allergy after egg and milk [48]. A recent population-based study in southeast Asia in children 14–16 years old has shown a prevalence of fish allergy of 2.29% in the Philippines, whereas in Singapore and Thailand it is 10 times lower [49].

Although there are no longitudinal studies assessing the loss of reactivity, fish allergy is considered to be a life-long food allergy.

Shellfish

Shellfish include crustacean and mollusks. Crustaceans are more widely consumed and induce more frequent allergic reactions: six times more than mollusks in a Spanish nationwide survey [50]. Shellfish allergy is more frequently found in adult patients than in children [4,10,12,13,50], and it is considered to be life-long [51]. The prevalence is lower in westernized countries than in southeast Asia or Colombia [4,10,37,52–56]. Interestingly, in the Spanish Canary Islands, an area with a subtropical climate and high seafood consumption, shellfish allergy is the most prevalent food allergy in adults [57].

The overall prevalence of shellfish allergy in studies performed in Europe and the US that combined symptoms and IgE sensitization was 0.6% [4]. The prevalence of challenge-confirmed shrimp allergy in Danish adults was 0.3% [27]. In random telephone surveys performed in the US and Canada, the prevalence in adults was 2.5% and 1.7%, respectively, whereas in Canadian children the estimate was 0.5% [31,56]. In a US national survey, the overall prevalence of probable shrimp allergy (symptoms and IgE sensitization) was 1%, but it was not found in children below 6 years of age [13].

In Asian countries, the overall prevalence of shellfish allergy is higher, and it is the most important food allergy in school-age children, adolescents, and adults [12,37,52–54]. In Singapore and the Philippines, the prevalence of shellfish allergy in school children of Asian origin was 1.2% in the 4–6-year-olds and increased to 5.2% in those 14–16 years old. In contrast, in the Western-born expatriates the prevalence for the same age groups was 0.55% and 0.96%, respectively [37].

Fruits and Vegetables

Allergy to fruits and vegetables has been mainly investigated in Europe, with very little information available from other areas. In Europe, the prevalence of fruit allergy confirmed by oral food challenges varied from 0.1% to 4.3%, and the fruits most commonly involved are those belonging to the *Rosaceae* family (apple, peach, cherry, etc.) [5,27,40,41,50,58–61]. Allergy to vegetables is less frequent, with a prevalence in the general population from 0.1% to 1.8%

[5,27]. The vegetables most commonly involved in allergic reactions are those from the *Apiaceae* family (celery, carrot) and tomato [5,27].

One of the main features of allergy to fruits and vegetables is the frequent association with pollen allergies (pollen-food syndrome). The primary sensitization is induced by pollen exposure, and the plant food allergy appears later as a result of cross-reactive IgE to allergens found in pollens and foods (i.e., Bet v 1 homologues, profilins). Allergies to tree nuts and peanut can also be linked to pollen allergy. For this reason, plant food allergies are more frequently found in pollen allergic patients [42]. In a study performed in Denmark, the prevalence of plant food allergies in (birch) pollen allergic patients was 19.2% for hazelnut, 16.7% for apple, 13.3% for kiwi, 7.6% for celery, and 5% for tomato [27].

Primary plant food allergies have been mainly described in Italy and Spain, and are linked to lipid transfer proteins (LTP), allergens found in plant foods and also in pollens. In this so-called LTP syndrome, the foods most frequently involved are *Rosaceae* fruits (mainly peach), tree nuts (mainly walnut and hazelnut), vegetables (mainly tomato and lettuce), and a long list of other plant foods including grape, kiwi, citrus fruits, cereals (especially corn and wheat), sunflower seed, peanut, etc. [60−62]. The prevalence in the general population has not yet been established.

Recently, within the EuroPrevall project, two epidemiological surveys were carried out in school children and adults from the general population and, additionally, more than 2,000 patients were evaluated in 12 allergy clinics across Europe (Athens, Greece; Lodz, Poland; Madrid, Spain; Manchester, UK; Milan, Italy; Prague, Czech Republic; Reykjavik, Iceland; Sofia, Bulgaria; Strasbourg, France; Utrecht, The Netherlands; Vilnius, Lithuania; Zürich, Switzerland) [63,64]. In the cross-sectional study in allergy clinics, the foods most frequently involved in children below 4 years of age were milk, egg, fish, and peanut. In children between 4 and 14 years of age, milk and egg allergies decreased, and there was an increase in peanut and tree nuts (especially hazelnut and walnut) and in apple allergic reactions. In patients over 14 years of age, the most common food allergies were of plant origin, mainly tree nuts, fruits (especially apple, peach, and kiwi), peanuts, and vegetables (carrot, celery, and tomato). Shrimp and fish allergy accounted for less than 7% and 4%, respectively, of all allergic reactions recorded in adult patients. Geographical differences were also observed within Europe. Hazelnut, apple, and celery allergies were more frequently found in central and northern Europe; shrimp and fish allergies in Madrid, Athens, and Reykjavik; peach and melon in Madrid and Milan (unpublished data) (Figures 2.1, 2.2).

SYMPTOMS AND SEVERITY OF FOOD ALLERGY

Food allergies may present clinically with an array of symptoms that can occur either in an isolated form or differently associated. Symptoms may involve different organ systems like the skin, the gastrointestinal tract, and the airways and may lead to anaphylaxis, the most dangerous of all allergic

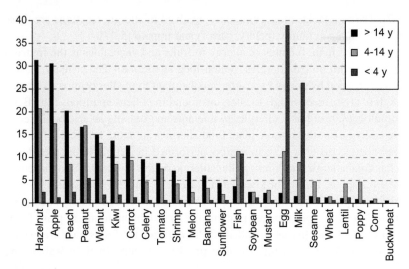

FIGURE 2.1 Foods involved in allergic reactions in different age groups across Europe. Results coming from the analysis of the cross-sectional study in allergy clinics of EuroPrevall (n=1671 patients). Food allergy was defined as a reported reaction within 2 hours of ingestion of the culprit food together with specific IgE to that food (either by skin prick test or serum IgE determination). Frequency is given as percentage of patients within the age group.

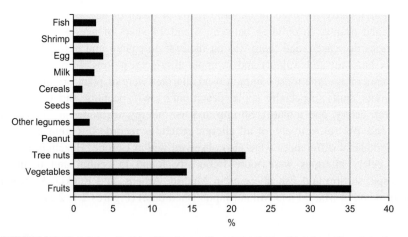

FIGURE 2.2 Foods involved in allergic reactions in patients older than 14 years of age across Europe. Results coming from the analysis of the cross-sectional study in allergy clinics of EuroPrevall (n=3108 allergic reactions). Food allergy was defined as a reported reaction within 2 hours of ingestion of the culprit food together with specific IgE to that food (either by skin prick test or serum IgE determination). Frequency is given as percentage of the total number of allergic reactions.

reactions, due to severe respiratory and/or cardiovascular involvement. Based on the pathogenesis, food allergies are divided into class 1 and class 2. The former are associated with sensitization to pepsin-stable allergens, which is generally believed to occur in the gastrointestinal tract; this type of food allergy may present with a spectrum of symptoms ranging from mild local reactions to very severe, potentially fatal, ones. Class 2 food allergies develop as a consequence of primary sensitization to seasonal airborne allergens due to the cross-reactivity between pollen proteins and homologous plant food allergens (pollen-food syndrome); the severity of this type of food allergy is in most cases limited, because of the heat- and pepsin-sensitivity of the causative allergens [65].

Symptoms of Food Allergy

Oral Allergy Syndrome (OAS)

OAS is a sort of contact urticaria that occurs within minutes and presents as itching, with or without angioedema (swelling), of lips, tongue, palate, throat, and/or ears and nose [66,67]. In most cases it is a mild and self-limited condition that disappears within 30–60 minutes, although in some cases the clinical course is more severe with potentially severe pharyngeal swelling or progression toward a generalized anaphylactic reaction [68]. OAS can be elicited by any food allergen, but it is the typical clinical presentation of the pollen-food syndrome. Due to the high prevalence of pollen allergy in adult patients, and its frequent association with plant food allergies, OAS is the most frequent symptom of food allergy in adults [27,40–42].

Gastrointestinal Symptoms

Food allergy may cause nausea, vomiting, gastric retention, intestinal hypermotility, abdominal pain due to colonic spasms, and diarrhea [69]. Symptoms usually develop within minutes to two hours after the ingestion of the offending food. Gastrointestinal symptoms are generally associated with sensitization to pepsin-stable allergen proteins; as a consequence they are generally associated with allergic symptoms in other target organs (skin, nose, lungs, and eyes).

Skin Symptoms

The skin is probably the most frequent target organ in IgE mediated food allergy [69]. Cutaneous symptoms may include pruritus, erythema, urticaria, and angioedema. Urticaria may follow the ingestion of the offending food by minutes, and may last for several hours. The offending food is often clear-cut. Some patients report contact urticaria after handling foods; in some instances they have reactions after ingestion as well; contact urticaria is particularly frequent in patients who are hypersensitive to peach lipid transfer protein [70], although raw meats, fish, and other fruits and vegetables may also induce such reactions [71].

Respiratory Symptoms

Respiratory symptoms (rhino-conjunctivitis and asthma) may occur in food allergic patients following the ingestion of the offending foods, but are rarely present in an isolated form [69]. Respiratory symptoms are not uncommon in food allergic patients following inhalation of food dusts or vapors.

Anaphylaxis

Anaphylaxis represents the most severe allergic reaction and is always a medical emergency. It is caused by a sudden and massive release of mediators from mast cells and/or basophils throughout the body and is characterized by a variable association of gastrointestinal (nausea, vomiting, abdominal cramps), skin (urticaria, erythema, itching, angioedema), and respiratory (asthma, rhinitis, dyspnea due to edema of the glottis) symptoms, in some cases associated with cardiovascular symptoms (hypotension, palpitations, collapse, and dysrhythmia). Anaphylaxis may start minutes or even seconds after contact with the offending food, and very small amounts may be sufficient to induce a fatal or near-fatal reaction [72—75]. In up to 20% of cases anaphylactic reactions are biphasic, with the appearance of a recurrent reaction 4 to 12 hours after the first one [76].

Factors Influencing the Occurrence and/or Severity of Symptoms

Several factors may influence the clinical presentation as well as the severity of a food allergy.

a) The physicochemical characteristics of the allergen. Allergen proteins that are able to resist thermal or other types of processing and, most importantly, pepsin digestion, may reach the gastrointestinal tract in an unmodified form and are therefore able to induce systemic reactions upon absorption [77—79].

b) The way the offending food is eaten. The fact that the offending food is ingested alone after fasting or associated with other foods (matrix effects, fat content) that may delay its absorption certainly plays a role in the occurrence and severity of the allergic reaction [78—81].

c) Dose (amount of food protein ingested).

d) Severe or uncontrolled asthma has been associated with severe anaphylaxis, especially in adolescents and adults [72,73,82].

e) The presence of co-factors such as aspirin and other non-steroidal anti-inflammatory drugs (NSAIDs), alcoholic beverages, use of antacids, and exercise has been associated with the onset and/or increased severity of food allergic reactions. Increased intestinal permeability may be the basis, but the mechanisms involved are still unclear [74,82]. The best characterized is wheat-dependent exercise-induced anaphylaxis in patients sensitized to omega-5-gliadin [83,84]. NSAIDs seem to enhance allergic reactions to plant foods in patients sensitized to LTPs [62,85].

f) The level of food-specific IgE to the food in question.

g) Other still poorly defined 'host' factors.

MANAGEMENT

At present the cornerstones of food allergy management are avoidance of the culprit food, and rescue medication to treat an eventual reaction. Obviously this does not cure food allergy. Different immunotherapeutic approaches aiming to develop a curative treatment for food allergy are currently under investigation.

Elimination Diet

Strict avoidance of the offending food is currently the only proven therapy for food allergy. To indicate an adequate diet, a correct diagnosis with an accurate identification of the culprit food(s) is absolutely essential [6,8,9,68]. Patients may also react to foods that share homologous proteins due to IgE cross-reactivity, and therefore, if the tolerance of closely related cross-reactive foods has not been assessed after a confirmed diagnosis to a member of the family, the patient should be instructed on the possible danger related to cross-reactivities [86].

Elimination diets, especially if a large number of foods are involved, may lead to eating disorders and malnutrition, and thus supervision by a dietician may be necessary, particularly in growing children [6].

In order to avoid accidental exposure to a hidden allergen in a processed food, patients (or their parents) should check the labels and should be aware of the different names the same food can be given in ingredients lists. Therefore, a correct avoidance diet needs constant vigilance, is a source of stress, and has a negative impact on the quality of life of food allergic patients and their families (discussed in chapter 1). Eating out of the home increases the risk of accidental exposure to allergens. Indeed, many of the most severe food allergic reactions occur outside the home. Meals at school, restaurants, or friends' and relatives' homes may be dangerous, and patients and their families restrict their social activities. For the safety of school children, it is essential that school staff are informed about the child's allergy and are able to recognize and treat allergic reactions [87–90].

Associations of food allergic patients provide extensive information and social support and are of great help. They provide advice to allergic individuals in different situations such as traveling or sending an allergic child to school. They may have information about food alerts (foods found to contain unlabelled allergens) or lists of currently 'safe' foods. The support of such groups helps allergic individuals to handle their food allergy safely and confidently, with less impact on their daily activities [87].

For infants and young children allergic to cow's milk, hypoallergenic formulas have been developed as a substitute to cow's milk until tolerance is developed. Extensively hydrolyzed casein or whey formula are preferred, and they are generally well tolerated by most allergic children. For those who do not tolerate them, an amino acid—based formula may be an adequate alternative [91—93]. These hypoallergenic milk substitutes are the only hypoallergenic foods currently available.

Rescue Medication

Given the difficulty of avoiding food allergens, patients are at risk of experiencing allergic reactions due to accidental food ingestion. All the patients (and their parents, school staff, or caregivers) should be trained in the early recognition and early treatment of reactions and given rescue medication.

A comprehensive management of food allergy should include adequate (written) indications to avoid the culprit food(s), and a written emergency action plan for accidental reactions that comprise a description of potential anaphylaxis symptoms, when to use self-injectable adrenaline, how to administer it intramuscularly in the lateral thigh, additional medications, and the recommendation to take the patient to a medical emergency room. The management should be personalized and reviewed with the patient on a regular basis. Within the emergency plan, patients should be given additional therapy, such as oral antihistamines and corticosteroids, and short acting bronchodilators, that have to be taken after the adrenaline injection, or as the only rescue medication when adrenaline is not indicated [6,8,9,94—96].

All the patients with a systemic reaction, even if they have responded favorably to the rescue medication, should be referred to an emergency medical facility where other medical therapies will be applied if needed, and where they must remain under close medical observation for possible biphasic reactions [6,8,9,94—96].

Immunotherapy

Specific allergen immunotherapy (IT) has proven to be an effective treatment in respiratory and insect venom allergies, and can be a therapeutic option in food allergy as well. Subcutaneous (SCIT), sublingual (SLIT), and oral (OIT) routes of administration have been investigated in clinical trials in patients with severe and persistent food allergies, who are at risk of health threatening reactions. The epicutaneous administration of food allergens on the intact skin as an alternative method for IT is currently under investigation [97,98].

In two studies of SCIT with peanut, an increase in peanut tolerance was shown, but systemic reactions were very frequently observed (13—39% of doses) [99,100]. The risk/benefit ratio was considered unacceptable. However, the administration. of recombinant food hypoallergens subcutaneously may have a better safety profile, and is currently under investigation [101].

OIT (or specific tolerance induction) for food allergy has been the subject of much research in recent years, especially in severe milk, egg, and peanut allergies. In OIT the allergen dose is immediately swallowed, and progressively increasing doses are given usually up to a normal serving dose. The results of studies show positive trends for short-term tolerance. After completing the trial, more than 50% of patients (up to 100% in some studies) are able to tolerate a normal food serving. Those who do not tolerate a food serving have in most cases an increased threshold that protects them for accidental exposure of hidden amounts of the food. However, when the daily intake of the food is stopped, tolerance is lost in more than 60% of patients. Systemic allergic reactions are observed in the majority of patients during the course of OIT, although some patients are able to complete the whole course of therapy without any [102−115].

Randomized controlled trials of SLIT have investigated its therapeutic potential in severe allergies to hazelnut, peach, peanuts, and milk. In SLIT the allergen is held under the tongue for 1 or 2 minutes before being swallowed. The safety profile was found to be better than in SCIT and OIT, with a low rate of systemic reactions (<3% of doses) all of mild intensity, and a variable rate of mild oropharyngeal local reactions (from 7% in hazelnut SLIT to 40% in peach SLIT patients). Although an increase in the food threshold was observed in the four SLIT clinical trials, together with some immunomodulatory effects, the efficacy seems to be lower than that of OIT [116−119].

Patients with pollen-food allergy syndrome are often reactive to a wide variety of plant foods, and extensive elimination diets are not uncommon. The issue in this clinical situation is less the health risk of an accidental exposure and more the impact on quality of life and the possible nutritional deficiencies if avoidance of a large number of (fresh) plant foods is needed. Therefore, there is an indication for IT. A number of studies have investigated the effect of birch pollen SCIT and SLIT on the associated food allergies, mainly apple and hazelnut [120−126]. Although some studies did not find a beneficial effect, others have shown a reduced clinical reactivity. The magnitude of the effect on apple tolerance differed among the studies and appeared to be transient. Altogether, the data available in the literature are controversial in regard to the beneficial effect of IT with birch pollen on the associated plant food allergies, and cannot be recommended currently as a therapeutic option. Well-conducted, double-blind, placebo-controlled clinical trials with food challenges to monitor efficacy are needed to establish the effect of pollen SCIT on related food allergies.

In summary, although further studies are needed to refine the safety and efficacy profile of food IT, current experience suggests that it is a promising therapy that can modify the patient's reactivity to the food, and that will therefore have a great impact in the current risk and risk management of food allergy.

REFERENCES

[1] Johansson SG, Bieber T, Dahl R, Friedmann PS, Lanier BQ, Lockey RF, et al. Revised nomenclature for allergy for global use: Report of the Nomenclature Review Committee of the World Allergy Organization, October 2003. J Allergy Clin Immunol 2004; 113:832−6.

[2] Hefle SL, Nordlee JA, Taylor SL. Allergenic foods. Crit Rev Food Sci Nutr 1996;36: S69−89.

[3] Food and Agriculture Organization of the United Nations. Rome, Italy: Report of the FAO Technical Committee on Food Allergies; 1995. November 13—14.

[4] Rona RJ, Keil T, Summers C, Gislason D, Zuidmeer L, Sodergren E, et al. The prevalence of food allergy: a meta-analysis. J Allergy Clin Immunol 2007;120:638—46.

[5] Zuidmeer L, Goldhahn K, Rona RJ, Gislason D, Madsen C, Summers C, et al. The prevalence of plant food allergies: a systematic review. J Allergy Clin Immunol 2008;121:1210—8.

[6] Fernández-Rivas M, Ballmer-Weber B. Food allergy: diagnosis and management. In: Mills ENC, Wichers H, Hoffmann-Sommergruber K, editors. Managing allergens in food. Cambridge, UK: Woodhead Publishing Limited; 2007. p. 3—28.

[7] Bernstein IL, Li JT, Bernstein DI, Hamilton R, Spector SL, Tan R, et al. Allergy diagnostic testing: an updated practice parameter. Ann Allergy Asthma Immunol 2008; 100(3 Suppl 3):S1—148.

[8] Chafen JJ, Newberry SJ, Riedl MA, Bravata DM, Maglione M, Suttorp MJ, et al. Diagnosing and managing common food allergies: a systematic review. JAMA 2010;303:1848—56.

[9] NIAID-Sponsored Expert Panel, Boyce JA, Assa'ad A, Burks AW, Jones SM, Sampson HA, et al. Guidelines for the diagnosis and management of food allergy in the United States: report of the NIAID-sponsored expert panel. J Allergy Clin Immunol 2010;126(6 Suppl):S1—58.

[10] Sicherer SH. Epidemiology of food allergy. J Allergy Clin Immunol 2011;127: 594—602.

[11] Fiocchi A, Brozek J, Schünemann J, Bahna SL, von Berg A, Beyer K, et al. World Allergy Organization (WAO) Diagnosis and Rationale for Action against Cow's Milk Allergy (DRACMA) Guidelines. World Allergy Organ J 2010;3:57—161.

[12] Urisu A, Ebisawa M, Mukoyama T, Morikawa A, Kondo N. Japanese Guideline for Food Allergy. Allergol Int 2011;60:221—36.

[13] Liu AH, Jaramillo R, Sicherer SH, Wood RA, Bock SA, Burks AW, et al. National prevalence and risk factors for food allergy and relationship to asthma: results from the National Health and Nutrition Examination Survey 2005—2006. J Allergy Clin Immunol 2010;126:798—806.

[14] Osborne NJ, Koplin JJ, Martin PE, Gurrin LC, Lowe AJ, Matheson MC, et al. Prevalence of challenge-proven IgE-mediated food allergy using population-based sampling and predetermined challenge criteria in infants. J Allergy Clin Immunol 2011; 127:668—76.

[15] Høst A, Halken S, Jacobsen HP, Christensen AE, Herskind AM, Plesner K. Clinical course of cow's milk protein allergy/intolerance and atopic diseases in childhood. Pediatr Allergy Immunol 2002;13(Suppl. 15):23—8.

[16] Saarinen KM, Pelkonen AS, Mäkelä MJ, Savilahti E. Clinical course and prognosis of cow's milk allergy are dependent on milk-specific IgE status. J Allergy Clin Immunol 2005;116:869—75.

[17] Martorell A, García Ara MC, Plaza AM, Boné J, Nevot S, Echeverria L, et al. The predictive value of specific immunoglobulin E levels in serum for the outcome of the development of tolerance in cow's milk allergy. Allergol Immunopathol (Madr) 2008;36:325—30.

[18] Skripak JM, Matsui EC, Mudd K, Wood RA. The natural history of IgE-mediated cow's milk allergy. J Allergy Clin Immunol 2007;120:1172—7.

[19] Steinke M, Fiocchi A, Kirchlechner V, Ballmer-Weber B, Brockow K, Hischenhuber C, et al. REDALL study consortium. Perceived food allergy in children in 10 European nations. A randomised telephone survey. Int Arch Allergy Immunol 2007;143:290—5.

[20] Burney P, Summers C, Chinn S, Hooper R, van Ree R, Lidholm J. Prevalence and distribution of sensitization to foods in the European Community Respiratory Health Survey: a EuroPrevall analysis. Allergy 2010;65:1182—8.

[21] Boyano-Martinez T, Garcia-Ara C, Diaz-Pena JM, Martín-Esteban M. Prediction of tolerance on the basis of quantification of egg white-specific IgE antibodies in children with egg allergy. J Allergy Clin Immunol 2002;110:304—9.

[22] Savage JH, Matsui EC, Skripak JM, Wood RA. The natural history of egg allergy. J Allergy Clin Immunol 2007;120:1413−7.

[23] Eigenmann PA. Anaphylactic reactions to raw eggs after negative challenges with cooked eggs. J Allergy Clin Immunol 2000;105:587−8.

[24] Des RA, Nguyen M, Paradis L, Primeau MN, Singer S. Tolerance to cooked egg in an egg allergic population. Allergy 2006;61:900−1.

[25] Lemon-Mulé H, Sampson HA, Sicherer SH, Shreffler WG, Noone S, Nowak-Wegrzyn A. Immunologic changes in children with egg allergy ingesting extensively heated egg. J Allergy Clin Immunol 2008;122:977−83.

[26] Eggesbø M, Botten G, Halvorsen R, Magnus P. The prevalence of allergy to egg: a population-based study in young children. Allergy 2001;56:403−11.

[27] Osterballe M, Hansen TK, Mortz CG, Host A, Bindslev-Jensen C. The prevalence of food hypersensitivity in an unselected population of children and adults. Pediatr Allergy Immunol 2005;16:567−73.

[28] Chen J, Hu Y, Allen KJ, Ho MHK, Li H. The prevalence of food allergy in infants in Chongqing, China. Pediatr Allergy Immunol 2011;22:356−60.

[29] Hourihane JO, Roberts SA, Warner JO. Resolution of peanut allergy: case control study. BMJ 1998;316:1271−5.

[30] Skolnick HS, Conover-Walker MK, Barnes Koerner C, Sampson HA, Burks W, Wood RA. The natural history of peanut allergy. J Allergy Clin Immunol 2001;107:367−74.

[31] Ben-Shoshan M, Harrington DW, Soller L, Fragapane J, Joseph L, St Pierre Y, et al. A population-based study on peanut, tree nut, fish, shellfish, and sesame allergy prevalence in Canada. J Allergy Clin Immunol 2010;125:1327−35.

[32] Pereira B, Venter C, Grundy J, Clayton CB, Arshad SH, Dean T. Prevalence of sensitization to food allergens, reported adverse reaction to foods, food avoidance, and food hypersensitivity among teenagers. J Allergy Clin Immunol 2005;116:884−92.

[33] Venter C, Pereira B, Voigt K, Grundy J, Clayton CB, Higgins B, et al. Prevalence and cumulative incidence of food hypersensitivity in the first 3 years of life. Allergy 2008; 63:354−9.

[34] Du Toit G, Katz Y, Sasieni P, Mesher D, Maleki SJ, Fisher HR, et al. Early consumption of peanuts in infancy is associated with a low prevalence of peanut allergy. J Allergy Clin Immunol 2008;122:984−91.

[35] Morisset M, Moneret-Vautrin DA, Kanny G, Allergo-Vigilance Network. Prevalence of peanut sensitization in a population of 4,737 subjects − an Allergo-Vigilance Network enquiry carried out in 2002. Eur Ann Allergy Clin Immunol 2005;37:54−7.

[36] Rancé F, Grandmottet X, Grandjean H. Prevalence and main characteristics of school-children diagnosed with food allergies in France. Clin Exp Allergy 2005;35:167−72.

[37] Shek LP, Cabrera-Morales EA, Soh SE, Gerez I, Zhing P, Yi FC, et al. A population-based questionnaire survey on the prevalence of peanut, tree nut, and shell-fish allergy in 2 Asian populations. J Allergy Clin Immunol 2010;126:324−31.

[38] Sicherer SH, Munoz-Furlong A, Godbold JH, Sampson HA. US prevalence of self-reported peanut, tree nut and sesame allergy: 11-year follow-up. J Allergy Clin Immunol 2010;125:1322−6.

[39] Fleischer DM, Conover-Walker MK, Matsui EC, Wood RA. The natural history of tree nut allergy. J Allergy Clin Immunol 2005;116:1087−93.

[40] Roehr CC, Edenharter G, Reimann S, Ehlersz I, Worm M, Zuberbier T, et al. Food allergy and non-allergic food hypersensitivity in children and adolescents. Clin Exp Allergy 2004;34:1534−41.

[41] Zuberbier T, Edenharter G, Worm M, Ehlers I, Reimann S, Hantke T, et al. Prevalence of adverse reactions to food in Germany - a population study. Allergy 2004;59: 338−45.

[42] Osterballe M, Hansen TK, Mortz CG, Bindslev-Jensen C. The clinical relevance of sensitization to pollen-related fruits and vegetables in unselected pollen-sensitized adults. Allergy 2005;60:218−25.

[43] Morita E, Chinuki Y, Takahashi H, Nabika T, Yamasaki M, Shiwaku K. Prevalence of wheat allergy in Japanese adults. Allergol Int 2012;61:101−5.

[44] Keet CA, Matsui EC, Dhillon G, Lenehan P, Paterakis M, Wood RA. The natural history of wheat allergy. Ann Allergy Asthma Immunol 2009;102:410−5.

[45] Savage JH, Kaeding AJ, Matsui EC, Wood RA. The natural history of soy allergy. J Allergy Clin Immunol 2010;125:683−6.

[46] Dalal I, Binson I, Reifen R, Amitai Z, Shohat T, Rahmani S, et al. Food allergy is a matter of geography after all: sesame as a major cause of severe IgE-mediated food allergic reactions among infants and young children in Israel. Allergy 2002;57:362−5.

[47] Aaronov D, Tasher D, Levine A, Somekh E, Serour F, Dalal I. Natural history of food allergy in infants and children in Israel. Ann Allergy Asthma Immunol 2008;101: 637−40.

[48] Crespo JF, Pascual C, Burks AW, Helm RM, Esteban MM. Frequency of food allergy in a pediatric population from Spain. Pediatr Allergy Immunol 1995;6:39−43.

[49] Connett GJ, Gerez I, Cabrera-Morales EA, Yuenyongviwat A, Ngamphaiboon J, Chatchatee P, et al. A population-based study of fish allergy in the Philippines, Singapore and Thailand. Int Arch Allergy Immunol 2012;159:384−90.

[50] Fernández Rivas M. Food allergy in Alergológica 2005. J Investig Allergol Clin Immunol 2009;19(Suppl. 2):37−44.

[51] Daul CB, Morgan JE, Lehrer SB. The natural history of shrimp hypersensitivity. J Allergy Clin Immunol 1990;86:88−93.

[52] Lee BW, Shek LP, Gerez IF, Soh SE, Van Bever HP. Food allergy-lessons from Asia. World Allergy Organ J 2008;1:129−33.

[53] Santadusit S, Atthapaisalsarudee S, Vichyanond P. Prevalence of adverse food reactions and food allergy among Thai children. J Med Assoc Thai 2005;88(Suppl. 8): S27−32.

[54] Wu TC, Tsai TC, Huang CF, Chang FY, Lin CC, Huang IF, et al. Prevalence of food allergy in Taiwan: a questionnaire-based survey. Intern Med J 2012;42:1310−5.

[55] Marrugo J, Hernández L, Villalba V. Prevalence of self-reported food allergy in Cartagena (Colombia) population. Allergol Immunopathol (Madr) 2008;36:320−4.

[56] Sicherer SH, Muñoz-Furlong A, Sampson HA. Prevalence of seafood allergy in the United States determined by a random digital telephone survey. J Allergy Clin Immunol 2004;114:159−65.

[57] Castillo R, Delgado J, Quiralte J, Blanco C, Carrillo T. Food hypersensitivity among adult patients: epidemiological and clinical aspects. Allergol Immunopathol (Madr) 1996;24:93−7.

[58] Kanny G, Moneret-Vautrin DA, Flabee J, Beaudouin E, Morisset M, Thevenin F. Population study of food allergy in France. J Allergy Clin Immunol 2001;108:133−40.

[59] Falcão H, Lunet N, Lopes C, Barros H. Food hypersensitivity in Portuguese adults. Eur J Clin Nutr 2004;58:1621−5.

[60] Asero R, Antonicelli L, Arena A, Bommarito L, Caruso B, Crivellaro M, et al. EpidemAAITO: features of food allergy in Italian adults attending allergy clinics: a multi-center study. Clin Exp Allergy 2009;39:547−55.

[61] Asero R, Antonicelli L, Arena A, Bommarito L, Caruso B, Colombo G, et al. Causes of food-induced anaphylaxis in Italian adults: a multi-center study. Int Arch Allergy Immunol 2009;150:271−7.

[62] Pascal M, Muñoz-Cano R, Reina Z, Palacín A, Vilella R, Picado C, et al. Lipid transfer protein syndrome: clinical pattern, cofactor effect and profile of molecular sensitization to plant-foods and pollens. Clin Exp Allergy 2012;42:1529−39.

[63] Mills EN, Mackie AR, Burney P, Beyer K, Frewer L, Madsen C, et al. The prevalence, cost and basis of food allergy across Europe. Allergy 2007;62:717−22.

[64] Kummeling I, Mills EN, Clausen M, Dubakiene R, Pérez CF, Fernández-Rivas M, et al. The EuroPrevall surveys on the prevalence of food allergies in children and adults: background and study methodology. Allergy 2009;64:1493−7.

[65] Breiteneder H, Ebner C. Molecular and biochemical classification of plant-derived food allergens. J Allergy Clin Immunol 2000;106:27−36.

[66] Amlot P, Kemeny DM, Zachary C, Parker P, Lessof MH. Oral allergy syndrome (OAS) symptoms of IgE-mediated hypersensitivity to foods. Clin Allergy 1987;17:33−42.

[67] Mari A, Ballmer-Weber BK, Vieths S. The oral allergy syndrome: improved diagnostic and treatment methods. Curr Opin Allergy Clin Immunol 2005;5:267−73.

[68] Sampson HA. Food allergy − accurately identifying clinical reactivity. Allergy 2005; 60(suppl. 79):19−24.

[69] Sampson HA. Food allergy. Part 1: immunopathogenesis and clinical disorders. J Allergy Clin Immunol 1999;103:717−28.

[70] Asero R. Peach-induced contact urticaria is associated with lipid transfer protein sensitization. Int Arch Allergy Immunol 2011;154:345−8.

[71] Winston G, Lewis C. Contact urticaria. Int J Dermatol 1982;21:573−8.

[72] Sampson HA, Muñoz-Furlong A, Bock SA, Schmitt C, Bass R, Chowdhury BA, et al. Symposium on the definition and management of anaphylaxis: summary report. J Allergy Clin Immunol 2005;115:584−91.

[73] Sampson HA, Mendelson L, Rosen JP. Fatal and near-fatal anaphylactic reactions to food in children and adolescents. N Engl J Med 1992;327:380−4.

[74] Hompes S, Köhli A, Nemat K, Scherer K, Lange L, Rueff F, et al. Provoking allergens and treatment of anaphylaxis in children and adolescents−data from the anaphylaxis registry of German-speaking countries. Pediatr Allergy Immunol 2011;22:568−74.

[75] Worm M, Edenharter G, Ruëff F, Scherer K, Pföhler C, Mahler V, et al. Symptom profile and risk factors of anaphylaxis in Central Europe. Allergy 2012;67:691−8.

[76] Tole JW, Lieberman P. Biphasic anaphylaxis: review of incidence, clinical predictors, and observation recommendations. Immunol Allergy Clin North Am 2007;27:309−26.

[77] Lehmann K, Schweimer K, Reese G, Randow S, Suhr M, Becker WM, et al. Structure and stability of 2S albumin-type peanut allergens: implications for the severity of peanut allergic reactions. Biochem J 2006;395:463−72.

[78] Nowak-Wegrzyn A, Fiocchi A. Rare, medium, or well done? The effect of heating and food matrix on food protein allergenicity. Curr Opin Allergy Clin Immunol 2009;9:234−7.

[79] Lepski S, Brockmeyer J. Impact of dietary factors and food processing on food allergy. Mol Nutr Food Res 2013;57:145−52.

[80] Grimshaw KE, King RM, Nordlee JA, Hefle SL, Warner JO, Hourihane JO. Presentation of allergen in different food preparations affects the nature of the allergic reaction−a case series. Clin Exp Allergy 2003;33:1581−5.

[81] Arena A. Anaphylaxis to apple: is fasting a risk factor for LTP-allergic patients? Eur Ann Allergy Clin Immunol 2010;42:155−8.

[82] Calvani M, Cardinale F, Martelli A, Muraro A, Pucci N, Savino F, et al., Italian Society of Pediatric Allergy and Immunology Anaphylaxis' Study Group. Risk factors for severe pediatric food anaphylaxis in Italy. Pediatr Allergy Immunol 2011;22:813−9.

[83] Du Toit G. Food-dependent exercise-induced anaphylaxis in childhood. Pediatr Allergy Immunol 2007;18:455−63.

[84] Aihara Y, Takahashi Y, Kotoyori T, Mitsuda T, Ito R, Aihara M, et al. Frequency of food dependent, exercise induced anaphylaxis in Japanese junior-highschool students. J Allergy Clin Immunol 2001;108:1035−9.

[85] Cardona V, Luengo O, Garriga T, Labrador-Horrillo M, Sala-Cunill A, Izquierdo A, et al. Co-factor-enhanced food allergy. Allergy 2012;67:1316−8.

[86] Sicherer S. Clinical implications of cross-reactive food allergens. J Allergy Clin Immunol 2001;108:881−90.

[87] Fernández Rivas M, Miles S. Plant food allergies. Clinical and psychosocial perspectives. In: Shewry PR, Mills ENC, editors. Plant food allergens. Oxford, UK: Blackwell Scientific Publications; 2004. p. 1−23.

[88] Kim JS, Sicherer SH. Living with food allergy: allergen avoidance. Pediatr Clin North Am 2011;58:459−70.

[89] Flokstra-de Blok BM, Dubois AE. Quality of life measures for food allergy. Clin Exp Allergy 2012;42:1014−120.

[90] Sicherer SH, Mahr T, American Academy of Pediatrics Section on Allergy and Immunology. Management of food allergy in the school setting. Pediatrics 2010;126: 1232−9.

[91] Host A, Halken S. Hypoallergenic formulas — when, to whom and how long: after more than 15 years we know the right indication! Allergy 2004;59(Suppl. 78):45−52.

[92] Niggemann B, von Berg A, Bollrath C, Berdel D, Schauer U, Rieger C, et al. Safety and efficacy of a new extensively hydrolyzed formula for infants with cow's milk protein allergy. Pediatr Allergy Immunol 2008;19:348−54.

[93] Hill DJ, Murch SH, Rafferty K, Wallis P, Green CJ. The efficacy of amino acid-based formulas in relieving the symptoms of cow's milk allergy: a systematic review. Clin Exp Allergy 2007;37:808−22.

[94] Ewan PW, Clark AT. Efficacy of a management plan based on severity assessment in longitudinal and case-controlled studies of 747 children with nut allergy: proposal for good practice. Clin Exp Allergy 2005;35:751−6.

[95] Sicherer SH, Simons FE. Quandaries in prescribing an emergency action plan and self-injectable epinephrine for first-aid management of anaphylaxis in the community. J Allergy Clin Immunol 2005;115:575−83.

[96] Muraro A, Roberts G, Clark A, Eigenmann PA, Halken S, Lack G, et al. The management of anaphylaxis in childhood: position paper of the European academy of allergology and clinical immunology. Allergy 2007;62:857−71.

[97] Mondoulet L, Dioszeghy V, Ligouis M, Dhelft V, Dupont C, Benhamou PH. Epicutaneous immunotherapy on intact skin using a new delivery system in a murine model of allergy. Clin Exp Allergy 2010;40:659−67.

[98] Dupont C, Kalach N, Soulaines P, Legoué-Morillon S, Piloquet H, Benhamou PH. Cow's milk epicutaneous immunotherapy in children: a pilot trial of safety, acceptability, and impact on allergic reactivity. J Allergy Clin Immunol 2010;125:1165−7.

[99] Oppenheimer JJ, Nelson HS, Bock SA, Christensen F, Leung DY. Treatment of peanut allergy with rush immunotherapy. J Allergy Clin Immunol 1992;90:256−62.

[100] Nelson HS, Lahr J, Rule R, Bock A, Leung D. Treatment of anaphylactic sensitivity to peanuts by immunotherapy with injections of aqueous peanut extract. J Allergy Clin Immunol 1997;99:744−51.

[101] Zuidmeer-Jongejan L, Fernandez-Rivas M, Poulsen LK, Neubauer A, Asturias J, Blom L, et al. FAST: Towards safe and effective subcutaneous immunotherapy of persistent life-threatening food allergies. Clin Transl Allergy 2012;2(1):5.

[102] Buchanan AD, Green TD, Jones SM, Scurlock AM, Christie L, Althage KA, et al. Egg oral immunotherapy in nonanaphylactic children with egg allergy. J Allergy Clin Immunol 2007;119:199−205.

[103] Morisset M, Moneret-Vautrin DA, Guenard L, Cuny JM, Frentz P, Hatahet R, et al. Oral desensitization in children with milk and egg allergies obtains recovery in a significant proportion of cases. A randomized study in 60 children with cow's milk allergy and 90 children with egg allergy. Eur Ann Allergy Clin Immunol 2007;39:12−9.

[104] Rolinck-Werninghaus C, Staden U, Mehl A, Hamelmann E, Beyer K, Niggemann B. Specific oral tolerance induction with food in children: transient or persistent effect on food allergy? Allergy 2005;60:1320−2,

[105] Staden U, Rolinck-Werninghaus C, Brewe F, Wahn U, Niggemann B, Beyer K. Specific oral tolerance induction in food allergy in children: efficacy and clinical patterns of reaction. Allergy 2007;62:1261−9.

[106] Martorell Aragonés A, Félix Toledo R, Cerdá Mir JC, Martorell Calatayud A. Oral rush desensitization to cow milk. Following of desensitized patients during three years. Allergol Immunopathol (Madr) 2007;35:174−6.

[107] Longo G, Barbi E, Berti I, Meneghetti R, Pittalis A, Ronfani L, et al. Specific oral tolerance induction in children with very severe cow's milk-induced reactions. J Allergy Clin Immunol 2008;121:343−7.

[108] Meglio P, Giampietro PG, Gianni S, Galli E. Oral desensitization in children with immunoglobulin E-mediated cow's milk allergy — follow-up at 4 yr and 8 months. Pediatr Allergy Immunol 2008;19:412−9.

[109] Skripak JM, Nash SD, Rowley H, Brereton NH, Oh S, Hamilton RG, et al. A randomized, double-blind, placebo-controlled study of milk oral immunotherapy for cow's milk allergy. J Allergy Clin Immunol 2008;122:1154−60.

[110] Hofmann AM, Scurlock AM, Jones SM, Palmer KP, Lokhnygina Y, Steele PH, et al. Safety of a peanut oral immunotherapy protocol in children with peanut allergy. J Allergy Clin Immunol 2009;124:286−91.

[111] Narisety SD, Skripak JM, Steele P, Hamilton RG, Matsui EC, Burks AW, et al. Open-label maintenance after milk oral immunotherapy for IgE-mediated cow's milk allergy. J Allergy Clin Immunol 2009;124:610−2.

[112] Varshney P, Steele PH, Vickery BP, Bird JA, Thyagarajan A, Scurlock AM, et al. Adverse reactions during peanut oral immunotherapy home dosing. J Allergy Clin Immunol 2009;124:1351−2.

[113] Jones SM, Pons L, Roberts JL, Scurlock AM, Perry TT, Kulis M, et al. Clinical efficacy and immune regulation with peanut oral immunotherapy. J Allergy Clin Immunol 2009;124:292−300.

[114] Blumchen K, Ulbricht H, Staden U, Dobberstein K, Beschorner J, de Oliveira LC, et al. Oral peanut immunotherapy in children with peanut anaphylaxis. J Allergy Clin Immunol 2010;126:83−91.

[115] Fisher HR, du Toit G, Lack G. Specific oral tolerance induction in food allergic children: is oral desensitisation more effective than allergen avoidance? A meta-analysis of published RCTs. Arch Dis Child 2011;96:259−64.

[116] Enrique E, Pineda F, Malek T, Bartra J, Basagaña M, Tella R, et al. Sublingual immunotherapy for hazelnut food allergy: a randomized, double-blind, placebo-controlled study with a standardized hazelnut extract. J Allergy Clin Immunol 2005;116:1073−9.

[117] Fernández-Rivas M, Garrido Fernández S, Nadal JA, Díaz de Durana MD, García BE, González-Mancebo E, et al. Randomized double-blind, placebo-controlled trial of sublingual immunotherapy with a Pru p 3 quantified peach extract. Allergy 2009; 64:876−83.

[118] Kim EH, Bird JA, Kulis M, Laubach S, Pons L, Shreffler W, et al. Sublingual immunotherapy for peanut allergy: clinical and immunologic evidence of desensitization. J Allergy Clin Immunol 2011;127:640−6.

[119] Keet CA, Frischmeyer-Guerrerio PA, Thyagarajan A, Schroeder JT, Hamilton RG, Boden S, et al. The safety and efficacy of sublingual and oral immunotherapy for milk allergy. J Allergy Clin Immunol 2012;129:448−55.

[120] Asero R. Effects of birch pollen specific immunotherapy on apple allergy in birch pollen hypersensitive patients. Clin Exp Allergy 1998;28:1368−73.

[121] Asero R. How long does the effect of birch pollen injection SIT on apple allergy last? Allergy 2003;58:435−8.

[122] Bucher X, Pichler WJ, Dahinden CA, Helbling A. Effect of tree pollen specific, subcutaneous immunotherapy on the oral allergy syndrome to apple and hazelnut. Allergy 2004;59:1272−6.

[123] Hansen KS, Khinchi MS, Skov PS, Bindslev-Jensen C, Poulsen LK, Malling HJ. Food allergy to apple and specific immunotherapy with birch pollen. Mol Nutr Food Res 2004;48:441−8.

[124] Bolhaar STHP, Tiemessen MM, Zuidmeer L, van Leeuwen A, Hoffmann-Sommergruber K, Bruijnzeel-Koomen CAFM, et al. Efficacy of birch pollen immunotherapy on cross-reactive food allergy confirmed by skin tests and food challenges. Clin Exp Allergy 2004;34:761−9.

[125] van Hoffen E, Peeters KA, van Neerven RJ, van der Tas CW, Zuidmeer L, van Ieperen-van Dijk AG, et al. Effect of birch pollen-specific immunotherapy on birch pollen-related hazelnut allergy. J Allergy Clin Immunol 2011;127:100−1.

[126] Mauro M, Russello M, Incorvaia C, Gazzola G, Frati F, Moingeon P, et al. Birch-apple syndrome treated with birch pollen immunotherapy. Int Arch Allergy Immunol 2011; 156:416−22.

Chapter | three

The Epidemiology of Food Allergy

Peter Burney[1], Thomas Keil[2], Linus Grabenhenrich[2], Gary Wong[3]

[1]*National Heart Lung Institute, Imperial College, London, UK*
[2]*Institute for Social Medicine, Epidemiology and Health Economics, Charité University Medical Center, Berlin, Germany*
[3]*Department of Paediatrics, Chinese University of Hong Kong, New Territories, Hong Kong, China*

CHAPTER OUTLINE

45

Risk Management for Food Allergy. http://dx.doi.org/10.1016/B978-0-12-381988-8.00003-8

PREVALENCE

There have been many studies of food allergy in clinical settings but fewer that have undertaken surveys in the general population (Figure 3.1). Two recent meta-analyses have summarized information on the prevalence of 'food allergy' according to different definitions as estimated from population-based surveys and have reported very variable prevalences [1,2].

Table 3.1 gives the range of estimates provided by the systematic review of Rona et al. Symptoms ascribed to food hypersensitivity ranged widely, from 3% to 35%, the prevalence of biological evidence for sensitization ranged rather less, from 4% to 17%, and the combination of symptoms

Table 3.1 Percent Prevalence in General Populations of 'Food Allergy' for Animal-Derived Foods and Peanut According to Different Methods

Food	Symptoms	Sensitization (IgE)	Sensitization (Skin Prick Test)	Symptoms and Sensitization	Food Challenge
Any Food	3–35	4–6	7–17	2–4.5	1–10.8
Milk	1.2–17	2–9	0.2–2.5	0–2	0–3
Egg	0.2–7	<1–9	0.5–5	0.5–2.5	0–1.7
Peanut	0–2	<1–6	1–6	0.5–2.5	0.2–1.6
Fish	0–2	~0	0–2	0–0.5	~0 - 1
Shellfish	0–10	–	~2.5	0–1.4	~0

From Rona et al., 2007 [1].

Table 3.2 Percent Prevalence of Perceived 'Food Allergy' to Fruits, Vegetables, and Tree Nuts by Age Group

	Age Group		
Food	0–6 years	6–16 years	> 16 years
Fruit	2.2–11	–	0.4–6.6
Vegetable	0.7–3.3	–	0.5–2.2
Tree Nut	0.03–0.2	0.2–2.3	0.4–1.4

From the systematic review of Zuidmeer et al. [2].

and sensitization ranged from 2% to 4.5%. Studies using food challenge provided estimates that varied from 1% to 10.8%. The food most commonly reported to give problems was milk, followed by egg and peanut. Fish and shellfish gave fewer problems [1].

Table 3.2 gives the percent prevalence of perceived food allergy reported for fruits, vegetables, and tree nuts as provided by Zuidmeer et al. in their systematic review. Fruits reportedly caused the most common problems, comparable to the prevalence of reports of milk hypersensitivity, followed by vegetables. Tree nuts were much less commonly reported to cause problems [2].

Both the systematic reviews raised major issues around the comparability of studies, both in their design and in the methods used to identify food allergy. Although all of the studies were of general populations, the protocols varied, in particular where it came to the more intensive tests such as double-blind challenge. Their response rates also varied considerably. Most importantly they varied in the method by which the outcome was measured and reported. Questionnaires varied from one or two questions to much more extensive information and were in some cases followed by more intensive interviews. 'Objective tests' such as specific IgE or skin prick tests varied according to the allergens used, the standardization of the extracts, and the cutoff points used to define a positive test. This suggests that there is real ambiguity about the definition of disease and that the answer to the question 'What is the prevalence of disease?' needs to be qualified. Comparative studies that have meaning and utility can, however, be undertaken providing they are well standardized, and such studies are now underway.

In 2003, a telephone survey was conducted, questioning adults over the age of 18 years about perceived food allergy in the youngest child of that adult. The survey covered 10 European countries, with results based on a total of 8,825

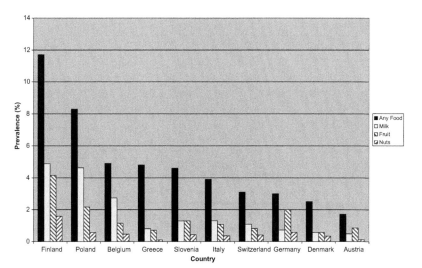

FIGURE 3.1 Prevalence of reported food hypersensitivity in children overall and to milk, fruit, and nuts. Adapted from Steinke et al. Arch Allergy Immunol 2006;143:290−295.

children [3]. The prevalence of perceived problems with any food varied widely, from 11.7% in Finland to 1.7% in Austria. These variations are clearly not due to differences in methods but could still be due to variations in the way that symptoms are interpreted in different countries. This is particularly likely to be the case, as the term 'allergy' was used throughout the survey questionnaire. Nevertheless there is some internal evidence that there are real differences in the prevalence of food hypersensitivity. Firstly, there is a very similar distribution of children who were reported by their parents to have a skin problem due to the food in question, and in all places skin problems were reported as the most common problem. Secondly, there is a similarity in the relative prevalence with which the main food was mentioned. So, milk was more commonly complained about than fruits, which were in turn more commonly reported than problems with nuts. The exception to this was in Germany and Austria, where fruits were more commonly reported than milk. This paper also reported a peak prevalence in 2—3 year olds with a smaller peak at age 7—12 years.

The other major international study to report on the relative prevalence of symptoms associated with food was the European Community Respiratory Health Survey, which reported prevalences mostly in Western Europe with some centers in Australia, New Zealand, and the US. Woods and colleagues published reported prevalence of food intolerance in 17,280 adults aged 20 to 44 years old, surveyed in 15 countries in 1991—3. They reported very variable prevalences, from 4.6% in Spain to 19.1% in Australia. They reported rash and itchy skin as the most common symptoms, most commonly associated with strawberries, apple, and hazelnut, closely followed by diarrhea and vomiting, which were most commonly associated with cow's milk, oysters, and chocolate [4].

In the second round of the European Community Respiratory Health Survey (ECRHS), the same people in many of the centers were reassessed 8 to 9 years later, and their serum samples were assessed for evidence of food-associated

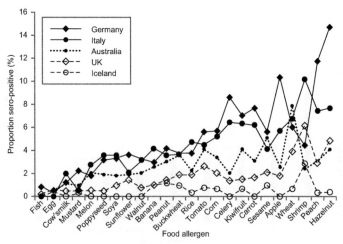

FIGURE 3.2 Prevalence of sensitization to different allergens in five countries of the European Community Respiratory Health Survey [5].

IgE. The prevalence of sensitization (specific IgE ≥ 0.35 kU/L) to any of 24 food allergens ranged from 7.7% in Reykjavik, Iceland, to 24.6% in Portland, OR, US (Figure 3.2). Despite this very wide variation in prevalence, the relative prevalence of the different food allergens was very similar in all centers. The most common sensitizations were to tree nuts and fruits such as apple and peach. The prevalence of sensitization to fish, milk, and egg was universally extremely low, in all cases less than 1% of the population, even though these were the most commonly reported causes of food-associated symptoms. Some of this may have been due to non-IgE-mediated responses, and it was noted that there was a relative excess of gastrointestinal symptoms associated with reported milk hypersensitivity, suggesting that some of the excess reports for milk might be due to lactase deficiency [5].

TIME TRENDS OF FOOD ALLERGY PREVALENCE AND HEALTH CARE UTILIZATION

Results from large population-based studies using standardized methodologies have shown that allergies, including asthma, rhinitis, and eczema, are the most common chronic diseases worldwide. Studies of time trends for these allergies confirmed the assumption that their rates have increased further during recent decades, not only in Europe, North America, Australia, and New Zealand but also in South America and many Asian countries [6]. These increases parallel other changes in sensitization to aeroallergens that have been shown to be associated with successive generations or birth cohorts [7,8], though no such changes have been directly demonstrated for food allergens.

In contrast, time trends of the prevalence of food allergy have rarely been examined. Very few population studies have been repeated in the same region to examine temporal changes. Trends in peanut allergy have been of particular interest, as this represents an increasing public health concern, reactions to peanuts being more often severe and life threatening than other food allergens and more likely to persist from childhood to adulthood [9].

Available published information on time trends of food allergy is limited and comes predominantly from the US, United Kingdom, and Australia.

UK, 1993—1999, Preschool Children, Peanut (Sensitization and Parent-Reported)

In 1993, a British study team examined allergic sensitization to peanut in 4-year-old children from the Isle of Wight and repeated the same investigation in children of a similar age from this island six years later. Parent-reported reactions to peanut, presumed to be allergic, doubled from 0.5% to 1.1% (p=0.2), whereas within this same relatively short time period, skin test reactivity to peanut increased threefold, from 1.1% to 3.3% (p=0.001). The prevalence of clinically relevant peanut allergy was assessed only during the latter study, when food challenge tests were performed to confirm suspected peanut allergy [10].

US, 1997−2002, Children and Adults, Peanut and Tree Nut (Self-Reported)

Sicherer et al. conducted two telephone surveys in the US, one in 1997 and the other in 2002, with 12,000 and 13,000 individuals, respectively. Although self-reported allergic reactions to peanut and tree nut had not increased in adults, the prevalence of parent-reported peanut allergy in children had doubled from 0.4% to 0.8% (p=0.05). The prevalence of parent-reported tree nut allergy in children remained the same at 0.2% at both time points. Other food items were not assessed [11].

US, 1997−2007, Children, Any Food (Self-Reported)

Over the 10 years from 1997 to 2007, the prevalence of parent-reported food allergy in children rose from 3.3% to 3.9% based on the National Health Interview Survey (NHIS) from representative samples of US households. Regarding the specific question on possible food allergy in their child, parents were not given guidance on what constituted '... any kind of food or digestive allergy'. Parental interpretations of this question might include conditions such as lactose intolerance (which is not an allergic condition) or celiac disease (which is not IgE mediated). The authors suggested that the increasing prevalence of parent-reported food allergy might indicate increasing recognition of food allergy among parents who had previously regarded symptoms as due to other causes [12].

Australia, 1995−2006, Young Children, Any Food (Referrals to Allergy Practice)

The changing demand for specialist food allergy services was examined by Mullins in Canberra, Australia. Patient records from 1,500 children aged 0 to 5 years who attended a community-based specialist allergy practice were evaluated retrospectively. For the time period from 1995 to 2006, the proportion of children with food allergy among all children seen increased from 20% to 58%, and the proportion of children with food anaphylaxis increased from 9% to 15%. Food allergy was diagnosed in children with a history of acute systemic allergic reactions after known food exposure and a positive skin prick test to the relevant food, but not with food challenge tests. Increased service demand is not necessarily due to increased prevalence. Interpretation of this single practice study is limited by the fact that only a subgroup of patients seeking medical assistance were examined, rather than a random sample of the entire local population. The practice's reputation for having an interest in these conditions, referral bias, increased awareness that food allergy may cause eczema, and the drift of patients from pediatrics or other specialties could be alternative explanations [13].

UK, 1990−2004, Children and Adults, Any Food (United Kingdom (UK) International Classification of Disease (ICD) Codes of Hospital Admissions)

A similar trend to that seen in the Australian study was found by Gupta et al. in an analysis of national, routinely collected health service data for England.

During the 14 years from 1990 to 2004, hospital admissions for food allergy showed continuing increases overall, from 5 to 26 per million, and particularly in children up to the age of 14 years which increased from 16 to 107 per million. Food allergy diagnosis in this study was based on ICD-9 code 693.1 and ICD-10 code 127.2 (both 'Dermatitis due to food') [14,15], and apart from true changes in the epidemiology of food allergy, alternative explanations for this dramatic increase in health care usage for presumed food allergy include changes in the coding behavior of doctors, changes in the awareness and behavior of patients, and changes in the health system.

China, 1999–2009, Infants, Any Food (Assessed by Food Challenge Tests)

Two cross-sectional Chinese studies from Chongqing showed a statistically significant increase in food allergy prevalence, from 3.5% in 1999 to 7.7% in 2009 among young children aged 0–24 months [16]. A particular strength of these investigations was the confirmation of food allergy by food challenge tests. Similarly, sensitization to food allergens assessed by positive skin prick tests almost doubled from 9.9% to 18% during the same 10-year period. As in the studies from Europe, America, and Australia, hen's egg and cow's milk were the most common foods causing allergic reactions in this age group.

In summary, we identified only a few studies, mostly from English-speaking countries, that analyzed time trends for food allergy. Most of these studies used sub-optimal methods for the accurate assessments of food allergy. The diagnostic gold standard, *double-blind placebo-controlled* food challenge tests, were not used in any of the time trend studies [17]. In particular, the dramatic increases in health care usage for food allergy and estimates

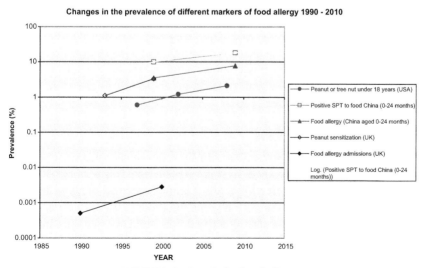

FIGURE 3.3 Trends in food allergy.

based on self- or parent-reported symptoms need to be interpreted with great caution; they are not necessarily equivalent to increased prevalence of food allergy.

This said, the evidence on time trends shows a consistent pattern of increase when plotted on a logarithmic scale (Figure 3.3), and the increase reflects the general increase in the prevalence of sensitization to inhalant allergens [7,8]. The best evidence for the evaluation of time trends of allergic reactions to food in children came from China. This investigation, using open food challenge tests to confirm suspected allergy, showed a two-fold increase over a period of ten years, mostly for reactions to hen's egg and cow's milk.

In conclusion, the evidence that food allergies have been increasing in a similar fashion to respiratory allergies is suggestive but not yet based on conclusive evidence. To examine time trends of food allergy properly we need population-based studies with representative samples and stringent methods including double-blind food challenge tests, the diagnostic gold standard for assessing food allergy.

RISK FACTORS FOR FOOD ALLERGY

Food allergy mediated by IgE requires both the development of IgE to food-related allergens and the development of disease. As sensitization without symptoms is common, both these processes are necessary. It is rare to have sensitivity to food allergens and to have no sensitivity to common aeroallergens. The initial questions are therefore why do people develop IgE-mediated sensitivity to any allergens, and then why do some of these develop IgE to foods? The secondary question is, why do some people sensitized to foods develop symptoms, while others do not?

There is considerable though indirect evidence that the prevalence of sensitization to allergens has increased substantially over the last 80 years at least, and that this increase is associated with successive birth cohorts [7, 8]. This, and the variety and ubiquity of some of the allergens involved, makes it unlikely that this change was related to increases in exposure to allergens. Such a conclusion is also supported by the observation that those exposed to large quantities of grass allergens while living on Alpine farms are less likely, not more likely, to become sensitized to grass [18].

The Hygiene Hypothesis

The most longstanding 'explanation' of why some people become sensitized and some do not is the so-called 'hygiene hypothesis'. This was initially proposed to explain why children from large families were less likely to develop rhinitis than children with no, or only one, brother or sister. The suggestion was that this might be because older children brought infections into the home and that these infections were in some way protective [19]. The initial observations on rhinitis have also been extended to sensitization [20], and other evidence has been used to support the general hypothesis. This includes the protective effect of early school attendance in children with no brothers or sisters [21] and the protective effect of living on a farm in early life, with its potential for heavy exposure to a wide variety of bacterial species [22]. A lack of sensitization has also been associated with positive serology for enteric pathogens with atopy in young recruits to the armed forces [23]. The hypothesis is further supported by the presence of considerable experimental evidence of a role for innate responses to infections in the regulation of the immune system [24].

Diet

Another line of inquiry has led to the suggestion that diet is important in determining the development of atopy. There is some evidence that vitamin D supplements [25] and high intake of vitamin A in mice [26] may be associated with a higher prevalence of sensitization, but much of the evidence on diet is related to the manifestation of clinical symptoms rather than sensitization, the development of specific IgE to allergens or positive skin prick tests, and cannot be interpreted in terms of the risk of sensitization. Speculation has, however, been most active around antioxidants and lipids.

Evidence that antioxidants affect sensitization has been found in some studies in relation to vitamin E, which has been associated with reduced sensitization in adults [27,28]. Although there has been some evidence that a high consumption of fruits is associated with lower levels of sensitization, there is little direct evidence that vitamin C is associated with lower levels of sensitization [28,29]. There is a high probability of uncontrolled confounding in any dietary survey, and the lack of successful intervention trials makes any conclusion difficult.

Most of the theoretical reasons for believing that fats might be important in allergic disease relate to the expression of disease, rather than to sensitization. Nevertheless there are studies that have suggested that dietary fats might be influential in sensitization. An early ecological analysis suggested that high consumption of mono-unsaturated fatty acids might be associated with a higher prevalence of sensitization [30]. This has been confirmed in studies of children [31], although one of these showed a significant effect only in those with a family history of allergic disease [32]. Similar to dietary lipids, there are some studies suggesting that sensitization may be related to consumption of foods rich in mono-unsaturated fatty acids, though the results often depend on looking at subgroups of the population [33]. The converse of this, a protective effect of whole milk consumption, has been reported from Australia [29]. Although an association has been reported between higher levels of serum or cell membrane eicoso-penta-enoic acid and γ-linolenic acid [34] and lower levels of sensitization, others have found no convincing associations with any specific fatty acids or groups of fatty acids [35,36].

Breast Feeding

Much research among children has focused on early feeding and weaning practices and, in particular, the ages at which potentially allergenic foods are introduced into the diet and how this might affect the immunological response to foods. The teaching has been that a long period of exclusive breast feeding and the delayed introduction of potentially allergenic foods is preferred and lowers risk. However, based on observations that early exposure in some cultures was associated with a low prevalence of allergy to those foods, and the potential importance of the gut-associated immune system in mediating tolerance to allergens, this orthodoxy has been challenged [37]. The experimental evidence to support a policy on weaning is currently being collected.

Cross-Reactions

Although most people with food sensitization are sensitized to other allergens, not everyone who is sensitized to aeroallergens becomes sensitized to foods. The issue is what determines which allergic people will develop food allergy, and what are the risk factors involved. Some people with 'food allergy' are thought to have a primary sensitization to an aeroallergen that cross-reacts with a food. This is most typical for people who have pollen allergy and have become sensitized to allergens that are common to both pollen and foods. The most common of these cross-reactions is between birch pollen, a very common allergy in Northern Europe, and tree nuts, apples, and other fruit and vegetables. For the most part these people have relatively minor symptoms, very often confined to itching and tingling of the lips and mouth on eating the food.

pH in the Stomach

It has been argued that the pH in the stomach affects the likelihood of sensitization to allergens that are ingested. For this there is both epidemiological evidence based on an increased level of sensitization in those who have been taking antacid treatment [38,39] and animal data [40]. One explanation given for the early occurrence of food allergy, and one of the arguments put forward for delaying introduction of potentially allergenic foods into the diet, has been that the pH of the stomach in infancy is higher than in later life and therefore less likely to denature the allergen during digestion [41].

Genes

Underlying these and other mechanisms there is variation in genetic susceptibility, and the most heavily researched area has been the genetic variation in the Toll-like receptors that mediate innate immune responses to microorganisms [42] and genes that regulate the barrier function of the skin and mucosa [43].

It is likely that the selection of allergens to which people become sensitized is also in some ways under genetic control, and the most likely link is with the

Human Leucocyte Antigen (HLA) system. There is relatively little information on this, though associations have been found between specific HLA alleles and tree nut allergy [44] and latex allergy [45].

The Link Between Sensitization and Allergy

Being sensitized to a food is not the same as developing clinical disease, and there are many people who are sensitized and have no symptoms. The factors that influence this probably include the likelihood of encountering the allergen after sensitization and a number of factors that influence the size of the skin wheal response to the allergen [46]. Although these factors have not been investigated specifically in relation to food allergy, there is some information in relation to aeroallergens. This includes the amount of the specific IgE [46] as well as a number of other factors that modulate the immune response. The most widely studied of these is the presence of parasites [47].

In very low income countries, there is evidence that low body mass is associated with normal levels of specific IgE to aeroallergens, but low levels of sensitization as shown by skin tests [48], and that re-feeding those with protein-calorie malnutrition may lead to the reappearance of positive skin tests [49]. There is no evidence for such an effect in developed market economics. In Africa, a 'rural' diet has been shown to be associated with a low level of positive skin prick tests in relation to serum-specific IgE compared with those on a more 'urban' diet, even after adjusting for the residence of the child, though the dietary component that is important has not been identified [50].

THE NATURAL HISTORY OF FOOD ALLERGIES

Prediction of the course of food allergy in a specific patient is best based on sound longitudinal data, tracing food allergic individuals prospectively. Unfortunately, most data available today on the natural history of the disease is based on cross-sectional investigations or on retrospective evaluation of patient records. Furthermore, the case definition is rarely confirmed by more accurate diagnostic tests such as food challenges.

Hen's Egg

Food allergy against hen's egg begins in infancy. Most cases are expected to resolve within the first years of the disease but may persist until early adulthood. Several population-based studies described development of tolerance in about two thirds of patients before entering primary school [51]. Other allergic diseases seem to play a major role as risk factors for persistence, since in a highly atopic sample of patients suffering egg allergy, as few as 12% had developed tolerance by 6 years of age. Only at 16 years of age had the expected two thirds lost their clinically manifest disease at a continuous rate [52]. High initial specific IgE has been suggested as a predictor of the course of hen's egg allergy [53].

Cow's Milk

Onset of cow's milk allergy occurs in the first two years of life, with the majority starting before the first birthday. Development of tolerance is expected in most cases, with a higher frequency in non-IgE-mediated intolerance. Similar to hen's egg, cow's milk allergy is expected to resolve in about 80% of children within the first six years of life [54]. In an atopic population, tolerance to cow's milk is less likely to develop, with only 12% tolerant by 6 years and 55% at 16 years [55]. High initial levels of specific IgE predict persistence. The presence of other allergic diseases such as asthma or allergic rhinitis and breast vs. formula feeding in the first months are also reported to predict the course of cow's milk allergy [55].

Peanut and Tree Nuts

Peanut allergy usually starts later in life than cow's milk and hen's egg allergy and tends to persist in a higher proportion of patients. Only about 10% to 20% outgrow their disease later in life [56]. Peanut allergy is severe and potentially fatal. Besides a rigorous avoidance of products containing traces of peanut, a thorough follow-up to identify development of tolerance is of major importance for patient management. Similar to peanuts, allergy against tree nuts tends to persist in most patients. Only about 10% outgrow their disease later in life [57].

Other Foods

Population-based data on the natural history of food allergy is scarce for other foods. Wheat and soy seem to follow a favorable pattern, similar to hen's egg and cow's milk. Fish and shellfish allergy resembles a picture traced in peanut or tree nut allergy, with late onset and a high proportion of persistence.

Specific IgE

Levels of specific IgE are commonly used to predict whether and when the disease is likely to disappear, with some uncertainty. For example, high initial specific IgE against hen's egg was shown to be associated with a high frequency of persistence [52], and a drop in specific IgE over the course of the disease is thought to indicate the development of tolerance. However, clinically apparent food allergy can persist while serologic markers decline, and specific IgE regularly remains high despite development of tolerance [57,58]. Change in specific skin reactivity also has low diagnostic value in detecting clinical tolerance in individuals.

In the absence of a conclusive history of accidental ingestion, food challenges — preferably double-blind and placebo-controlled — are needed to identify progression towards tolerance.

On a molecular level, characteristics of the epitope not only predispose certain food items to be common triggers for allergic reactions but may influence the natural history, including development of tolerance. Linear epitopes, seen for example in allergenic proteins of peanut, are associated with a persistent type of food allergy, where as conformational epitopes are commonly seen to be responsible for transient disease, for instance in cow's milk allergy [59].

The age of onset, duration, expected age at which clinically manifest food allergy will be outgrown, and the likelihood of severe anaphylactic reactions all vary greatly between different food items. Cow's milk and hen's egg allergy — observed mainly in infancy — are expected to disappear by school age in the majority of children. Peanut, tree nut, and fish allergy on the other hand — with onset at a higher age — are found to last longer and more frequently cause severe anaphylaxis. Recurrence of food allergy after development of clinical tolerance is documented for peanut allergy [60] but is not expected for other food items.

THE CONSEQUENCES OF FOOD ALLERGY

The clinical consequences of reactions associated with food allergy range from a minor itch of the skin or a runny nose to a life-threatening reaction of anaphylaxis characterized by alarming symptoms or signs such as a significant drop of blood pressure and difficulty with breathing. The majority of food allergic reactions are mediated through the presence of specific IgE, while others are mediated by cellular mechanisms. Depending on the mechanisms involved, the timing and presentation of symptoms will be different. Although most food-induced allergic reactions may be relatively mild, anaphylaxis remains an important condition that frequently results in visits to accident and emergency departments. Allergic reactions resulting in different forms of skin changes and oral allergy syndrome are the most common manifestations of food allergy [61]. Of the various types of skin reaction, the rapid development of urticaria or hives shortly after ingestion of the food allergen is the most common presentation. Other skin manifestations include flushing and itching of the skin and other forms of rash. Angioedema is a variant form of urticaria in which the deeper instead of the superficial layer of the skin is affected [62]. These immediate types of reactions usually occur within minutes of ingesting the food allergen, and they are mediated through the presence of existing IgE directed against the food allergen in affected individuals. Many of these reactions can be treated by self-administered medications, and the patients may not need to visit their doctors if the symptoms are mild and they have previous experience of treating the reactions themselves. Other delayed types of reactions such as worsening of eczema are usually not mediated by IgE and may occur hours or days after ingestion of the offending food [63]. Diagnosis of these reactions is difficult because of the variable nature of presentation and timing after the ingestion of the food allergen. Nevertheless, in patients with severe atopic eczema, food allergy should be considered. The common food allergens associated with this type of reaction include milk, peanut, egg, and seafood. An appropriate elimination diet will result in improvement of the eczema.

Food allergic reactions affecting the respiratory tract are highly variable, ranging from very mild symptoms of the upper respiratory tract to very severe breathing difficulty such as an asthma attack. The typical nasal symptoms are sneezing, runny nose, and congestion. Patients with nasal symptoms frequently complain of eye symptoms at the same time [64]. Eye symptoms may include redness, tearing, and swelling of the eyelids. The nasal symptoms are very similar to the symptoms in patients with seasonal nasal allergy. When the lower airways are also affected, patients will experience chest tightness, breathing difficulty, wheezing attacks, or even death due to inability to breathe because of severe constriction of the airways. Food allergic reactions in patients with preexisting unstable asthma are at high risk of causing severe respiratory symptoms [65]. These severe reactions will result in visits to accident and emergency departments or even hospitalizations. One uncommon form of reaction to food allergen is by inhalation. This type of reaction occurs in highly sensitive individuals and results in a reaction after the inhalation of airborne food allergens such as the smell of fish or shellfish in a seafood restaurant or seafood market [66].

Subjective symptoms affecting the gastrointestinal tract pose a particular difficulty for accurate diagnosis of food allergy. These symptoms include nausea, vomiting, abdominal pain, and diarrhea. Young children may not be able to describe their symptoms. Patients with food poisoning or intolerance to lactose may develop symptoms similar to allergic reactions to food. Although patients with food allergy usually present with symptoms within two hours of ingestion, some patients may present later with severe diarrhea due to problems further down in the gut. The oral allergy syndrome is a special form of food allergy. Patients are initially sensitized to pollens such as birch, resulting in the production of IgE, which may then cross-react with other foods such as apples and pears [67]. Patients who have ragweed allergy may develop symptoms after following contact with melons and bananas. The typical symptoms are swelling and itchiness of the lips, tongue, and throat minutes after ingestion of the fresh fruits. These reactions may also occur when the affected subjects are eating the food for the first time.

Other forms of allergic reactions affecting the gastrointestinal tract are mediated by cellular mechanism. Patients with these conditions have specific cells of the immune system that may react with specific food allergens, resulting in manifestations affecting the gut. These conditions include celiac disease and food protein-induced enteropathy. Celiac disease is also known as gluten-sensitive enteropathy. The majority of patients with this disease have one of the specific genetic markers [68]. Testing for these markers will help in making the diagnosis. The common sources of gluten in the human diet include wheat, barley, rye, and malt. When exposed to foods with gluten, immune-mediated damage will occur in the mucosal lining of the gut. Patients with untreated celiac disease will present with diarrhea, abdominal pain, bloating, and weight loss. Affected young children frequently present with failure to thrive (see also chapter 7). Food protein-induced enteropathy typically occurs in young children presenting with intractable diarrhea in the first few months of life. The other symptoms include vomiting and failure to gain weight appropriately. The common food proteins associated with this illness are cow's milk, wheat, soybean, and egg.

Anaphylaxis is the most severe form of allergic reaction. It has been estimated that food-induced anaphylaxis accounts for up to 50% of all anaphylactic reactions treated in emergency departments [69], and each year more than 150 patients die because of food-induced anaphylaxis in the US [70]. In developed countries, the common foods resulting in anaphylactic reactions are peanuts, tree nuts, shellfish, and fish. These reactions usually progress rapidly over 10 to 20 minutes shortly after ingestion of the food. Patients usually start with symptoms affecting the mouth and throat such as itching and swelling. The skin is affected in the early stage of the reaction resulting in redness, swelling, itch, and development of hives. Left untreated, the patients may develop wheezing, difficulty with breathing, drop in blood pressure, fainting, and collapse. Not infrequently, severe reactions occur in subjects with known food allergies in situations of accidental exposure, such as eating out in a restaurant. Patients with a history of or who are at risk of anaphylaxis should be given a self-injector of adrenaline. Prompt and early treatment with adrenaline injection is life saving. Reports analyzing fatal or near-fatal reactions have revealed that early use of adrenaline injection might have prevented these fatalities [71]. Occasionally, anaphylactic reactions may have a biphasic pattern. The patients may be treated and the symptoms improve. However, their symptoms will recur within a few hours after the initial presentation. One unusual form of anaphylaxis is termed food-dependent exercise-induced anaphylaxis. This form of reaction occurs when the patients have ingested a particular food coupled with exercise within 2−4 hours. Ingestion of the food or exercise alone would not precipitate the reaction. The common foods associated with this form of anaphylaxis are grains such as wheat and buckwheat, seafood, and vegetables and fruits such as celery and grape [72]. Accurate diagnosis can be difficult, as these reactions are difficult to replicate in the clinic. Although the consequence of food allergy in most patients may be relatively minor and easily treated with oral medication, the fear and psychological burden of carrying an auto-injector for adrenaline in case of severe reactions should not be underestimated.

CONCLUSIONS

We need to distinguish between 'sensitization to foods', the development of specific IgE to allergens contained in foods, and 'food allergy', the development of clinical symptoms as a consequence. Considerable advances have been made in recent years in mapping the distribution of sensitization rates to foods. We now have internationally standardized surveys that show wide variations in prevalence that are broadly consistent with the prevalence of sensitization to aeroallergens.

Most people sensitized to food allergens are also sensitized to aeroallergens, but what makes sensitization to food allergens more likely in some people, other than the presence of a general tendency to respond with IgE, is so far unknown. At least in later life, this seems unlikely to be related solely to exposure to the food. The factors that determine who becomes tolerant to allergens during childhood are also largely unknown.

The relative prevalence of food allergy has been much more difficult to study. Double-blind, placebo-controlled challenge tests, still the gold standard for identifying true food allergy, are very time-consuming and the extensive use of such tests in the general population is extremely difficult. It is, however, clear that the link between complaints about individual foods and serological evidence of sensitization to those foods is not strong. This makes the interpretation of simple surveys of either symptoms or markers of sensitization difficult to interpret in relation to true food allergy.

Studies have been undertaken to find less demanding methods to approximate the answers from double-blind, placebo-controlled challenge tests and these have shown some promise [73]. Nevertheless, the ability of these methods to replicate the results of double-blind, placebo-controlled food challenge tests across different environments and different age groups is so far unknown.

The EuroPrevall surveys [74], and their partnerships in middle- and low-income countries [75], have yet to publish their full results but should make an important contribution to answering some of these questions.

REFERENCES

[1] Rona RJ, Keil T, Summers C, Gislason D, Zuidmeer L, Sodergren E, et al. The prevalence of food allergy: A meta-analysis. J Allergy Clin Immunol 2007;120(3):638.

[2] Zuidmeer L, Goldhahn K, Rona R, Gislason D, Madsen C, Summers C, et al. The prevalence of plant food allergies: A systematic review. J Allergy Clin Immunol 2008; 121(5):1210.

[3] Steinke M, Fiocchi A, Kirchlechner V, Ballmer Weber B, Brockow K, Hischenhuber C, et al. Perceived food allergy in children in 10 European nations. A randomised telephone survey. Int Arch Allergy Immunol 2007;143(4):290.

[4] Woods RK, Abramson M, Bailey M, Walters EH. International prevalences of reported food allergies and intolerances. Comparisons arising from the European Community Respiratory Health Survey (ECRHS) 1991–1994. Eur J Clin Nutr 2001 04;55(4): 298–304.

[5] Burney P, Summers C, Chinn S, Hooper R, van Ree R, Lidholm J. Prevalence and distribution of sensitization to foods in the European Community Respiratory Health Survey: A EuroPrevall analysis. Allergy 2010.

[6] Asher MI, Montefort S, Björkstén B, Lai CKW, Strachan D, Weiland S, et al. Worldwide time trends in the prevalence of symptoms of asthma, allergic rhinoconjunctivitis, and eczema in childhood: ISAAC phases one and three repeat multicountry cross-sectional surveys. Lancet 2006;368(9537):733–43.

[7] Jarvis D, Luczynska C, Chinn S, Potts J, Sunyer J, Janson C, et al. Change in prevalence of IgE sensitization and mean total IgE with age and cohort. J Allergy Clin Immunol 2005 09;116(3):675–82.

[8] Law M, Morris JK, Wald N, Luczynska C, Burney P. Changes in atopy over a quarter of a century, based on cross sectional data at three time periods. BMJ 2005;330(7501):1187–8.

[9] Sicherer S, Munoz-Furlong A, Godbold J, Sampson H. US prevalence of self-reported peanut, tree nut, and sesame allergy: 11-year follow-up. J Allergy Clin Immunol 2010; 125(6):1322.

[10] Grundy J, Matthews S, Bateman B, Dean T, Arshad SH. Rising prevalence of allergy to peanut in children: Data from 2 sequential cohorts. J Allergy Clin Immunol 2002 11; 110(5):784−9.

[11] Sicherer S, Munoz-Furlong A, Sampson HA. Prevalence of peanut and tree nut allergy in the United States determined by means of a random digit dial telephone survey: A 5-year follow-up study. J Allergy Clin Immunol 2003;112(6):1203−7.

[12] Branum A, Lukacs S. Food allergy among children in the United States. Pediatrics 2009;124(6):1549.

[13] Mullins R. Paediatric food allergy trends in a community-based specialist allergy practice, 1995−2006. Med J Aust 2007;186(12):618.

[14] Gupta R, Sheikh A, Strachan D, Anderson HR. Increasing hospital admissions for systemic allergic disorders in England: Analysis of national admissions data. BMJ 2003 11/15;327(7424):1142−3.

[15] Gupta R, Sheikh A, Strachan DP, Anderson HR. Time trends in allergic disorders in the UK. Thorax 2007;62(1):91.

[16] Hu Y, Chen J, Li H. Comparison of food allergy prevalence among Chinese infants in Chongqing, 2009 versus 1999. Pediatr Int 2010;52(5):820−4.

[17] Woods RK, Stoney RM, Raven J, Walters EH, Abramson M, Thien FC. Reported adverse food reactions overestimate true food allergy in the community. Eur J Clin Nutr 2002;56(1):31−6.

[18] Braun-Fahrlander C, Gassner M, Grize L, Neu U, Sennhauser FH, Varonier HS, et al. Prevalence of hay fever and allergic sensitization in farmer's children and their peers living in the same rural community. Clin Exp Allergy 1999;29:28−34.

[19] Strachan D. Hay fever, hygiene, and household size. BMJ 1989;299(6710):1259−60.

[20] Svanes C, Jarvis D, Chinn S, Burney P. Childhood environment and adult atopy: Results from the European community respiratory health survey. J Allergy Clin Immunol 1999 Mar;103(3 Pt 1):415−20.

[21] Kramer U, Heinrich J, Wjst M, Wichmann H. Age of entry to day nursery and allergy in later childhood. Lancet 1999;353:450−4.

[22] Braun Fahrlander C, Riedler J, Herz U, Eder W, Waser M, Grize L, et al. Environmental exposure to endotoxin and its relation to asthma in school-age children. N Engl J Med 2002;347(12):869.

[23] Matricardi PM, Rosmini F, Ferrigno L, Nisini R, Rapicetta M, Chionne P, et al. Cross sectional retrospective study of prevalence of atopy among Italian military students with antibodies against hepatitis A virus. BMJ 1997;314:999−1003.

[24] Hammad H, Lambrecht BN. Dendritic cells and airway epithelial cells at the interface between innate and adaptive immune responses. Allergy 2011;66(5):579.

[25] Hypponen E, Sovio U, Wjst M, Patel S, Pekkanen J, Hartikainen A, et al. Infant vitamin D supplementation and allergic conditions in adulthood: Northern Finland birth cohort 1966. Ann N Y Acad Sci 2004;1037:84.

[26] Ruhl R, Hanel A, Garcia A, Dahten A, Herz U, Schweigert F, et al. Role of vitamin A elimination or supplementation diets during postnatal development on the allergic sensitisation in mice. Mol Nutr Food Res 2007;51(9):1173−81.

[27] Fogarty A, Lewis S, Weiss S, Britton J. Dietary vitamin E, IgE concentrations, and atopy. Lancet 2000;356(9241):1573.

[28] McKeever TM, Lewis SA, Smit H, Burney P, Britton J, Cassano PA. Serum nutrient markers and skin prick testing using data from the third National Health and Nutrition Examination Survey. J Allergy Clin Immunol 2004 Dec; 114(6):1398−402.

[29] Woods R, Walters EH, Raven J, Wolfe R, Ireland P, Thien FCK, et al. Food and nutrient intakes and asthma risk in young adults. Am J Clin Nutr 2003;78(3):414.

[30] Heinrich J, Hoelscher B, Bolte G, Winkler G. Allergic sensitization and diet: Ecological analysis in selected European cities. Eur Respir J 2001;17(3):395.

[31] Bolte G, Frye C, Hoelscher B, Meyer I, Wjst M, Heinrich J. Margarine consumption and allergy in children. Am J Respir Crit Care Med 2001;163(1):277.

[32] Sausenthaler S, Kompauer I, Borte M, Herbarth O, Schaaf B, von Berg A, et al. Margarine and butter consumption, eczema and allergic sensitization in children. The LISA birth cohort study. Pediatr Allergy Immunol 2006;17(2):85.

[33] Trak Fellermeier MA, Brasche S, Winkler G, Koletzko B, Heinrich J. Food and fatty acid intake and atopic disease in adults. Eur Respir J 2004;23(4):575.

[34] Hoff S, Seiler H, Heinrich J, Kompauer I, Nieters A, Becker N, et al. Allergic sensitisation and allergic rhinitis are associated with n-3 polyunsaturated fatty acids in the diet and in red blood cell membranes. Eur J Clin Nutr 2005;59(9):1071.

[35] Woods RK, Raven JM, Walters EH, Abramson MJ, Thien FCK. Fatty acid levels and risk of asthma in young adults. Thorax 2004;59(2):105.

[36] Kompauer I, Demmelmair H, Koletzko B, Bolte G, Linseisen J, Heinrich J. Association of fatty acids in serum phospholipids with hay fever, specific and total immunoglobulin E. Br J Nutr 2005;93(4):529.

[37] Du Toit G, Katz Y, Sasieni P, Mesher D, Maleki S, Fisher H, et al. Early consumption of peanuts in infancy is associated with a low prevalence of peanut allergy. J Allergy Clin Immunol 2008;122(5):984.

[38] Untersmayr E, Bakos N, Scholl I, Kundi M, Roth-Walter F, Szalai K, et al. Anti-ulcer drugs promote IgE formation toward dietary antigens in adult patients. FASEB J 2005 04;19(6):656−8.

[39] Scholl I, Untersmayr E, Bakos N, Roth-Walter F, Gleiss A, Boltz-Nitulescu G, et al. Antiulcer drugs promote oral sensitization and hypersensitivity to hazelnut allergens in BALB/c mice and humans. Am J Clin Nutr 2005 01;81(1):154−60.

[40] Pali Scholl I, Herzog R, Wallmann J, Szalai K, Brunner R, Lukschal A, et al. Antacids and dietary supplements with an influence on the gastric pH increase the risk for food sensitization. Clin Exp Allergy 2010;40(7):1091.

[41] Walker WA. Allergen absorption in the intestine: Implication for food allergy in infants. J Allergy Clin Immunol 1986 11;78(5 pt 2):1003−9.

[42] Vercelli D, Baldini M, Stern D, Lohman C, Halonen M, Martinez F. CD14: A bridge between innate immunity and adaptive IgE responses. J Endotoxin Res 2001;7(1):45−8.

[43] Blumenthal MN, Langefeld CD, Barnes KC, Ober C, Meyers DA, King RA, et al. Rich SS and for the Collaborative Study on the Genetics of Asthma. A genome-wide search for quantitative trait loci contributing to variation in seasonal pollen reactivity. J Allergy Clin Immunol 2006;117(1):79.

[44] Hand S, Darke C, Thompson J, Stingl C, Rolf S, Jones KP, et al. Human leucocyte antigen polymorphisms in nut-allergic patients in South Wales. Clin Exp Allergy 2004 05;34(5):720−4.

[45] Blanco C, Sanchez-Garcia F, Torres-Galvan MJ, Dumpierrez AG, Almeida L, Figueroa J, et al. Genetic basis of the latex-fruit syndrome: Association with HLA class II alleles in a spanish population. J Allergy Clin Immunol 2004 11;114(5):1070−6.

[46] Sporik R, Hill DJ, Hosking CS. Specificity of allergen skin testing in predicting positive open food challenges to milk, egg and peanut in children. Clin Exp Allergy 2000;30(11):1540.

[47] van den Biggelaar AHJ, van Ree R, Rodrigues LC, Lell B, Deelder AM, Kremsner PG, et al. Decreased atopy in children infected with Schistosoma haematobium: A role for parasite-induced interleukin-10. Lancet 2000;356:1723−7.

[48] Calvert J, Burney P. Effect of body mass on exercise-induced bronchospasm and atopy in African children. J Allergy Clin Immunol 2005;114(3):531−7.

[49] Abbassy AS, Badr E, Hassan AI, Aref GH, Hammad SA. Studies of cell mediated immunity and allergy in protein energy malnutrition. II immediate hypersensitivity. J Trop Med Hyg 1974;77(1):18−21.

[50] Hooper R, Calvert J, Thompson RL, Deetlefs ME, Burney P. Urban/rural differences in diet and atopy in South Africa. Allergy [Internet] 2008;63(4):425.

[51] Boyano Martinez T, Garcia Ara C, Diaz Pena J, Martin Esteban M. Prediction of toler-ance on the basis of quantification of egg white-specific IgE antibodies in children with egg allergy. J Allergy Clin Immunol 2002;110(2):304.

[52] Savage J, Matsui E, Skripak J, Wood R. The natural history of egg allergy. J Allergy Clin Immunol 2007;120(6):1413.

[53] Tey D, Heine RGa. Egg allergy in childhood: An update. Curr Opin Allergy Clin Immunol 2009;9(3):244−50.

[54] Saarinen K, Pelkonen A, Makala M, Savilahti E. Clinical course and prognosis of cow's milk allergy are dependent on milk-specific IgE status. J Allergy Clin Immunol 2009; 116(4):869.

[55] Skripak J, Matsui E, Mudd K, Wood R. The natural history of IgE-mediated cow's milk allergy. J Allergy Clin Immunol 2007;120(5):1172.

[56] Fleischer D, Conover Walker M, Christie L, Burks AW, Wood R. The natural progres-sion of peanut allergy: Resolution and the possibility of recurrence. J Allergy Clin Immunol 2003;112(1):183.

[57] Hill DJ, Firer MA, Ball G, Hosking CS. Natural history of cows' milk allergy in children: Immunological outcome over 2 years. Clin Exp Allergy 1993; 23(2):124.

[58] Sampson HAM. Update on food allergy. J Allergy Clin Immunol 2004;113(5): 805−19.

[59] Vila L, Beyer K, Jarvinen K-, Chatchatee P, Bardina L, Sampson HA. Role of confor-mational and linear epitopes in the achievement of tolerance in cow's milk allergy. Clin Exp Allergy 2001;31(10):1599−606.

[60] Fleischer D, Conover Walker M, Matsui E, Wood R. The natural history of tree nut allergy. J Allergy Clin Immunol 2005;116(5):1087.

[61] Mansoor DK, Sharma HP. Clinical presentations of food allergy. Pediatr Clin N Am 2011;58:315−26.

[62] Krishnamurthy A, Naguwa SM, Gershwin ME. Pediatric angioedema. Clin Rev Allergy Immunol 2008;34:250−9.

[63] Sicherer SH, Sampson HA. Food hypersensitivity and atopic dermatitis: Pathophysi-ology, epidemiology, diagnosis, and management. J Allergy Clin Immunol 1999;104: s114−22.

[64] Sicherer SH, Sampson HA. Food allergy. J Allergy Clin Immunol 2010;125: s116−25.

[65] NIAID-Sponsored Expert Panel, Boyce JA, Assa'ad A, Burks AW, Jones SM, Sampson HA, et al. Guidelines for the diagnosis and management of food allergy in the United States: Report of the NIAID-sponsored expert panel. J Allergy Clin Immunol 2010;126:s1−58.

[66] Jeebhay MF, Robins TG, Lehrer SB, Lopata AL. Occupational seafood allergy: A review. Occup Environ Med 2001;58:553−62.

[67] Katelaris CH. Food allergy and oral allergy or pollen-food syndrome. Curr Opin Allergy Clin Immunol 2010;10:246−51.

[68] Volta U, Villanacci V. Celiac disease: Diagnostic criteria in progress. Cell Mol Immunol 2011;8:96−102.

[69] Yocum MW, Butterfield JH, Klein JS, Volcheck GW, Schroeder DR, Silverstein MD. Epidemiology of anaphylaxis in Olmsted county: A population-based study. J Allergy Clin Immunol 1999;104:452−6. 1999;104:452−6.

[70] Sampson HA. Anaphylaxis and emergency treatment. Pediatrics 2003;111:1601−8.

[71] Bock SA, Munoz-Furlong A, Sampson H,AJ. Fatalities due to anaphylactic reactions to foods. J Allergy Clin Immunol 2001;107:101−3.

[72] Robson-Ansley P, Toit GD. Pathophysiology, diagnosis and management of exercise-induced anaphylaxis. Curr Opin Allergy Clin Immunol 2010;10:312−7.

[73] DunnGalvin AP, Daly DR, Cullinane CR, Stenke EM, Keeton DR, Erlewyn-Lajeunesse MMD, et al. Highly accurate prediction of food challenge outcome using routinely available clinical data. J Allergy Clin Immunol 2011; 127(3):633−9.

[74] Kummeling I, Mills ENC, Clausen M, Dubakiene R, Pérez CF, Fernández-Rivas M, et al. The EuroPrevall surveys on the prevalence of food allergies in children and adults: Background and study methodology. Allergy 2009;64(10):1493.

[75] Wong GWK, Mahesh PA, Ogorodova L, Leung TF, Fedorova O, Holla AD, et al. The EuroPrevall-INCO surveys on the prevalence of food allergies in children from China, India and Russia: The study methodology. Allergy 2010;65(3):385.

Allergen Thresholds and Risk Assessment

Chapter four

How to Determine Thresholds Clinically

Barbara K. Ballmer-Weber[1], André C. Knulst[2], Jonathan O'B. Hourihane[3]

[1]*Allergy Unit, Department of Dermatology, University Hospital Zürich, Switzerland*
[2]*Department of Dermatology/Allergology, University Medical Center, Utrecht, The Netherlands*
[3]*Department of Paediatrics and Child Health, Clinical Investigations Unit, Cork University Hospital, Wilton, Cork, Ireland*

CHAPTER OUTLINE

INTRODUCTION AND DEFINITIONS

The double-blind placebo-controlled food challenge (DBPCFC) is a technique that has been applied since the 1970s. Common diagnostic methods used in allergy investigations assess the presence or absence of allergen-specific sensitization. Sensitization is a prerequisite for IgE-mediated allergy but is not always accompanied by manifest allergy symptoms. Thus, whereas the absence of sensitization can be often used to exclude allergy as a cause of the observed symptoms, its presence is not sufficient for a positive diagnosis of allergy. To date, no *in vitro* or *in vivo* test exists which shows full correlation with clinical food allergy. Food challenges, in particular the DBPCFC, still represent the most reliable and the only scientifically accepted way to establish

67

Risk Management for Food Allergy. http://dx.doi.org/10.1016/B978-0-12-381988-8.00004-X

or rule out an allergy to a food, though they are neither 100% sensitive nor 100% specific [1−3]. Therefore the DBPCFC is to date the recommended way to definitively assess food allergy. The earlier challenge protocols were developed simply to establish a diagnosis of food allergy, without consideration of a threshold or minimum eliciting dose (MED). Accordingly, the starting doses in the early studies would now be considered relatively high. In those studies many patients reacted to the first dose, and therefore no hard conclusions could be drawn on doses of food below which no allergic subject was likely to experience symptoms. These tests only confirmed a food allergy, but nothing about a threshold could be inferred except that a threshold or MED had been exceeded. In recent years, the focus of interest in food challenge studies has moved away from the pure intention to produce a simple yes/no answer toward an assessment of a dose-response-relation.

A threshold is usually defined as "a limit below which a stimulus causes no reaction"; i.e., the minimal amount of an allergenic food that does elicit an allergic response [4]. Threshold doses can be determined at an individual and a population level. Individual thresholds can be defined in an experimental setting, i.e., by titrated challenges. Population thresholds, however, have to be estimated by statistical approaches based on individual threshold results. Another concept (adapted/derived from pharmacotoxicology studies) deals with the terms No Observed Adverse Effect Level (NOAEL) and Lowest Observed Adverse Effect Levels (LOAEL). The NOAEL can be defined as the highest dose of a substance observed in a study not to produce any adverse effect, essentially the adverse effect threshold. Similarly, the LOAEL can be defined as the lowest dose that is observed to produce an adverse effect [4]. In the following chapter we will focus on the procedure of titrated challenges to establish individual threshold doses.

MEAL, SOURCE MATERIAL, AND THE MATRIX ISSUE

A suitable challenge meal for DBPCFC has to fulfill a couple of requirements. It should adequately blind the food under investigation. A number of foods might be hidden in the same recipe on separate occasions, but some foods might require specific adaptations, e.g., fish and shrimp. Adequate blinding involves taste, palatability, texture, and smell. Blinding is usually easy at lower doses, but may be very difficult at higher doses (in the gram range), especially for foods with a typical, strong taste or smell. Since NOAEL and LOAEL usually are in the milligram range, they can generally be easily determined.

Furthermore, the portions should be tasty, and of an adequate volume and energy content to allow the patient to ingest all of it without (gastrointestinal) problems. Children having DBPCFC should have the final serving adapted to a normal daily helping for their age.

Another important issue is the source material. In practice it is not easy to find a supplier that can guarantee that the food delivered as source material is free from all other allergens. Nevertheless this is a prerequisite for an adequate recipe. The other ingredients used in the recipe should not be other well-known allergenic foods.

Challenge meals used are usually low in fat content. While this might not be representative of all situations in daily life, since allergens can also occur in products with a high fat content, e.g., chocolate, certain bakery and meat products, the single clinical study available so far showed a higher fat content that resulted in an increase of the threshold (symptoms starting at a higher dose). More importantly, more severe symptoms were elicited at that threshold dose compared to a lower fat recipe. This might be due to both delayed emptying of the stomach by the higher fat recipe and lack of early allergic symptoms (which would have terminated the challenge) as the allergen was bound to the fatty food matrix and was possibly less 'bioavailable' [5].

FACTORS AFFECTING THE OUTCOME OF CHALLENGES

The outcome of a challenge is related to several different factors, such as the age of the patient. Younger children tend to show more frequently objective symptoms, possibly due to the fact that they cannot verbally report subjective symptoms adequately. The type of protocol used may influence the outcome as well. Short intervals between the doses may make it difficult to interpret which dose was responsible for the symptoms. Theoretically, previous doses can influence the reaction of following doses. Both enhancing and decreasing effects are possible. The clinical impact of this with the currently used protocols seems to be low. Unrecognized unstable disease (rhinoconjunctivitis/asthma or intercurrent disease) can aggravate the reactions. Therefore, it is important to take good care of the patient in the period prior to the challenge, and to instruct the patient to report any changes in health or medication status. Ideally adult patients are challenged starting in a fasting state to avoid interaction with any food already ingested. Sometimes this is impossible, especially with children, so a light, low fat meal such as sandwiches and marmalade or sugar can be allowed.

When evaluating symptoms reported by a patient, especially but not exclusively children, it is important to avoid suggestive or leading questions ('How are you feeling now?' is better than 'Are you feeling sick after that dose?'). It is prudent to not over-interpret minimal subjective symptoms, since experience with low dose DBPCFC shows that these usually disappear spontaneously and the challenge can continue. Otherwise these symptoms might be over-reported and might even lead to stopping the challenge prematurely.

INCLUSION AND EXCLUSION CRITERIA FOR PATIENTS TO UNDERGO TITRATED CHALLENGES

Patients of any age with a case history of an adverse reaction to food can be included in titrated food challenges. However, especially in case of a history of very severe reaction, the benefit of a challenge has to be weighed against

Table 4.1 Drugs Contraindicated for Challenges

Antihistamines in the previous 3 days, except hydroxyzine and dexclorfeniramina in the previous 10 days

Systemic corticosteroids in the 2 previous weeks

Tricyclic antidepressants in the previous 5 days

Immunosuppressive treatment

Monoamine oxidase (MAO) inhibitors in the previous 2 weeks

Beta-blocking agents in the previous 24 hours

Angiotensin-converting enzyme (ACE) inhibitors in the previous 2 days

the risk. Participants in low dose challenges have informally reported increased confidence and reassurance about their ability to deal with future low dose exposures in the community, but this has not been studied formally. When the culprit food is unknown, for instance, because of ingestion of a meal with many different food allergens, it is essential to determine the food allergen causing the severe reaction in order to give adequate and informed advice about future safe eating. In cases of several challenges to identify the culprit food each starting dose must be low (microgram or low milligram range).

Exclusion criteria are pregnancy, continuous intake of essential medications which might prevent or aggravate the allergic reaction or which might interfere with the treatment of a challenge induced allergic reaction (Table 4.1) as well as any contraindications for the administration of adrenaline (e.g., ischemic heart disease). In addition, patients with ongoing disease which might either hamper the interpretation of the challenge outcome (chronic urticaria, seasonal allergy rhinitis, asthma, active eczema) or aggravate the severity of the allergic reaction (acute viral infection, mastocytosis) are usually not considered for challenges. Seasonal influences (rhinitis) can be overcome by rescheduling until out of the relevant season and unstable asthma can be controlled by review of medication, smoking cessation advice, etc.

SAFETY ASPECTS OF TITRATED DBPCFC

As in all types of food challenges, the possibility of a severe allergic reaction occurring under provocation has to be taken into account. Therefore, a patient has to be under continuous experienced nursing and medical supervision during the procedure. Usually this is primarily done by nurses with medical support at hand. The nurses should be specifically trained in the early recognition of symptoms that can precede severe reactions, e.g., rhinoconjunctivitis, dry cough, itching in axillae and groin, and also in the administration of appropriate treatment, usually in close collaboration with a medical doctor. The supervision of children is a particular skill set, as non-verbal and preschool children cannot always explain how they feel. It has been widely observed that children who are relaxed and playing happily before and during the early

Table 4.2 Basic Clinical Requirements for the Implementation of Food Challenges

Medical doctor well-trained in the treatment of allergic diseases/particular anaphylaxis

Anesthesiology team (or equivalent team particularly trained in resuscitation) on call; at hand within 5 minutes, possibility for hospitalization and longer observation

Laryngoscope, intubation tube, ventilation bag, O_2 at hand

Heart defibrillator at hand

Peak flow-meter, spirometry apparatus at hand

High skills in inserting infusion lines warranted

Infusion line, infusion fluid at hand

Inhalative beta-2 mimeticum and corticosteroid inhaler at hand

Epinephrine inhaler at hand (e.g., Priamatene Mist or equivalent product) or 1 mg epinephrine in 2 mL NaCl to use in a nebulizer

Antihistamines and corticosteroids p.o. and i.v. at hand

Epinephrine i.m. (i.v.) at hand

challenge steps may become less active and quieter when they start to experience allergic symptoms that they cannot explain. They often revert to cuddling their parent. Experienced challenge nurses take great care in observing them for escalating reactions when these changes of mood and behavior are seen. Protocols of how and when to use the different medications should be in place. Basic requirements in terms of medical skills and equipment are listed in Table 4.2.

PRE- AND POST-CHALLENGE ASSESSMENTS OF PATIENTS

Patients have to be carefully examined before undergoing DBPCFC. Pulse rate, blood pressure, lung auscultation, basic peak flow value (PEF) and − in subjects with a case history of asthma or a bronchospasm as a consequence of an allergic reaction to the food under investigation − FEV_1 are recommended to be assessed. Furthermore, the oral cavity and the skin have to be carefully inspected. In case of a positive reaction to the challenged food, or at the end of the DBPCFC session, all parameters need to be reassessed. Intravenous access is recommended to be established before low dose DBPCFC in patients with history of systemic reactions to the study food or according to the judgment of the responsible physician. After intake of the last challenge dose the patients are usually kept under observation for at least 2 hours. In case of a severe allergic reaction, the post-challenge observation time has to be extended to at least 4 hours after complete recovery, taking into account the possibility of a biphasic reaction which might rarely occur up to 4 hours after the primary allergic response [6]. At discharge it is recommended to provide the patients with emergency drugs (antihistamines and corticosteroids), written information on how and when to use them, and contact details of the hospital in case symptoms develop after discharge. They need to be advised to report any late onset reactions occurring within 48 hours after the challenge. It is routine practice in some centers for study staff to contact the participant the day after the challenge to ensure all is well.

WHICH SYMPTOM DEFINES THE INDIVIDUAL THRESHOLD DOSE?

As the dose increases in titrated challenges in patients who have systemic (objective) reactions to the challenged food, the allergic patients' first allergic manifestation at a discrete dose is usually a subjective symptom. As the dose increases further, objective symptoms occur, often at an amount of the allergenic food that is substantially higher than the dose which provoked the subjective response. Table 4.3 summarizes the subjective symptoms that might be observed under challenge, and the corresponding objective symptoms which might occur at a later stage of the challenge. This dose-response relation has been shown in many different studies. In peanut allergy the LOAEL for subjective symptoms was 0.1 mg of peanut protein and 2 mg for objective symptoms in one titrated challenge study [7], or 10 mg and 100 mg, respectively, in another investigation [8]. In soy allergy, however, the LOEAL for subjective symptoms was 5.3 mg soy protein, and for objective symptoms of 241 mg, which is clearly much higher than for peanut [9]. The reasons for these differences in threshold doses between even closely related allergenic foods are not known at present.

A subjective symptom is an important warning signal for the allergic patient and also for the physician performing titrated challenges, since it might be indicative of a following objective, systemic, and more severe reaction. However, taking into account the challenge situation and the fact that patients under challenge are often under mental stress since they are expecting the allergic reaction and are afraid of what might occur, subjective reactions tend

Table 4.3 Organ-Related Subjective and Corresponding Objective Symptoms That Might Be Observed Under Challenge

Organ	Subjective Symptoms	Objective Symptoms
Skin	Itch	Flush, urticaria, angioedema
Oral mucosa	Itch	Blisters, redness, swelling
Gastrointestinal tract	Nausea, pain, cramps	Diarrhea, vomiting
Eyes/nose	Itch	Rhinitis, conjunctivitis
Lung	Tightness, chest pain, dyspnea	Hoarseness, wheezing, reduction of lung function, stridor
Cardiovascular system	Dizziness, vertigo	Tachycardia, drop of blood pressure

to be overemphasized and sometimes misinterpreted as significant challenge-related allergic symptoms, e.g., cholinergic urticaria, headache, irritation of the throat. Therefore, objective symptoms are more reliable. Within the framework of an international consensus conference, LOAEL has been defined as the dose of an allergenic food that induces mild objective symptoms in highly sensitive individuals [10]. This is, however, an academic statement and does often not reflect what occurs in actual titrated challenges. Patients with histories of systemic food-induced reactions often respond repetitively to the lower doses of the titrated challenges with pure subjective symptoms and might suddenly develop a severe allergic reaction at the next (higher) challenge dose. Therefore, the physician in charge has to be fully competent to treat a patient with severe or even anaphylactic reactions. To have the best possible read-out of a titrated challenge, all reactions, whether reported by the patient or observed by the physician in charge of the challenge have to be carefully recorded to allow the NOAEL and LOAEL to be determined for subjective and objective symptoms. In terms of the individual risk assessment, and as a basis to adequately advise the patient about future elimination diets, the assessment of thresholds for both subjective and objective symptoms is important.

CHALLENGE PROTOCOLS AND DOSING

Few reports have been published dealing with the standardization of DBPCFC protocols. In 2004, a position paper of the European Academy of Allergy and Clinical Immunology [2] was the first to provide general guidelines for the safe conduct of DBPCFC investigations. A consensus conference developed a standardized low dose challenge protocol [10–11]. These reports suggested starting doses of the order of 10 μg of the suspected food and a dose progression ranging from doubling to half or full-log intervals. Taking into account these past experiences, the EuroPrevall project designed a protocol for titrated challenges that has been successfully applied in a multicenter approach. Major recommendations have been recently published and are summarized in Table 4.4 [4]. The dosing schedule applied started with 3 μg food protein and went up to a discrete dose of 3 g, cumulatively 4.4 g protein as listed in Table 4.5. Using this regimen, few first dose responses have been observed for subjective reactions, but the NOAEL for objective symptoms could be determined for most but not all foods (unpublished data). It is essential to reach a cumulative top dose equal to a normal daily serving to assess the reactivity of a patient to a food. A 'non-reactivity' to small doses of the allergenic food is only exploitable in a patient who responds at a higher dose with allergic symptoms confirming that he or she is truly food allergic. As we learned from the EuroPrevall project, in foods for which 4.4 g proteins does not equal a normal serving, the challenge does not reliably confirm or exclude food allergy. In such foods, higher cumulative doses have to be respected in future threshold dose studies. For instance, 4.4 g of fish proteins equals an amount of 18 g native fish, a dose too low to elicit symptoms in a subgroup of fish allergic subjects.

Table 4.4 General Recommendations for Conduct of Challenge Studies According to the EuroPrevall Experiences [4]

1. Individual patient thresholds should be determined using DBPCFCs starting with very low doses of the suspected food, i.e., at or below 10 mcg/3 mcg protein, to ensure that no-one reacts at the lowest dose, and therefore an NOAEL can be established for the study.
2. To ensure adequate statistical power at least 29 patients with a food allergy previously confirmed by DBPCFC and not reacting to a discrete low dose must be used for the determination of NOAELs. The allergic status of these patients should be fully characterized so that they can be related to the overall allergic population.
3. Clinically relevant food allergy should be confirmed for the determination of a patient's LOAEL by increasing the doses until the first convincing (preferably objective) allergic reaction occurs or until a full daily serving has been ingested to exclude clinically relevant food allergy.
4. The challenge matrix should be low fat in content.
5. Sensory testing by trained testers should be used to confirm that participants cannot tell which preparation contains allergen.
6. The dose of allergen present in the prepared challenge materials should be verified in a representative sample by allergen detection techniques.
7. The challenge matrix should contain the investigated food in its most allergenic form, if known, with due regard for patient safety, e.g., raw egg would not be used due to possible contamination with *Salmonella* spp.
8. Time interval between the discrete doses should be preferably at least 20–30 min.
9. Placebo and active challenges should preferably be performed on separate days.
10. All reactions, whether reported by the patient (subjective) or observed by study personnel (objective) must be recorded in detail to allow for determination of NOAEL and LOAEL for each type of symptom and sign.

Table 4.5 Recommendations for a Dosing Regimen According to the EuroPrevall Project

Dose	Amount Protein
1	3 μg
2	30 μg
3	300 μg
4	3 mg
5	30 mg
6	100 mg
7	300 mg
8	1000 mg
9	3 g
Total (cumulative dose)	4.4 g

Active and placebo challenges are performed on two different days, for instance, about 1 week apart, with doses usually administered at an interval of 20 to 30 min. The sequence of the administration of DBPCFC meals, e.g., placebo and active meal, is determined by an independent co-worker not involved in the challenge process.

Challenges are discontinued after the dose leading to the first objective allergic symptoms is given or after ingestion of the whole meal. Objective symptoms considered for discontinuation of the process are alterations of the oral mucosa, such as blisters, swelling, or intense reddening; skin symptoms such as flush; urticaria; angioedema; rhinitis; conjunctivitis; drop of blood pressure of at least 20 mmHg, drop of FEV1 > 12% or PEF of at least 20%; laryngeal edema; diarrhea; or emesis. Severe, persistent, moderate to subjective symptoms lasting for more than 45 minutes, such as severe itching of palms, soles, and head, severe nausea, or gastric/abdominal pain, may be taken into account.

AN OPEN FOOD CHALLENGE ALWAYS HAS TO FOLLOW A NEGATIVE DBPCFC

In case of a negative DBPCFC, it is essential to perform an open food challenge to confirm the tolerance of the patient to the investigated food. False negative DBPCFCs do occur, due to degradation of the responsible allergens during preparation of the challenge meal, decreased releasability of the allergens from the matrix or a reduced allergenicity of the selected source material. Patients have to undergo an open food challenge with a normal helping (e.g., one hen's egg, one apple, 20 shelled raw hazelnuts, etc. preferably of the same source as used in DBPCFC). The starting dose is selected according the patient's history in regard to clinical sensitivity to the investigated food. One straightforward protocol is to administer three doses of the food in its most allergenic form, i.e., 1/8, 1/4, and 5/8 of the serving, at intervals of 30 minutes. The same pre- and post-challenge assessments and criteria for positivity as described for DBPCFC should be applied.

REFERENCES

[1] Asero R, Ballmer-Weber BK, Beyer K, Conti A, Dubakiene R, Fernandez- Rivas M, et al. IgE-mediated food allergy diagnosis: current status and new perspectives. Mol Nutr Food Res 2007;51:135−1347.
[2] Bindslev-Jensen C, Ballmer-Weber BK, Bengtsson U, Blanco C, Ebner C, Hourihane J, et al. European Academy of Allergology and Clinical Immunology. Standardization of food challenges in patients with immediate reactions to foods − position paper from the European Academy of Allergology and Clinical Immunology. Allergy 2004;59:690−7.
[3] Lidholm J, Ballmer-Weber BK, Mari A, Vieths S. Component-resolved diagnostics in food allergy. Curr Opin Allergy Clin Immunol 2006;6:234−40.
[4] Crevel RW, Ballmer-Weber BK, Holzhauser T, Hourihane JO, Knulst AC, Mackie AR, et al. Thresholds for food allergens and their value to different stakeholders. Allergy 2008;63:597−609.

[5] Grimshaw KE, King RM, Nordlee JA, Hefle SL, Warner JO, Hourihane JO. Presentation of allergen in different food preparations affects the nature of the allergic reaction — a case series. Clin Exp Allergy 2003;33:1581—5.

[6] Järvinen KM, Amalanayagam S, Shreffler WG, Noone S, Sicherer SH, Sampson HA, et al. Epinephrine treatment is infrequent and biphasic reactions are rare in food-induced reactions during oral food challenges in children. J Allergy Clin Immunol 2009;124:1267—72.

[7] Hourihane JO'B, Kilburn SA, Nordlee JA, Hefle SL, Taylor SL, Warner JO. An evaluation of the sensitivity of subjects with peanut allergy to very low doses of peanut protein: a randomized, double-blind, placebo-controlled food challenge study. J Allergy Clin Immunol 1997;100:596—600.

[8] Flintermann AE, Pasmans SG, Hoekstra MO, Meijer Y, van Hoffen E, Knol EF, et al. Determination of no-observed-adverse-effect levels and eliciting doses in a representative group of peanut-sensitized children. J Allergy Clin Immunol 2006;117:448—54.

[9] Ballmer-Weber BK, Holzhauser T, Scibilia J, Mittag D, Zisa G, Ortolani C, et al. Clinical characteristics of soybean allergy in Europe: a double-blind, placebo-controlled food challenge study. J Allergy Clin Immunol 2007;119:1489—96.

[10] Taylor SL, Hefle SL, Bindslev-Jensen C, Bock SA, Burks AW, Christie L, et al. Factors affecting the determination of threshold doses for allergenic foods: how much is too much? J Allergy Clin Immunol 2002;109:24—30.

[11] Taylor SL, Hefle SL, Bindslev-Jensen C, Atkins FM, Andre C, Bruijnzeel- Koomen C, et al. A consensus protocol for the determination of the threshold doses for allergenic foods: how much is too much? Clin Exp Allergy 2004;34:689—95.

Chapter | five

Thresholds or 'How Much Is Too Much?'

René W.R. Crevel[1], Barbara K. Ballmer-Weber[2], Steve L. Taylor[3], Geert Houben[4], Clare Mills[5]

[1]*Safety and Environmental Assurance Center, Unilever, Sharnbrook, Bedfordshire, UK*
[2]*Allergy Unit, Department of Dermatology, University Hospital Zürich, Switzerland*
[3]*Food Allergy Research & Resource Program, University of Nebraska, Lincoln, NE, US*
[4]*Food & Nutrition, TNO, Zeist, The Netherlands*
[5]*Institute of Inflammation and Repair, Manchester Academic Health Science Centre, Manchester Institute of Biotechnology, University of Manchester, UK*

CHAPTER OUTLINE

Risk Management for Food Allergy. http://dx.doi.org/10.1016/B978-0-12-381988-8.00005-1

INTRODUCTION

Since the emergence of food allergy as a public health issue in the early to mid 1990s, the question 'how much is too much' has been at the forefront of the mind of risk assessors and regulators, as well as allergic consumers and clinicians. Initial impressions from anecdotal reports suggested that thresholds were extremely low, although with hindsight such reports inevitably presented a biased picture focused on the more interesting cases. It was soon recognized that they formed a poor basis for risk assessment, and efforts began to generate clinical data [1]. In parallel, initiatives were set up to systematically gather available data, most of which were unpublished [2]. The latter effort has since been updated by the US FDA's Threshold Working Group [3,4]. This work revealed very significant data gaps but also highlighted some early conclusions about the difficulties in determining population thresholds, which are critical to the public health dimension. They thus also spurred new lines of investigation into methods to use these data effectively, while also highlighting considerations that were critical to data quality and usability.

The EuroPrevall project, which ran from 2005 to 2009, built on these earlier observations to deliver one of its core objectives, namely data and tools to improve food allergy and food allergen management. Actual data on thresholds were a critical element of these data, but just as important was the application and further improvement of new methodologies to analyze such data at the population level. A strong emphasis on high quality of data ran through the strategy of the project, delivered through rigorously defined protocols, applied to a consistent standard and with a high degree of resolution within the data. This chapter describes the unique features of the EuroPrevall strategy, linking it to the pre existing data and knowledge. It also considers their contribution to delivering the objectives of the project and the way that they will thereby help to improve the management of allergens from a public health perspective.

WHAT IS MEANT BY THRESHOLDS IN THE CONTEXT OF FOOD ALLERGY AND ALLERGENS?

The Concise Oxford English Dictionary (ninth edition) defines threshold (physiology) as 'a limit below which a stimulus causes no reaction', which operationally translates to a dose at, or below which, a response is not seen in an experimental setting [5].

Individual clinical thresholds as determined in a challenge study lie between the highest dose observed not to produce any adverse effect (No Observed Adverse Effect Level [NOAEL]) and the lowest dose to produce an adverse effect (the Lowest Observed Adverse Effect Level [LOAEL]). In food allergy, the term 'threshold' has often been approximated to the LOAEL, although the accuracy of this approximation depends on dose spacing. Allergic people respond over a very wide range of doses, and this, together with the limitations inherent in studies of human beings, makes the prospect of obtaining absolute experimental thresholds for food allergens for human populations a remote possibility.

Allergic responses, in common with other immune responses, consist of two phases: sensitization and elicitation. Thresholds probably apply to both phases. However, little is known about thresholds of sensitization to food proteins in human beings, and in practice the term 'threshold' in food allergy is largely used in relation to the elicitation phase. This chapter therefore only addresses thresholds of elicitation and furthermore limits itself to IgE-mediated reactions, which are those that can produce the most acutely life-threatening manifestations.

Thresholds exist at both an individual and a population level. Individual thresholds can be estimated experimentally, but this does not hold in practice for population thresholds. The term 'threshold' is also invested with different meanings in different contexts (e.g., regulatory thresholds and analytical thresholds), and the term 'minimum eliciting dose' is therefore preferred [6]. In modeling the distribution of minimum eliciting doses for any given allergenic food, the term Eliciting Dose (EDp) can thus be used to designate the amount of allergen predicted to produce a reaction in a defined proportion (for instance 0.5, 1, or 5%: ED0.5, ED01, or ED05) of the allergic population, to distinguish it from experimentally determined thresholds. The EDp can be considered as a threshold for a defined proportion of the allergic population.

THRESHOLDS BEFORE EUROPREVALL: WHAT DATA WERE AVAILABLE AND HOW USEFUL WERE THEY FOR RISK ASSESSMENT?
What Data Existed on Thresholds?

Case reports and series show that exposure to small quantities of an offending food can sometimes elicit a severe allergic reaction in a sensitized individual [7,8]. However, these studies provide little quantitative information. Diagnostic, double-blind, placebo-controlled food challenges (DBPCFC), in use since the 1970s [9,10], have generated more quantitative information on thresholds of reactivity. However, the design of these studies resulted in a high proportion of first dose reactors, which made them unsatisfactory for modeling the distribution of minimum eliciting doses and more generally for risk assessment [11]. Taylor et al. [2], in an analysis of data produced up to the late 1990s, found that several hundred patients had been challenged at lower doses with cows' milk [n=598], egg [n=782], and peanuts [n=663], as well as smaller numbers with other allergenic foods. However because these data were often obtained by means of different protocols, the estimation of a threshold dose was very difficult. Studies designed specifically to establish low dose reactivity did not appear until the late 1990s [1].

The most reliable and plentiful data on minimum eliciting doses (MEDs) result from challenge studies performed in peanut allergic patients. These data originated from a range of studies, including diagnostic challenge series using a low dose challenge methodology, low dose challenges designed to determine MEDs, but also immunotherapy studies. Data from these various sources, together with previously unpublished data, covering altogether over 450 patients, proved suitable for dose distribution modeling [12,13]. These analyses revealed ED10s (i.e., the doses estimated to give a reaction in 10% of peanut allergic individuals) on the order of 4 mg of peanut protein for the populations in question. Very recently, an extensive analysis of published and unpublished low dose challenge data on 13 allergenic foods was conducted by an Expert Panel convened by the Australian Allergen Bureau to review the action levels used in their Voluntary Incidental Trace Allergen Labeling (VITAL) scheme, which is described in further detail below.

The VITAL Scientific Expert Panel and Thresholds

The VITAL scheme is a comprehensive system for allergen management developed by the Allergen Bureau of Australia. It was first introduced in 2007 and was recently the subject of an extensive review and overhaul. It is beyond the scope of this chapter even to give an overview of the system. However, thresholds for labeling have been a critical and integral component of the system from the start and were therefore included in the recent review. Unlike other elements of the system, the Allergen Bureau decided that this review should be conducted by a panel of independent, internationally recognized experts.

In 2011, an extensive analysis of published and unpublished low dose challenge data on 13 allergenic foods was conducted by the VITAL Scientific Expert Panel for the Australian-New Zealand Allergen Bureau [14]. The VITAL Scientific Expert Panel convened by the Allergen Bureau is founded on a collaboration between the Food Allergy Research and Resource Program (FARRP, University of Nebraska, US) and the Netherlands Organization for Applied Scientific Research (TNO, Zeist, The Netherlands) together with other experts. The panel had access to and analyzed threshold data from published literature, unpublished clinical records in the Netherlands and Germany, and partially completed FARRP studies and concluded that sufficient data exist for most major allergenic foods of concern for the distribution of MEDs in the various populations of individuals who had undergone food challenges to be modeled statistically. The resulting dose distribution curves enable the establishment of an eliciting dose for each allergenic food (EDp) at which a certain proportion of the allergic population (p) would be likely to react. This approach was used to establish Eliciting Dose (ED) values to be used as reference doses for guiding decision making regarding the use of precautionary labeling ('may contain' labeling), which warns of the possible presence of small amounts of unintended allergen.

MED distributions based on both discrete and cumulative doses were modeled using three different statistical models (log normal, log logistic, and Weibull). ED values for all three models were determined, with preference being given to the model with the best fit at low doses, as determined by statistical and visual examination. Where sufficient data existed, in addition to the combined data, dose distributions were modeled separately for infants and children versus adults, in addition to the whole dataset. The challenge doses were normalized in all cases to mg of protein from the allergenic food.

Sufficient data from the available studies existed to allow dose distribution modeling for 11 major allergenic foods (see Tables 5.1 and 5.2). For four allergens, the number of data points was sufficiently abundant (good to excellent data set) to define ED01 values reliably (i.e., without recourse to low dose extrapolation beyond the experimental data set). For seven allergens with a dataset based on fewer individual MEDs, but still sufficient for statistical modeling, ED01 values sometimes might be less reliable, and the lower confidence interval of the ED05 was also considered as the basis for the

Table 5.1 Summary of VITAL Scientific Expert Panel Recommendations

Allergen	Reference Dose (mg Protein)	Basis of Reference Dose	Quality of Database[**]
Peanut	0.2	ED01	Excellent
Milk	0.1	ED01	Excellent
Egg	0.03	ED01 and ED05 95% lci[*]	Excellent
Hazelnut	0.1	ED01 and ED05 95% lci	Good
Soy	1	ED05 95% lci Note: this level may not completely protect certain individuals sensitive to soy milk	Sufficient
Wheat	1	ED05 95% lci Note: wheat-allergic consumers would be protected by foods containing < 20 ppm gluten	Sufficient
Cashew	2	ED05 95% lci	Sufficient
Mustard	0.05	ED05 95% lci	Sufficient
Lupin	4	ED05 95% lci	Sufficient
Sesame	0.2	ED05 95% lci	Marginally sufficient
Shrimp	10	ED05 95% lci	Marginally sufficient
Celery	n/a		Insufficient
Fish	n/a		Insufficient
Other tree nuts (walnut, pecan, almond, pistachio, brazil nut, macadamia, pine nut)			Insufficient

[*] Lower confidence interval
[**] The classification of quality reflects the abundance of data and its distribution across the dose range
(Allergen Bureau 2011)

Table 5.2 Reference Values for Various Allergenic Foods Recommended by the Vital Scientific Expert Panel and Example Action Levels Based on Two Food Intake Examples (5 and 50 g)*

Allergen	Reference Dose (mg Protein)	Action Level (ppm) for 5 g Serving Size: [Action Level in 2007 Version of VITAL]		Action Level (ppm) for 50 g Serving Size:
Peanut	0.20	40	[2]	4.0
Milk	0.10	20	[5]	2.0
Egg	0.03	6	[2]	0.6
Hazelnut	0.10	20	[2 – tree nuts]	2.0
Soy	1.00	200	[10]	20.0
Wheat	1.00	200	[20 – gluten]	20.0
Cashew	2.00	400		40.0
Mustard	0.05	10		1.0
Lupin	4.00	800		80.0
Sesame	0.20	40	[2]	4.0
Shrimp	10.00	2000	[2 – crustacea]	200.0
Celery	Insufficient data			
Fish	Insufficient data			

*Recommended health-based reference doses based upon the Eliciting Dose (ED)01 or Lower Confidence Interval (LCI) of ED05 for objective symptoms expressed as mg total protein from allergenic source to be used in VITAL calculation tool to calculate concentration action levels for precautionary (advisory) labeling (Allergen Bureau (2011)).

establishment of a reference dose. Taking into consideration the conservatism of some of the choices made (e.g., selection of the model giving the lowest EDp in many cases), reference doses were established for 11 allergens that will be protective for more than 95% to 99% of the allergic population. Based on the description of clinical symptoms from exposure to low doses of allergen, incidental effects that may occur with allergen exposures at or below these reference doses will generally be mild and transitory and require no medical intervention. Exquisitely sensitive allergic consumers may not be fully protected by the reference doses, but these consumers are generally recommended not to consume pre-packaged processed foods and can be given meaningful advice related to the reference doses. Other allergic consumers can rely on the safety of the reference doses (meaning that intakes at or below these doses will not induce severe reactions) provided they are in a stable state (no active infections, no unstable asthma, etc.). Meaningful advice on managing their condition when in a potentially unstable state can be provided by a physician. No safety factor was used in

Table 5.3 Protocols Used in Low Dose Challenges with Peanuts

Authors	Starting Dose	Last Dose	Peanut Material and Matrix	Interval Between Doses (Min)	Stop Criteria	Duration of Challenge
Peeters et al. 2007 [49]	10 µg	3 g	Partially defatted peanut flour in whole wheat instant cereals	15–30	3 times subjective symptoms or 45 mn duration otherwise objective symptoms	1 d
Flintermann 2006 [43]	5 µg	1.5 g	Partially defatted peanut flour in whole wheat instant cereals	15–30	3 times subjective symptoms otherwise objective symptoms	1 d
Hourihane 2005 [20] Grimshaw 2003 [32]	1 mg (first dose reactors)	4 g	Roasted and part. defatted peanut flour chocolate	15–30	Not defined	1 d
Wensing 2002 [19]	30 µg	1 g	Roasted peanut meal (Runner) Mashed potato cereal	30	subjective symptoms lasting > 1 h otherwise objective symptoms	1 d 2nd day for higher doses
Moneret-Vautrin et al. 1998 [53]	5 mg	20 g	Peanut seeds in mashed potato or apple puree	not defined	abdominal pain otherwise objective symptoms	1 d 2nd day for higher doses
Hourihane 1997) [1]	10 µg	50 mg	Peanut flour In wholegrain wheat flour	10–15	objective symptoms	1 d

From: Crevel et al. (2008) Reference [50]

the establishment of reference doses because of the built-in conservatism of the assumptions used both in the derivation of the reference doses (e.g., use of worst case statistical model) and in their subsequent application (e.g., it will be assumed that 100% of the produced batches of food will be contaminated and that all contamination is at the action level). Furthermore, the Expert Panel aimed at the best possible quantification of risks based on scientific data and probabilistic risk assessment principles, the latter considered to be the most suitable methodology for food allergen risk assessment for risk management purposes. The proposed reference doses are expected to result in an optimal balance between safety and practical value (feasibility, enforceability, monitoring possibilities, and meaningful labeling).

Action levels in the 2007 VITAL grid were expressed as concentrations based on a 5 g serving size. The panel recommended that action levels in the revised VITAL guidance should take into consideration differences in amounts of food consumed. An approach was recommended in which action levels are calculated from the reference doses using the following formula:

$$\text{Action Level (in mg/kg or ppm)} = [\text{reference Dose (in mg)} / \text{Intake (in g)}] \times 1000$$

The choice of intake figure significantly affects the calculated action level and hence the level of risk for sensitive consumers when using the action level as a cutoff for deciding on the use of a precautionary warning. The panel, therefore, recommends that the intake figure is determined by the use of accurate relevant dietary consumption data or internal company data, taking into account mean consumption and the 95th percentile consumption.

HOW HAVE THRESHOLD DATA BEEN GENERATED? PROTOCOLS AND THEIR EVOLUTION

The DBPCFC remains the 'gold standard' for confirming food allergy to this day, despite improvements in the predictive value of other diagnostic procedures, as acknowledged in guidelines [15,16]. In the earliest studies, starting doses, typically in the range of 250–500 mg of the food for the most sensitive subjects, were chosen to produce an objective but mild reaction [2,17]. Studies also differed in critical details, such as challenge procedures including timing of doses and whether placebo doses were interspersed with active ones [1], the form of the food used [11,18], the matrix in which it was presented [19–21], and the weight accorded to subjective and objective manifestations [21,22,23].

In the last 10 years, the DBPCFC has been adapted to generate threshold data, as the value of such data both for clinical management of food allergy, and subsequently for public health purposes, has become increasingly apparent. Several clinical trials, with doses ranging from micrograms to grams, have been specifically designed to determine MEDs for various allergenic foods (Table 5.2) [19,20,22,24–26]. The most recent have adhered fairly closely to a consensus low dose clinical challenge protocol, facilitating data analyses within and across studies, formulated at a roundtable conference under the aegis of the Food Allergy Research and Resource Program of the University of Nebraska, in 2002 [27]. Together with the 1999 conference on current knowledge of thresholds [2], it galvanized the clinical and regulatory community into recognizing the potential of low dose challenges to maximize the information from such procedures for the benefit of the individual patient but also for wider public health (see also chapter 4).

FACTORS AFFECTING THE OUTCOME OF CHALLENGE STUDIES AND THE TYPE OF DATA GENERATED

A wide variety of DBPCFC protocols, differing in potentially significant ways, have been used [2]. This has affected both the type of data generated, and therefore its value for specific purposes, and the extent to which studies could be compared, even when they nominally used the same outcome measures. The factors in challenge studies that can be controlled fall into three main categories: the challenge procedure itself, the selection of patients, and the challenge materials (summarized in Table 5.4). In addition, the data recorded about the response to the challenge material vary in their degree of detail and therefore the resolution with which symptom severity can be described.

Challenge Procedure

Conduct of the challenge determines the type of data generated and therefore its suitability for different purposes. The main factors influencing the precision of any threshold estimate include starting dose, dose progression, the time interval between doses, and the way in which placebo and active doses are randomized.

Table 5.4 Factors Affecting the Outcome of Challenge Studies and Type and Quality of Data Generated

Challenge Procedure

- Starting dose
- Dose progression
- Time interval between doses
- Placebo placement in sequence
- Scoring criteria
- Stop criteria

Patient-Related Criteria

- Individual thresholds
- Benefit
- Safety
- Thresholds for risk assessment studies (in addition to above)
- Documented reactivity to food
- Good patient characterization
- Include patients with previous severe reactions if safe

Challenge Materials

- Most allergenic form of food if known
- 'Real food' blinding matrix
- Dose and allergenic activity verification in matrix
- Sensory evaluation of blinding — taste, texture, smell

In some studies these have been interspersed [1], but in most recent studies active and placebo have been given on different days [28], which is also the recommendation of the Consensus conference [27]. Scoring of reactions and stop criteria, and therefore experience in the conduct of challenge studies, particularly in the case of subjective reactions, will also affect any threshold estimate. Recent recommendations on challenge protocols suggest starting doses of the order of 10 μg of the suspected food, dose progression ranging from doubling to half or full-log intervals, and a time interval between doses of 15−30 minutes, largely for practical reasons (Table 5.4) [27]. There is a recognition, however, that this type of dose progression resembles protocols for inducing tolerance in immunotherapy studies, which has led to proposals that 'one shot' studies should be conducted at a suitably low dose to validate dose distributions. In such study, a group of unselected allergic individuals, sufficiently large to give the study adequate statistical power, should be challenged at one single dose, for instance the ED5, to verify whether indeed approximately the expected proportion of the individuals (5% in the example) will show a reaction at that dose. Good practice and participant safety dictate that participants should discontinue medications likely to interfere with the outcome (or otherwise be excluded) and, if suffering from asthma, that their condition should be stable. This makes the allergen encounter during a challenge quite different from what might occur in the community, where the subjects' health and medication use, etc. may vary considerably [20].

Patient-Related Criteria

Inclusion and exclusion criteria for low dose challenge studies will differ according to the purpose for which the data are being generated. Where individual thresholds are being estimated, the primary concerns are benefit to and safety of the individual. Population studies on thresholds aim to generate data and test hypotheses that can be generalized to the relevant population and, in addition to the patient safety criteria, participant selection must reflect these needs. Participants must therefore be well characterized in relation to the allergic population, in particular in terms of their reactivity.

A key requirement in threshold studies is that the subjects are demonstrably still allergic to the food being tested. In several studies, the development of tolerance in previously allergic subjects has been demonstrated, in particular in children with milk and egg allergy [29]. Even up to 20% of peanut allergic patients may outgrow their food allergy [30]. This requirement becomes particularly important where interval-censoring survival analysis is used, since these data can be included.

Participant inclusion for risk assessment studies ideally needs to reflect population variability as far as possible and relevant. If this proves impracticable, the people tested should be characterized in as much detail as possible, such that the test group can be mapped onto the overall allergic population and generalizable conclusions can be drawn. Population variability encompasses both inter-individual variation within an otherwise homogeneous population and the possible existence of subpopulations with a different distribution of reactivity (e.g., children vs. adults). The most recent studies indicate that individual MEDs can span at least six orders of magnitude. However, statistical analysis of most studies suggests these figures represent the extremes of a continuous distribution, rather than two (or more) populations with distinct characteristics. With a few exceptions (e.g., [31]), MEDs have not been determined in random samples of the allergic population but instead using groups of food-allergic patients referred to specialist allergy clinics in tertiary care centers. The challenged population will therefore contain individuals who are more reactive than the general allergic population, although it may in many cases exclude any prior known severe reactors in the referred population.

No study exists to indicate what proportion of patients have been excluded from challenges on grounds of a previous severe reaction, so it is difficult to know to what extent their exclusion affects dose distributions and parameters derived from them. Patients with a history of moderate-to-severe reactions to peanut have been reported to have significantly lower MEDs than patients reporting mild reactions [19]. However, the difference was modest compared to the orders of magnitude differences between individuals, and most challenges were scored on subjective reactions, which could affect interpretation. A retrospective analysis of diagnostic challenges performed with milk, egg, peanut, soy, and wheat revealed that patients experiencing more severe reactions tended to react at a lower median percentage (15%) of the maximum dose (4 g of protein) than those experiencing milder reactions [11]. However, this still corresponds to a dose of 600 mg of food protein, nearly five orders of magnitude greater than the proposed starting doses in a low dose challenge protocol. This finding accords with the analysis of Sicherer et al. [10] that the majority of food-allergic patients do not even react to the first dose (400−500 mg) in the typical diagnostic challenge. Subjects with lower thresholds than those tested to date have been documented in a few case reports [23]. Thus, those undergoing low dose challenges are not automatically representative of the entire group with a specific food allergy. This may not necessarily occur solely, or even principally, because of the intentional exclusion of these highly sensitive individuals but could simply reflect their small number and consequently the smaller probability that they will be incorporated in a study. It should also be noted that the same phenomenon probably takes place at the other (high) end of the spectrum, for the same reasons. In fact, it could be argued that such individuals are less likely to come to the attention of health systems because they can manage their condition with relative ease. Overall, it remains unknown whether and to what extend these phenomena significantly influence the overall threshold distributions. Data are lacking, unfortunately, to quantify the relationship between the challenged population and the overall allergic population, although the EuroPrevall community studies should provide some insights.

Data are scarce about the existence of subpopulations with different thresholds. A few published challenge trials have evaluated both infants and adults for peanuts, but more often such studies have covered only certain populations for specific allergens. For instance, milk and eggs have been investigated largely in children, where these allergies occur most frequently. Taylor et al. [12] found differences between groups of challenge patients recruited for different purposes. However it is questionable whether these represent true subpopulations. Data have therefore generally proved too limited to conduct a systematic analysis of differences in threshold doses between infants/young children and adults. One study that looked at possible subpopulation differences in relation to asthma status did not confirm their existence [20].

Challenge Materials and Their Delivery

The key to success in the DBPCFC is the accurate delivery of a range of doses of the relevant allergenic food in a form that is unrecognizable to the patient. The test material must therefore be well characterized and its taste, smell, color, and texture must be masked.

Foods are consumed in various forms and following different types of processing. Ideally, in order to provide the greatest margin of safety, the threshold should be determined using the most allergenic and relevant form of the food, but in practice this can only be determined by challenge. Many different forms of food have been used in clinical DBPCFCs [2], ranging for instance from full fat to defatted foods (peanut flour, milk). Grimshaw et al. [32] showed the relevance of this observation, inasmuch as a higher-fat matrix resulted in a significantly higher MED in some subjects and more severe symptoms.

Processing can also influence the allergenicity of foods, but its effect is difficult to predict [33]. In pollen-related food allergy, cooked food is often better tolerated than raw food owing to the destruction of heat-labile proteins [34]. In contrast, in peanut allergy the challenge material and most of the allergenic proteins are much more stable and roasting may even increase allergenicity [30,35], although this remains to be confirmed clinically. Differences in allergen content among apple varieties correlate with reactivity in DBPCFC [36,37], but only minor differences have been noted between peanut cultivars [38], and there significance is still unconfirmed *in vivo*. Allergic people often differ in their reactivity to individual proteins in foods [39,40], but characterization of challenge materials with regard to their content and profile of allergenic proteins has received little attention.

The simplest way of masking a food is to deliver it in capsules. This resolves the sensory issues but makes it difficult to deliver relatively high doses. This route also by-passes the oral cavity as a site where symptoms may occur, which may bias the outcome towards more severe symptoms. For these reasons, recent recommendations discourage it [15]. The only alternative to capsules is a blinding matrix that is a 'real' food. These systems need not have an active and placebo that are indistinguishable; the patients must merely be unable to tell which is which. Taste and smell can be masked by a stronger taste and smell [41,42], although allergenic materials with strong smells or flavors can still pose problems and nose clips can be beneficial for particularly pungent foods. Pre-treatments of the food such as freeze drying or defatting may help reduce smell or taste, but these can only be used with due regard to their effect on the allergens. Similar handling considerations apply to labile allergens, such as the Bet v1 homologues in fruits.

Masking should always aim to maximize the amount of active compound in the matrix, thereby minimizing the amount of material that the patient is required to consume and the probability of non-specific gastrointestinal symptoms, which could decrease the sensitivity of the test. A close mimic to the active food in terms of the type of flavor could also be used as the placebo where applicable, although no instances have yet been reported.

As well as taste and smell, the other main sensory attribute that needs to be masked is texture, which can be achieved by adding material of a granular nature such as oatmeal, which can mask the texture of peanut flour [42]. Again mimics can be used effectively for masking texture in some foods. Starch-based thickeners can be very good for controlling the thickness of a challenge food, but the level of granularity can still be an issue. Color also needs to be masked, but this is generally straightforward as the addition of relatively small amounts of highly colored ingredients can be sufficient. Failing that, lighting can be controlled in the area the challenge is administered, as is standard practice in sensory testing. The effectiveness of masking should be verified by testing the recipes using a trained sensory panel and standard sensory testing protocols (e.g., triangle tests [43]).

The availability of allergen (releasability) in different matrices is also critical to the interpretation of challenge studies, and the food should ideally be presented in a form and matrix that assures maximum availability. As recently shown in peanut allergy [32,42] for instance, the fat content of a challenge vehicle can have a profound effect on the kinetics of the clinical reaction. Other constituents (e.g., polyphenols) may also reduce availability. These considerations highlight the need for a more thorough assessment of availability, as well as confirmation of a selection of the doses administered.

TOOLS FOR THE ANALYSIS OF THRESHOLD DATA

In parallel with the generation of threshold data, methodologies have been investigated to analyze these data effectively, particularly in the context of hazard characterization and subsequently risk assessment. Initial considerations driving these developments were the difficulty of identifying population thresholds owing to the wide range of reactivity of allergic patients but also the exclusion from challenges of individuals who had experienced severe reactions and were therefore assumed, although not proven, to be more sensitive. These considerations were coupled with the observation that a classical risk assessment approach, based on applying uncertainty factors to a NOAEL or

LOAEL, might result in such low regulatory thresholds that they would be impossible to apply in practice and would therefore drive the proliferation of precautionary labeling that research into risk assessment for food allergens sought to avoid. These fears were confirmed by a limited number of conference publications illustrating the approach for peanuts. Early work, under the aegis of the ILSI-Europe food allergy task force, proposed dose distribution modeling as a useful approach to characterize the risk from allergenic foods [44]. Conceptually, this approach attempts to build a cumulative distribution of minimum eliciting doses based on food challenge data and fit the resulting distribution of MEDs statistically. This enables prediction of the frequency of reaction to any given amount of allergen, although the limitations of the data and modeling tools must be clearly understood in order for sound conclusions to be drawn [6]. In particular, predictions based on extrapolation well beyond the area for which actual data exist are highly dependent on the model used and fraught with considerable uncertainty. Dose distribution modeling has benefited from the large amount of data that has become available for some allergenic foods and from development of the technique. Interval censoring survival analysis is one example of a methodology that has recently been applied, enabling large amounts of additional data to be used through the inclusion of first dose reactors and those who, although allergic, did not react at the highest dose used in a particular study. Data from different peanut allergic populations have now been effectively modeled [12,13] and now form a sound basis for assessing risk from small amounts of this most notorious allergenic food [45].

However, risk assessment implies probability and, therefore, involves a consideration of a number of factors. Thus, at its simplest, the probability of a reaction to an allergen will depend on the amount present, the probability that an allergic person will consume the product (in sufficient amount), and that the person is sensitive enough to respond to the amount of allergen consumed [46,47]. Both the probability that an allergic person will consume the product and the amount present can be influenced by other factors. For instance, a 'may contain' label could reduce the probability but may be disregarded if the consumer has eaten similar products without reacting. Unlike ingredients, which are added in known quantities, the amounts present by cross-contact can vary from absent to significant across a batch or production run, and this distribution will vary from product to product, often bearing no relationship to the absence, presence, or wording of a precautionary label [48]. This explains, of course, why the consumer mentioned above has eaten the product safely on previous occasions. The same estimate of cross-contact for products made according to different processes, using worst case assumptions, may also mask completely different allergen distributions and consequently different risk profiles. The probabilistic approaches described by Spanjersberg et al. [48] and Rimbaud et al. [49] open the possibility not only for more refined quantitative assessment of the global public health risk from a category of products. They can also be applied to specific processes and actively help to improve management of allergens by enabling the investigation of particular scenarios to see which are most effective in minimizing risk (see chapter 6).

EuroPrevall and the Development of Threshold Data

The EuroPrevall project was designed to address many of the drawbacks of available MED data, both from a qualitative and a quantitative perspective. The project's strategy was to develop a high quality dataset on allergenic foods of public health importance within the European Union of sufficient size to allow comparison and integration with existing data as well as to provide a sound basis for public health decisions on allergen management. Provision of high quality data focused on protocol and clinical data recording, both prior to and during the challenge, in order to characterize each challenged individual as closely as possible. This element also required training of those administering the challenges in order to ensure that similar descriptors reflected similar reactions in the challenged individuals.

Low Dose Challenge Protocol

The EuroPrevall low dose challenge protocol evolved from the protocol proposed at the Second Threshold Conference [27] and has been described in detail in other publications [50,51] and in chapter 4. Key features of the protocol itself include a dose range spanning six orders of magnitude (from 3 µg to 3 grams of protein), chosen in order to avoid or greatly minimize the number of first dose reactors as well as those failing to react at the highest dose despite meeting all the inclusion criteria for allergic reactivity. This approach enhanced confidence in the MEDs obtained in that region critical to risk assessment and from a statistical standpoint minimized right and left censoring when using interval censoring survival analysis.

Another feature of the EuroPrevall challenge protocol is the use of a universal matrix for delivery of allergen in bio-available form. This matrix was developed in the form of a real food product, specifically a dessert, in order to improve palatability and therefore acceptability to those were challenged but also to provide a more realistic vehicle compared with those usually used. Importantly, this matrix was tested among a small group of patients in order to demonstrate the bioavailability of a number of allergens. The matrix was, however, only suitable for children above the age of 5, but where it could not be used, other standardized matrices were used, thus maintaining the principle of the use of a well-defined challenge vehicle. The standard dessert matrix worked for most foods used, although specifically prepared vehicles had to be employed with some foods such as shrimp, apple, and peach.

Thorough characterization of test materials in terms of source and composition, together with verifying actual concentration in the matrix, has been an established feature in toxicological studies for many years. EuroPrevall grounded its approach in this principle. Well-defined forms of the allergens with relevance to what is normally consumed were used, usually in the form of ingredients supplied to food manufacturers. The doses attained in the matrix were also verified, as was the homogeneity of dilutions when prepared according to the detailed instructions supplied with the materials. These enabled people to be challenged in a highly reproducible fashion.

The nature of double-blind food challenges implies that blinding of the subjects as well as the operators to the material being administered is critical to the integrity of the procedure. As described, most foods (those with labile allergens being the exception) were tested by trained sensory panelists in triangle tests to ensure that the test material could not be identified, even though there might be a difference between test and control [51].

Development of a high quality dataset for food challenge studies requires more than accurate characterization of the allergenic foods and challenge materials. It also needs standardized conditions for conduct of the study and data recording, both prior to and during the challenge. Only under those circumstances can the results of a food challenge study be accurately and unequivocally interpreted in the context of the population tested. EuroPrevall put in place a number of procedures to achieve this high level of data quality. Firstly, a complete and detailed allergological history was taken from each patient, which subsequently enabled a detailed description of the profile of the challenged population, making it available for comparison with other populations challenged with the same allergenic food. Similar attention was paid to recording up to 20 different signs and symptoms during challenges, both subjective and objective, on a comprehensive case report form (Table 5.5). This will permit detailed comparisons between different groups of allergic individuals, in particular those recruited through the community surveys and who should be more representative of the distribution of reactivity in the allergic population at large than individuals recruited through attendance at allergy clinics. Finally, those clinicians administering challenges underwent training by clinicians expert in conducting food challenges to ensure adherence to the protocol and integrity of data recording.

Dose Distribution Analysis Tools

Development of tools to analyze dose distribution data is also essential to making the most effective use of the data generated. The type and quality of data generated as part of the project were predicated on the needs of the dose distribution modeling approach. Very specifically, as mentioned previously, as well as starting at a very low dose, the protocols encompassed a very wide range of doses in order to avoid first dose reactors, but they also included a sufficient range of doses to provide a good resolution for the dose distribution curves. Initial thoughts focused on modeling MEDs (i.e., LOAELs), but interval censoring survival analysis [12] was later introduced such that the curves more accurately reflected the distribution of actual thresholds. Dose distribution analyses described in the published literature focus entirely on the doses that provoked the reactions during the challenges, without any consideration of the response observed during the reaction. Severity is of course a

Table 5.5 Symptom Scoring in EuroPrevall Food Challenges

Symptom	
Short Code	Definition
OAS (S)	Oral Allergy Syndrome — itching and tingling sensation on lips, oral cavity, auditory canal, throat
B (O)	Blisters of the oral mucosa
Lpru (S)	localized pruritus
Gpru (S)	Generalized pruritus
R (O)	Rhinitis
C (O)	Conjunctivitis
Dph (S)	Dysphagia
N (S)	Nausea
F (O)	Flush
U (O)	Urticaria
Co (O)	Cough
D (S WP1.3, O WP1.1)	Dyspnea
AE (O)	Angioedema
G (S)	Gastric pain and/or burning, abdominal pain [worse children]
E (O)	Emesis [worse adults]
Di (O)	Diarrhea
L (O)	Larynx-edema
BS (O)	Bronchospasm: positive lung auscultation and/or significant decrease of basal FEV1 (> 12%) or PEF (> 20%)
BP (O)	Blood pressure drop (at least 20 mmHg)
S (O)	Shock

Note: (S) and (O) denote Subjective and Objective symptoms, respectively.

critical component of the public health impact of allergic reactions. The Euro-Prevall team examined ways in which symptom severity could be taken into account, although it acknowledged that only a limited range of severity would normally be encountered during properly conducted food challenges where stop criteria included any mild objective reaction. A severity visualization tool, originally developed to check data entries, showed promise as a way to visualize severity patterns among allergens, with the possibility of developing it further as a help to decision making.

EuroPrevall also drove the development of tools for the application of challenge data to risk assessment for public health. Madsen et al. [45] described three possible methods for conducting such analyses, dependent on the quantity and quality of data available. The safety assessment approach was the closest to classical toxicological evaluation of chemicals with a safety or uncertainty factor applied to the highest dose not to provoke a reaction. This approach, while highly conservative, makes use of only a single data point in dose distribution. In contrast, an approach based on the Benchmark dose method (BMDL) builds a dose distribution from the whole dataset, from which an appropriate point of departure (e.g., 95% lower confidence interval of the ED10 or ED5) is selected for calculation of the margin of exposure. Neither method, however, permits the elaboration of a quantitative risk assessment. The report thus strongly recommends the probabilistic modeling methodology. This takes into account not only the distribution of the MEDs but also the distribution of residual allergen contamination in the product population, as well as the extent to which the product is eaten (for a more detailed description see chapter 6). However, issues remain to be resolved. For instance, current models seem to predict a higher number of allergic reactions to foods compared with the numbers reported to disease registries or encountered in medical practice. It is unknown whether such differences are due to conservatism in the risk assessment models or to underreporting of actual allergic incidents, or indeed a combination of both factors. Resolution of these issues will require the outcomes of the 'one shot' studies, as well as more comprehensive adoption and analysis of allergic reaction registries, such as those proposed by Worm et al. [52]. At the time of writing, a prospective study enumerating and characterizing all the allergic reaction experienced by a cohort of allergic individuals is underway in the Netherlands. Another is planned in a different population as one of the objectives of a new EU-funded project on food allergy and allergen management.

Preliminary Observations on Thresholds from EuroPrevall

Initial analyses of the challenge data have already provided important insights (Tables 5.6a, b). While confirming previous observations that subjective symptoms occur at lower doses than objective ones, EuroPrevall data have begun to quantify the differences between the two types of symptom. Thus, ED10s based on subjective symptoms were approximately two orders of magnitude lower than ED10s based on objective symptoms, irrespective of the allergenic food tested.

Important observations were also made on individual allergenic foods. Peanut, well-documented in many other studies and long considered to be the most potent allergenic food, yielded an ED10 for objective responses of the same order (4 mg peanut protein) as previously reported [12,13]. However, this value was close to that of the hazelnut ED10, as well as the ED10s for fish and celery, consistent with previous reports of severe reactions to the latter. Shrimp, in contrast, had an ED10 two orders of magnitude higher than those allergenic foods. In contrast, milk and egg, which are predominantly allergens of infants and young children, showed exceptionally low ED10s in

Table 5.6a Preliminary Results of Dose Distribution Analysis in EuroPrevall (Subjective)

Food	Numbers Challenged	Numbers Positive (%) (total)	Left Censored (subjective)	ED10 (Subjective) (mg Protein)		
				Lognormal	Loglogistic	Weibull
Peanut	135	60 (44.4)	10	0.007	0.006	0.003
Hazelnut	132	91 (68.9)	8	0.009	0.01	0.004
Celeriac	64	41 (64.1)	7	0.002	0.003	0.001
Fish	50	34 (68.0)	2	0.2	0.5	0.3
Shrimp	55	30 (54.0)	3	9.3	85.5	63.9
Egg < 3.5	278	162 (58.3)	0			
Egg > 3.5	36	21 (58.3)	0	0.2	0.2	0.04
Milk < 3.5	382	133 (34.8)	0			
Milk > 3.5	36	14 (38.9)	0	0.02	0.02	0.01

Table 5.6b Preliminary Results of Dose Distribution Analysis in EuroPrevall (Objective)

Food	Numbers Challenged	Numbers Positive (%) (total)	Left Censored (objective)	ED10 (Objective) (mg Protein)		
				Lognormal	Loglogistic	Weibull
Peanut	135	62 (45.9)	2	2.8	6.6	5.2
Hazelnut	132	91 (68.9)	0	8.5	9.9	10.1
Celeriac	64	41 (64.1)	0	1.6	2.8	2.6
Fish	50	34 (68.0)	1	27.3	32.6	25.8
Shrimp	55	30 (54.0)	0	2504	2574	2532
Egg < 3.5	278	162 (58.6)	4	0.6	1.3	1.0
Egg > 3.5	36	21 (58.3)	0	27.1	26.4	20.4
Milk < 3.5	382	133 (34.8)	4	0.1	0.2	0.1
Milk > 3.5	36	14 (38.9)	0	5.3	7.6	6.6

dose distributions including participants less than 3.5 years old. Data from a limited number of older children and adults indicated, however, that as those allergies resolve in most of the infants and young children, the remainder evolve to a less sensitive state, with ED10s of the same order as peanut in the case of milk and one order of magnitude higher for egg. Other data on a small number of participants challenged more than once suggest that maintenance of a milk or egg allergy into later life is associated with initially greater sensitivity (lower MEDs).

CONCLUSIONS

Thresholds of elicitation for allergenic foods constitute a critical piece of data for the characterization of the hazard posed by those foods to allergic individuals and therefore also play a crucial role in assessing the risk from those foods. It is unsurprising therefore that much effort has gone into generating data on them and developing tools to analyze such data for the purposes of risk assessment, with the result that quantitative risk assessments are now possible for many allergenic foods. Data and knowledge to improve the management of allergens and thereby address the public health impact of food allergy were at the heart of the concept of the EuroPrevall project. For thresholds, this translated not only into over 1,000 challenges on nine priority foods according to a detailed and well-defined protocol with high resolution symptom recording. It also involved the development of new tools to analyze such data and maximize the value obtained from them, in particular with incorporation of a severity dimension, as well as the further elaboration and wider application of methods deployed in other studies during the lifetime of the project. While much of the data remain to be analyzed in detail, findings so far have begun to quantify the relationship between subjective and objective reactions, as well as provide insights into quantitative differences in reactivity to egg and milk between infants and pre-school children and older children and adults. Findings for other allergenic foods such as peanut and hazelnut have also largely confirmed other studies.

REFERENCES

[1] Hourihane JO'B, Kilburn SA, Nordlee JA, Hefle SL, Taylor SL, Warner JO. An evaluation of the sensitivity of subjects with peanut allergy to very low doses of peanut protein: a randomized, double-blind, placebo-controlled food challenge study. J Allergy Clin Immunol 1997;100:596−600.

[2] Taylor SL, Hefle SL, Bindslev-Jensen C, Bock SA, Burks AW, Christie L, et al. Factors affecting the determination of threshold doses for allergenic foods: how much is too much? J Allergy Clin Immunol 2002;109:24−30. US.

[3] FDA/CFSAN. Report prepared by the Threshold Working Group. Approaches to establish thresholds for major food allergens and for gluten in food. http://www.fda.gov/downloads/Food/IngredientsPackagingLabeling/UCM192048.pdf; 2006. Accessed 27 August 2013.

[4] US FDA Threshold Working Group. Approaches to establish thresholds for major food allergens and for gluten in food. J Food Prot 2008;71(No. 5):1043−88.

[5] Kroes R, Galli C, Munro I, Schilter B, Würtzen G. Threshold of toxicological concern for chemical substances present in the diet: a practical tool for assessing the need for toxicity testing. Food Chem Toxicol 2000;38:255−312.

[6] Crevel RWR, Briggs D, Hefle SL, Knulst AC, Taylor SL. Hazard characterization in food allergen risk assessment: the application of statistical approaches and the use of clinical data. Food Chem Toxicol 2007;45(5):691−701.

[7] Gern JE, Yang E, Evrard HM, Sampson HA. Allergic reactions to milk-contaminated 'non-dairy' products. N Engl J Med 1991;324:976−9.

[8] Laoprasert N, Wallen ND, Jones RT, Hefle SL, Taylor SL, Yunginger JW. Anaphylaxis in a milk-allergic child following ingestion of lemon sorbet containing trace quantities of milk. J Food Prot 1998;61:1522−4.

[9] Bock SA, Sampson HA, Atkins FM, Zeiger RS, Lehrer S, Sachs M, et al. Double-blind, placebo-controlled food challenge (DBPCFC) as an office procedure: a manual. J Allergy Clin Immunol 1988;82:986−97.

[10] Sicherer SH, Morrow EH, Sampson HA. Dose-response in double-blind, placebo-controlled oral food challenge in children with atopic dermatitis. J Allergy Clin Immunol 2000;105:582−6.

[11] Perry TT, Matsui EC, Conocer-Walker MK, Wood RA. Risk of oral food challenges. J Allergy Clin Immunol 2004;114:1164−8.

[12] Taylor SL, Crevel RWR, Sheffield D, Kabourek J, Baumert J. Threshold dose for peanut: risk characterization based upon published results from challenges of peanut-allergic individuals. Food Chem Toxicol 2009;47:1198−204.

[13] Taylor SL, Moneret-Vautrin DA, Crevel RW, Sheffield D, Morisset M, Dumont P, et al. Threshold dose for peanut: risk characterization based upon diagnostic oral challenge of a series of 286 peanut-allergic individuals. Food Chem Toxicol 2010; 48:814−9.

[14] Allergen Bureau. Summary of the VITAL Scientific Expert Panel Recommendations. http://allergenbureau.net/downloads/vital/VSEP-Summary-Report-Oct-2011.pdf; 2011. Accessed 27 August 2013.

[15] Bindslev-Jensen C, Ballmer-Weber BK, Bengtsson U, Blanco C, Ebner C, Hourihane J, et al. European Academy of Allergology and Clinical Immunology. Standardization of food challenges in patients with immediate reactions to foods − position paper from the European Academy of Allergology and Clinical Immunology. Allergy 2004;59:690−7.

[16] NIAID-Sponsored Expert Panel. Guidelines for the diagnosis and management of food allergy in the United States: report of the NIAID-Sponsored Expert Panel. J Allergy Clin Immunol 2010;126(6). Supplement S1−S58.

[17] Oppenheimer JJ, Nelson HS, Bock SA, Christensen F, Leung DYM. Treatment of peanut allergy with rush immunotherapy. J Allergy Clin Immunol 1992;90:256−562.

[18] Sampson HA, Ho DG. Relationship between food-specific IgE concentrations and the risk of positive food challenges in children and adolescents. J Allergy Clin Immunol 1997;100(4):444−51.

[19] Wensing M, Penninks AH, Hefle SL, Koppelman SJ, Bruijnzeel-Koomen CAFM, Knulst AC. The distribution of individual threshold doses eliciting allergic reactions in a population with peanut allergy. J Allergy Clin Immunol 2002;110:915−20.

[20] Hourihane JO, Knulst AC. Thresholds of allergenic proteins in foods. Toxicol Appl Pharmacol 2005;207(Suppl. 2):152−6.

[21] Osterballe M, Bindslev-Jensen C. Threshold levels in food challenge and specific IgE in patients with egg allergy: is there a relationship? J Allergy Clin Immunol 2003;112: 196−201.

[22] Wensing M, Penninks AH, Hefle SL, Akkerdaas JH, van Ree R, Koppelman SJ, et al. The range of minimum provoking doses in hazelnut-allergic patients as determined by double-blind, placebo-controlled food challenges. Clin Exp Allergy 2002b;2002(32): 1757−62.

[23] Bullock RJ, Barnett D, Howden ME. Immunologic and clinical response to parenteral immunotherapy in peanut anaphylaxis − a study using IgE and IgG4 immunoblot monitoring. Allergol Immunpathol (Madr) 2005;33:250−6.

[24] Morisset M, Moneret-Vautrin DA, Kanny G, Guenard L, Beaudouin E, Flabbee J, et al. Thresholds of clinical reactivity to milk, egg, peanut and sesame in immunoglobulin E-dependent allergies: evaluation by double-blind or single-blind placebo-controlled oral challenges. Clin Exp Allergy 2003;33:1046−51.

[25] Flintermann AE, Pasmans SG, Hoekstra MO, Meijer Y, van Hoffen E, Knol EF, et al. Determination of no-observed-adverse-effect levels and eliciting doses in a representative group of peanut-sensitized children. J Allergy Clin Immunol 2006;117: 448−54.

[26] Scibilia J, Pastorello EA, Zisa G, Ottolenghi A, Bindslev-Jensen C, Pravettoni V, et al. Wheat allergy: a double-blind, placebo-controlled study in adults. J Allergy Clin Immunol 2006;117:433−9.

[27] Taylor SL, Hefle SL, Bindslev-Jensen C, Atkins FM, Andre C, Bruijnzeel-Koomen C, et al. A consensus protocol for the determination of the threshold doses for allergenic foods: how much is too much? Clin Exp Allergy 2004;34:689−95.

[28] Ballmer-Weber BK, Holzhauser T, Scibilia J, Mittag D, Zisa G, Ortolani C, et al. Clinical characteristics of soybean allergy in Europe: a double-blind, placebo-controlled food challenge study. J Allergy Clin Immunol 2007;119:1489−96.

[29] Host A. Cow's milk protein allergy and intolerance in infancy. Some clinical, epidemiological and immunological aspects. Pediatr Allergy Immunol 1994;5(Suppl. 5):1−36.

[30] Skolnick HS, Conocer-Walker MK, Koerner CB, Sampson HA, Burks W, Wood RA. The natural history of peanut allergy. J Allergy Clin Immunol 2001;107:367−74.

[31] Zuberbier T, Edenharter G, Worm M, Ehlers I, Reimann S, Hantke T, et al. Prevalence of adverse reactions to food in Germany − a population study. Allergy 2004;59:338−45.

[32] Grimshaw KEC, King RM, Nordlee JA, Hefle SL, Warner JO, Hourihane JOB. Presentation of allergen in different food preparations affects the nature of the allergic reaction − a case series. Clin Exp Allergy 2003;33:1581−5.

[33] Thomas K, Herouet-Guicheney C, Ladics G, Bannon G, Cockburn A, Crevel R, et al. Evaluating the effect of food processing on the potential human allergenicity of novel proteins: international workshop report. Food Chem Toxicol 2007;45:1116−22.

[34] Ballmer-Weber BK, Scheurer S, Fritsche P, Enrique E, Cistero-Bahima A, Haase T, et al. Component resolved diagnosis using recombinant allergens in cherry allergic patients. J Allergy Clin Immunol 2002;110:167−73.

[35] Maleki SJ, Chung SY, Champagne ET, Raufman JP. The effects of roasting on the allergenic properties of peanut proteins. J Allergy Clin Immunol 2000;106:763−8.

[36] Vieths S, Jankiewicz A, Schoning B, Aulepp H. Apple allergy: the IgE-binding potency of apple strains is related to the occurrence of the 18-kDa allergen. Allergy 1994;49:262−71.

[37] Bolhaar STHP, van de Weg WE, van Ree R, Gonzalez-Mancebo E, Zuidmeer L, Bruijnzeel-Koomen CAFM, et al. *In vivo* assessment with prick-to-prick testing and double-blind, placebo-controlled food challenge of allergenicity of apple cultivars. J Allergy Clin Immunol 2005;116:1080−6.

[38] Koppelman SJ, Vlooswijk RA, Knippels LM, Hessing M, Knol EF, van Reijsen FC, et al. Quantification of major peanut allergens Ara h 1 and Ara h 2 in the peanut varieties Runner, Spanish, Virginia, and Valencia, bred in different parts of the world. Allergy 2001;56(2):132−7.

[39] Peeters KA, Koppelman SJ, Van Hoffen E, van der Tas CW, Hartog Jager CF, Penninks AH, et al. Does skin prick test reactivity to purified allergens correlate with clinical severity of peanut allergy? Clin Exp Allergy 2007;37(1):108−15.

[40] Mittag D, Vieths S, Vogel L, Becker WM, Rihs HP, Helbling A, et al. Soybean allergy in patients allergic to birch pollen: clinical investigation and molecular characterization of allergens. J Allergy Clin Immunol 2004;113:148−54.

[41] Huijbers GB, Colen AAM, Jansen JJN, Kardinaal AFM, Vlieg-Boerstra BJ, Martens BPM. Masking foods for food challenge: practical aspects of masking foods for a double-blind, placebo-controlled food challenge. J Am Diet Assoc 1994;94(6):645−9.

[42] van Odijk J, Ahlstedt S, Bengtsson U, Borres MP, Hulthen L. Double-blind placebo-controlled challenges for peanut allergy the efficiency of blinding procedures and the allergenic activity of peanut availability in the recipes. Allergy 2005;60(5):602−5.

[43] Vlieg-Boerstra BJ, Bijleveld CMA, van der Heide S, Beusekamp BJ, Wolt-Plompen SAA, Kukler J, et al. Development and validation of challenge materials for double-blind, placebo-controlled food challenges in children. J Allergy Clin Immunol 2004;113:341−6.

[44] Bindslev-Jensen C, Briggs D, Osterballe M. Can we determine a threshold level for allergenic foods by statistical analysis of published data in the literature? Allergy 2002;57:741−6.

[45] Madsen CB, Hattersley S, Buck J, Gendel SM, Houben GF, Hourihane JO'B, et al. Approaches to risk assessment in food allergy: report from a workshop 'developing a framework for assessing the risk from allergenic foods'. Food Chem Toxicol 2009; 47(2):480−9.

[46] Spanjersberg MQ, Kruizinga AG, Rennen MA, Houben GF. Risk assessment and food allergy: the probabilistic model applied to allergens. Food Chem Toxicol 2007;45(1): 49−54.

[47] Kruizinga AG, Briggs D, Crevel RW, Knulst AC, van den Bosch LM, Houben GF. Probabilistic risk assessment model for allergens in food: sensitivity analysis of the minimum eliciting dose and food consumption. Food Chem Toxicol 2008;46: 1437−43. Epub 2007 Oct 1.

[48] Spanjersberg MQ, Knulst AC, Kruizinga AG, Van Duijn G, Houben GF. Concentrations of undeclared allergens in food products can reach levels that are relevant for public health. Food Addit Contam Part A Chem Anal Control Expo Risk Assess 2010;27: 169−74.

[49] Rimbaud L, Heraud F, La Vieille S, Leblanc JC, Crepet A. Quantitative risk assessment relating to adventitious presence of allergens in food: a probabilistic model applied to peanut in chocolate. Risk Anal 2010;(1). 2010 Jan;30(1):7-19. Epub 2009 Dec 11.

[50] Crevel RWR, Ballmer-Weber BK, Holzhauser T, Hourihane JO, Knulst AC, Mackie AR, et al. Thresholds for food allergens and their value to different stakeholders. Allergy 2008;63(5):597−609.

[51] Cochrane SA, Salt LJ, Wantling E, Rogers A, Coutts J, Ballmer-Weber BK, et al. Development of a standardized low-dose double-blind placebo-controlled challenge vehicle for the EuroPrevall Project. Allergy 2012;67:107−13.

[52] Worm M, Timmermans F, Moneret-Vautrin A, Muraro A, Malmheden-Yman II, et al. Towards a European registry of severe allergic reactions: current status of national registries and future needs. Allergy 2010;65(6):671−80.

[53] Moneret-Vautrin DA, Rance F, Kanny G, Olsewski A, Gueant JL, Dutau G, Guerin L, et al. Food allergy to peanuts in France–evaluation of 142 observations. Clin Exp Allergy 1998 Sep;28(9):1113−9.

Chapter six

From Hazard to Risk — Assessing the Risk

Charlotte Bernhard Madsen[1], Geert Houben[2],
Sue Hattersley[3], René W.R. Crevel[4],
Ben C. Remington[5], Joseph L. Baumert[5]

[1]*DVM Research Leader Division of Toxicology and Risk Assessment,
National Food Institute, Technical University of Denmark*
[2]*Food & Nutrition, TNO, Zeist, The Netherlands*
[3]*Food Standards Agency, London, UK*
[4]*Safety and Environmental Assurance Center, Unilever, Sharnbrook, Bedfordshire, UK*
[5]*Department of Food Science & Technology and Food Allergy Research &
Resource Program, University of Nebraska, Lincoln, NE, US*

CHAPTER OUTLINE

101

Risk Management for Food Allergy. http://dx.doi.org/10.1016/B978-0-12-381988-8.00006-3

INTRODUCTION

It is well described that foods such as milk, egg, peanut, shrimp, etc. constitute a hazard to individuals who are allergic to these foods. To go from hazard (is this dangerous and to whom?) to risk (what is the probability that a sensitive person will meet this food in a sufficient amount to cause a reaction, and how serious will that reaction be?) is a procedure that demands detailed knowledge of levels and frequencies at which allergenic material is present in foods, amounts of food consumed, and data on the doses that elicit allergic reactions of certain types (e.g., oral allergy syndrome, skin effects, etc.) and how this reactivity is distributed in the allergic population. This level of detail is not always available, so it may be necessary to make assessments based on incomplete data. With incomplete data it may not be possible to produce a fully quantitative estimate of the risk, but rather judgment can be made, for instance, about whether the concentration of an allergen is likely to be unsafe or not.

Safety assessment and risk assessment are part of the risk analysis concept, which also includes risk management and risk communication. These elements are separate tasks often performed by different players, but they should be part of an interactive and iterative process [1]. Ideally, the safety assessment or risk assessment of allergenic foods is a purely scientific process that utilizes expertise in food allergy, toxicology, and food intake assessment.

Risk assessment of food allergens differs from most other assessments of food-borne hazards because only a small proportion of the population is at risk. In addition, the allergenic food that may be lethal to consume for the food allergic person is often an important nutrient for the rest of the population.

The attempt to estimate the risk from intake of hazardous chemicals is a classic toxicological discipline that has been in existence for many years. Most toxicological risk assessments are not able to determine a quantitative risk but establish a level that is judged to be safe, often translated into an acceptable daily intake (ADI). Assessing the risk from contamination with hazardous micro-organisms is also a well-recognized discipline, and advanced mathematical modeling has been developed, allowing an actual quantitative estimate of such risk. These probabilistic models are now also used in toxicology. Risk assessment in food allergy relies on the methods developed in toxicology and microbiology. As in the other disciplines, food allergy safety or risk assessment can be conducted using different methods, depending on the scope of the assessment and the data available. In this chapter we will present two safety assessment methods and one risk assessment method in food allergy, with examples of their use [2].

WHY AND WHEN IS IT NECESSARY TO ESTIMATE THE RISK FROM ALLERGENIC FOOD?

For many chemical substances, acceptable or tolerable levels in foods are defined in regulation (e.g., food additives, pesticides, mycotoxins). This means that the public and industrial risk managers can use these regulatory thresholds to decide whether a content or level of contamination is acceptable or not. As the levels are included in the legislation, they can be used and discussed and will be the same in products A and B, and often also the same in countries X and Y.

In contrast, regulatory thresholds for allergenic foods have not yet been developed. Current European, US, and other legislation on allergenic food ingredients define which allergenic foods must always be declared on a product label, regardless of the level of use. Except for Switzerland, this only applies to ingredients deliberately added to a food according to a recipe. This legislation does not set any specific thresholds for labeling of these allergenic foods. In reviews conducted several years ago, regulatory authorities generally concluded that data were inadequate to define safe thresholds for food allergens, although they accepted that such thresholds do exist [3,4]. Most legislation has not directly addressed the issue of allergen cross contamination. While there is some voluntary guidance that includes qualitative advice for industry on how to assess and manage risk from allergenic foods [5], there is currently no advice from regulatory bodies or compliance authorities on levels of allergen cross contamination above which precautionary (advisory) labeling (such as 'May Contain Nuts') should be used.

Because of the current absence of agreed upon defined thresholds, food producers as well as enforcement authorities have to decide what level of allergenic food in a given product constitutes a health risk and therefore requires action to manage and/or communicate the risk. The basis for this decision is a safety or risk assessment.

A safety or risk assessment for an allergenic food can be needed for many different reasons. However, food allergen risk assessment has gained most attention in relation to understanding the risk arising from the unintended presence of an allergen in a product (e.g., through cross contamination). This refers to situations where allergens are unintentionally present in food products, for instance due to practical issues with cleaning production facilities between production runs (for example, water cannot be used when cleaning chocolate production facilities) or due to residues of raw materials arising at any point in the supply chain (harvest, storage, transport, etc.). This chapter and the examples in it will mainly focus on risk assessment for such cross contamination scenarios. The risk assessment principles and methodologies, however, can be applied generally.

The approaches described are applicable to foods containing allergens in non-particulate distributions and cannot directly be used to assess the risk from sporadic contamination with particles such as whole seeds, pieces of nuts, or clots of dough, for which a different approach will be required. However, the probabilistic approach in risk assessment can also be used to deal with particular contamination scenarios, and this will be addressed in this chapter as well.

SAFETY ASSESSMENT IN FOOD ALLERGY USING ONE DATA POINT (NOAEL OR LOAEL) AND AN ESTIMATED FOOD CONSUMPTION

In traditional toxicological risk assessment approaches, data from animal experiments are typically used. The NOAEL (no observed adverse effect level) is typically divided by an uncertainly factor of 10 to allow for differences in sensitivity between animals and humans, and then divided by another uncertainty factor of 10 to account for inter-individual variation among humans. If the LOAEL (lowest observed adverse effect level) is used instead of the NOAEL, an additional uncertainly factor is applied. An allergic individual's LOAEL is equivalent to their minimal eliciting dose (MED) for an allergic reaction. The terms LOAEL and MED have been used interchangeably in previous texts, but the LOAEL will be used for the remainder of this chapter. In food allergen risk assessment it is neither relevant nor necessary to use data from animal experiments, as human data are available from diagnostic and other clinical food challenges. Furthermore, a reliable and predictive animal model for human allergic reactions to food does not currently exist. The most relevant information used for food allergen risk assessment is threshold data from food allergic individuals who have undergone clinical low dose challenge trials (see chapters 4 and 5). In most instances, individual NOAEL and LOAEL values can be derived from those clinical threshold studies. However, in some challenge trials a small fraction of the allergic patients may experience reactions even at the lowest dose administered, so NOAELs cannot be determined. It is also impossible to say with certainty that the most sensitive food allergic individual has been seen in these low dose challenge trials (or indeed any other food challenge studies).

As the data used in the food allergy assessment are from studies in humans, it is not relevant to use the first uncertainty factor of 10. Depending on the quality of the study and the inclusion criteria for the patients, it may be relevant to include an uncertainty factor that takes into account the uncertainties arising from the establishment of the NOAEL and the possible exclusion of a sensitive fraction of the allergic population.

The US Food and Drug Administration (FDA) Threshold Working Group outlined one example of how this approach might be used for food allergens [4], but there has been no overall consensus on how NOAEL or LOAEL data from clinical challenge trials, with or without the use of uncertainty factors, should be used in food allergy safety assessment. It is likely that one of the reasons for this is that when using NOAELs or LOAELs and an uncertainty factor of 10 or more, the numbers derived are so low that they are below the level that can be reasonably attained in production of food for normal consumption and below the limit of detection of analytical assays for food allergens — and hence not very useful for risk management. Furthermore, reliance on only one data point (or two if using separate data for adults and children) places heavy emphasis on the quality of study design (e.g., dose spacing) and introduces further uncertainty regarding the degree to which the threshold derived is representative for the whole population in question (more about uncertainty factors on page 20).

Despite the above, NOAELs or LOAELs from challenge studies may be used for an initial first assessment or for *the* assessment if this proves sufficient for a sound decision on the level of risk (e.g., the exposure dose of the allergen would be sufficiently high to pose an allergenic risk for the affected population) or if no more data are available.

Example: Spice Mix with Undeclared Wheat Flour as Carrier

A sauce has 10 g spice mixture/kg as an ingredient. It is found that the spice mixture contains (an unknown amount of) wheat flour as carrier. The wheat flour does not appear on the list of ingredients. The question to the risk assessor is, could the undeclared wheat flour be a risk to people with a wheat allergy?

NOAEL/LOAEL APPROACH

As the amount of wheat flour in the spice mix is unknown, it is assumed that 50% of the spice mix is wheat flour. The protein content of wheat flour is 10%. The serving size of the sauce is estimated to be 150 g. This gives a dose of 75 mg wheat protein per serving. Based on a literature search, the LOAEL for wheat protein in children based on objective symptoms is 2.6 mg wheat protein [6]. The dose of 75 mg wheat protein is significantly higher than the LOAEL for objective symptoms and could present a health risk to individuals with wheat allergy. Additionally, individuals that suffer from celiac disease are at risk from this level of wheat protein. So the simple answer is yes, the undeclared wheat flour can be a risk to people with wheat allergy (as well as those with celiac disease).

The challenge with using the NOAEL/LOAEL approach is that it does not take into consideration the population distribution of wheat allergic individuals, so an estimated number of individuals that would be predicted to react in the wheat allergic population cannot be determined. A risk assessor may want to consider what percentage of the population (e.g., 1% of the population) is predicted to be at risk in order to determine perhaps internally what level of acceptable risk he or she may be willing to consider. This approach also does not take into consideration the probable health risk that would be predicted in the wheat allergic population at this exposure level, and it does not take into consideration the number of units that may be out in the retail market (e.g., 1,000 units vs. 1 million units), which also contributes to the absolute risk (probability of a reaction occurring) involved with the product. All of these factors are very important to risk assessors when considering the overall risk from a product and whether it needs to be mitigated.

SAFETY ASSESSMENT IN FOOD ALLERGY BASED ON ALL AVAILABLE CHALLENGE DATA AND AN ESTIMATED INTAKE FOOD CONSUMPTION (BENCHMARK DOSE/MARGIN OF EXPOSURE APPROACH)

The NOAEL/LOAEL approach only uses one data point from one study to derive a reference point to use in the risk analysis. This is not an optimal way of using data from food allergy challenge studies. As described in chapter 5, it is possible to combine data from different challenge studies and, depending of the number of challenges, to estimate with reasonable confidence and accuracy a dose that may elicit a reaction in, for example, 1, 5, or 10% of patients allergic to the specific food (also called, respectively, the eliciting dose or ED1, ED5, ED10). Estimating a reference point using all available data is a method that is increasingly used in risk assessment of chemicals [7,8] in the form of the Benchmark Dose (BMD) approach.

In 2009 the European Food Safety Authority (EFSA) Scientific Committee concluded:

> *the BMD approach is a scientifically more advanced method to the NOAEL approach for deriving a Reference Point, since it makes extended use of available dose-response data and it provides a quantification of the uncertainties in the dose-response data. Using the BMD approach also results in a more consistent Reference Point, as a consequence of the specified benchmark response.*

The BMD approach is applicable to all chemicals in food, irrespective of their category or origin (e.g., pesticides, additives, or contaminants), and potentially to all situations where data are sufficient to describe some relationship between dose and response. The BMD approach is of particular value for:
  i) situations where the identification of a NOAEL is uncertain,
  ii) providing a reference point for the Margin of Exposure in case of substances that are both genotoxic and carcinogenic, and

iii) dose-response assessment of observational epidemiological data.

In the short term, the EFSA Scientific Panels and Units are strongly encouraged to adopt the BMD approach in situations such as those described above [8].

The steps involved in identifying the BMD for a particular study are:

- Specification of a low but measurable response level. For quantal (yes/no) data, EFSA (2009) recommend using a 10% increase in response compared with the background response. This is called the Bench Mark Response (BMR).

- Fitting a set of dose-response models, and calculation of the BMD and the Bench Mark Low Dose (BMDL) (lower 95% confidence interval of the BMD) for those models that describe the data according to statistical criteria.

Ideally, the BMR would reflect an effect size that is negligible or non-adverse. However the BMR chosen should not be too small, to avoid having to estimate a BMD by extrapolation outside the range of observation, such that the BMDL would then depend heavily on the model used. The default BMR may be modified based on statistical or toxicological considerations.

Although the BMR should reflect a response that is negligible, it is nonetheless based on a level where a response is expected. According to EFSA [8], the BMDL is comparable to the NOAEL. For more information on the statistical background to this discussion see EFSA [8]. The consequence is that the same uncertainty factors applied to the NOAEL should be used for the BMDL. For quantal data, this would then be the BMDL at 10% 'effect', also designated $BMDL_{10}$. This again is based on toxicological data where there is a background response. As this is not the case with food allergy challenge data, another approach needs to be taken. Chapter 5 describes how combining challenge data makes it possible to estimate a *Reference Dose,* a parallel to the BMD. Depending on the quality of the data the Reference Doses suggested in chapter 5 are based on a positive challenge reaction in 1% of the food allergic population where large numbers of individual NOAEL/LOAEL values exist (i.e., greater than 200 individual data points) or based on the lower confidence interval of the ED5. The process for deriving the *Reference Doses* in chapter 5 utilized a hybrid approach based on an initial BMD analysis, followed by analysis with probabilistic modeling (discussed in detail on page 9), which was used to validate the results of the BMD analysis. Multiple distributions were used in the BMD analysis, and distributions determined to fit the data best were used in the probabilistic modeling for each allergen. It is still at matter of debate if and how uncertainty factors should be applied to the results. For a detailed discussion of uncertainty factors see page 20.

The BMDL can be divided by the estimated intake of the chemical in the population, resulting in a Margin of Exposure (MoE). The size of the MoE is used to decide whether risk reduction measures should be taken [1]. The MoE may also be referred to as the Margin of Safety. An MoE=1 indicates a risk equal to the risk connected to the Reference Dose. An MoE < 1 indicates a risk higher than the risk connected to the Reference Dose. An MoE > 1 indicates a risk lower than the risk connected to the Reference Dose (e.g., an MoE$=100$ is analogous to an uncertainty factor of 100).

EXPOSURE ASSESSMENT

To be able to calculate an MoE, it is necessary to estimate the exposure to the allergenic food. Contamination with allergenic food may either be estimated if no data are available or be a result of allergen analyses. In a (IgE-mediated) food allergy, exposure during one meal or eating occasion is the relevant measure of exposure. In celiac disease (gluten intolerance), the intake per day should be used. In the spice mix example, the portion size was estimated using common knowledge of eating habits. A more scientific approach is to use dietary data from food surveys. These can be used in different ways. From dietary surveys it is possible to derive data on the mean consumption of a product on the basis of the whole population. This figure is not very useful in food allergy risk assessment as it says little about the portion size. As an example, the mean daily consumption of wheat bread in the Danish adult population is 71 g, but this figure also includes all the people who do not eat bread. A more relevant figure is the mean consumption based on the persons eating the product (eaters only). The consumption may also be expressed as a percentile (e.g., the 95th percentile) intake in eaters only. In a worst case scenario, the maximum intake of the food may be used. Using data from the Danish National Food Survey, the mean intake per meal of wheat bread is 78 g. The 95th percentile is 165 g and the maximum per meal is 260 g. It is apparent that the intake varies considerably depending on the data chosen. All the presented data are valid descriptions of the wheat bread consumption, but the outcome of the risk assessment heavily relies on which intake data are chosen. This is exemplified below in the 'bread' example.

In conclusion, calculating the Margin of Exposure provides data to describe the safety or lack of safety of a given level of allergen carryover and may inform the discussion on setting acceptable levels of contamination: the larger the MoE, the less reason for concern. Once a BMDL for a specific food is established it is easy, in risk management situations, to calculate the MoE for different intake and contamination situations or scenarios.

The as yet unresolved discussion of the use of and size of uncertainty factors is also relevant in determining the acceptable size of the MoE.

There has been much emphasis on getting and using the best possible food challenge data as a basis on which to determine action levels. The choice of food consumption data also influences the MoE.

At the operational level, the VITAL approach described later uses in essence an MoE, approach where an MoE < 1 is a trigger for precautionary labeling.

Example: Spice Mix with Undeclared Wheat Flour as Carrier — Revisited: Risk Analysis Using the BMD Approach

As outlined in the NOAEL/LOAEL risk analysis example above, the intake of wheat protein will again be 75 mg per meal in the BMD risk analysis example.

The *Reference Dose* suggested in chapter 5 for wheat is 1.0 mg wheat protein. This Reference Dose is based on the 95% lower confidence interval of the dose giving reactions in 5% of 40 wheat allergic patients who underwent clinical low doses challenge trials (EDLow$_5$). Translated into the terminology used for bench mark dose risk analysis described above, the EDLow$_5$ would be the Bench Mark Dose Low$_5$. The Reference Dose (BMDL$_5$) can be divided by the estimated exposure (1.0 mg/75 mg $= 0.01$). The Margin of Exposure or Margin of Safety would therefore be 0.01 in this example. It is clear that the MoE is $<< 1$ even if the content of wheat flour was grossly overestimated, indicating that there is a high risk of an allergic reaction.

Conclusion: the dose of 75 mg wheat protein per serving can be a risk to wheat allergic patients. It can also cause adverse reactions in persons with celiac disease (see chapter 7).

RISK ASSESSMENT IN FOOD ALLERGY BASED ON THE DISTRIBUTION OF INPUT DATA (CHALLENGE, CONTAMINATION, CONSUMPTION) (PROBABILISTIC APPROACH)

In contrast to the two safety assessment approaches above, the probabilistic approach is not based on a single value (no effect level or minimum eliciting dose) to use as Reference Dose. In addition, it does not rely on a single figure for consumption or a single figure for the concentration of the allergen in a food. The characteristic of this method is that it uses distributions. The input variables are the 'allergen exposure distribution' and the 'dose distribution curve' from challenge data. The 'allergen exposure distribution' is the combination of the 'distribution of the amount of food consumed' and the 'distribution of concentration of allergen' in the food product under investigation (Figure 6.1).

The outcome of the probabilistic risk assessment is the probability of an allergic reaction occurring upon consumption of the food product in question. The probability is a numerical value that estimates the magnitude of the risk. The advantage of this method is that it makes the basis of the value judgment that the risk manager must make very explicit — if this specific risk is acceptable or not, or what concentration is acceptable. The disadvantage is that, compared to the two safety assessment methods, it requires more detailed data as well as substantial mathematical/statistical skills.

The use of the probabilistic method in food allergy was first described by Spanjersberg et al. [9]. A further discussion and description of the statistics behind the method can be found in Kruizinga et al. [10] and Rimbaud et al. [11].

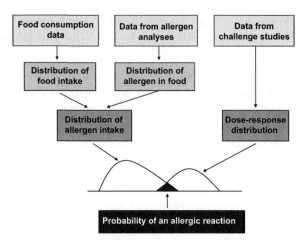

FIGURE 6.1 The figure illustrates the concept of probabilistic risk assessment in food allergy. Modified from Spanjersberg et al. [9].

The probabilistic risk assessment methodology can be used to estimate the proportion of an allergic population that may suffer from a reaction due to the presence of a certain level of (e.g., undeclared) allergen in a food product or, vice versa, to calculate a maximum allergen concentration in food products allowed on the basis of a desired level of safety (i.e., acceptable residual risk). The probabilistic approach can also be used in other risk analysis processes, for instance in assessing the potential reduction in risk brought about by more intense cleaning procedures. It can also be used to assess the value of the development and use of more sensitive analytical methodologies or in establishing the analytical sensitivity needed for a certain protection level.

When conducting probabilistic risk assessments the probability of an allergic reaction occurring can primarily be expressed in three different ways. The numerical results (e.g., 1 reaction per 1,000 individuals per eating occasion) will appear to differ significantly depending on the way the risk is expressed, so it is very important to understand the terminology. In reality the risk is not that different between the three methods of risk expression. The first way of expressing the risk is in terms of 'Allergic User Risk', which assumes that all individuals are allergic to the food allergen of interest and all individuals consume the food product in question. This is the most conservative way to express the probable risk of an allergic reaction occurring, as not everyone would be allergic to the particular food allergen of interest and not everyone would consume the particular food product. The expression of risk in terms of Allergic User Risk does have its advantages, as it requires the fewest assumptions about the allergic population and the consumption patterns of the population. A more practical way of thinking about the Allergic User Risk is that it is the risk that an allergic reaction would occur if the allergic individual purchased the food product in question and consumed the product.

The second way of expressing the risk is in terms of the 'Allergic Population Risk', which again assumes that everyone is allergic to the food allergen of interest, but in this risk assessment the percentage of the population that consumes the product of interest is taken into account. This percentage is obtained from national dietary surveys. One assumption that must be made in this case is that the allergic population would consume the particular product of interest in the same quantities and at the same frequency as the non-allergic population. Few data are currently available on the differences between consumption patterns of allergic and non-allergic individuals. This is one data gap that must be addressed in the future. The third way of expressing the risk is in terms of the 'Overall Population Risk', which is a more traditional expression of risk commonly used in toxicological risk assessments. Expression of allergic risk in the overall population assumes that a certain percentage of the population consumes the food product of interest and also assumes that a certain percentage of the population is allergic to the particular food allergen of interest. Prevalences of food allergies have been estimated using a variety of approaches throughout the world (e.g., random digit telephone surveys or more formal estimates based on the allergic population for a particular clinic or national population). For the purposes of the following probabilistic risk assessment examples, we will limit our analysis and expression of risk to the Allergic User Risk and the Allergic Population Risk.

Example: Spice Mix with Undeclared Wheat Flour as Carrier − Revisited: Probabilistic Approach

A 75 mg dose of wheat protein per serving equates to a protein level of 500 mg/kg in the sauce. The model uses consumption data based on the tomato sauce portion of spaghetti in the 2003−2008 United States National Health and Nutrition Examination Survey. Spaghetti sauce was chosen as it could still be consumed by wheat allergic and celiac individuals seeking a wheat free or gluten free noodle dish. Intake data were conservatively based on a per day basis and an estimation that 2.8% of the US population consumes tomato sauce. The wheat allergic thresholds used in the probabilistic risk assessment were those used to derive the suggested *Reference Dose* for wheat in chapter 5. From the inputs above, we can estimate that the risk of an objective reaction among wheat allergic individuals who consume tomato sauce is 41.4%. This means that if 1,000 representative wheat allergic individuals were to consume the sauce, a total of 414 reactions would be predicted. But as only 2.8% of the population eats tomato sauce, the risk in the wheat allergic population as a whole is 1.2%. These predicted reactions are based on the random pairing of an exposure dose (based on a certain consumption of tomato sauce and a corresponding concentration of undeclared wheat in the tomato sauce) with a random threshold value from the population threshold distribution of wheat allergic individuals. If the exposure dose of the simulated event is over the threshold of the simulated individual, a reaction is predicted to occur.

The probabilistic risk assessment model runs these random simulations hundreds of thousands to millions of times to determine the predicted number of reactions that could occur upon consumption of the tomato sauce product containing undeclared wheat. It is important to take a close look at the risk assessment data to determine where the predicted allergic reactions are occurring in the simulation. Due to the asymptotic nature of the distributions, the risk assessment program will continue to statistically select random individuals from the threshold distribution curve that have predicted thresholds well below the most sensitive individual in the actual data set (nano and picogram levels) or to select individuals that consume the food product of interest at quantities that are physically impossible to achieve in a single sitting. Expert interpretation of the risk assessment results is often needed to ensure that the estimates are realistic. When studying the predicted reactions in our wheat in tomato sauce example, 97.3% of predicted reactors had a threshold over the proposed action level of 1.0 mg wheat protein and fell within the observed range of clinical thresholds (Figure 6.2). Additionally, 52.5% had thresholds over 20 mg wheat protein These percentages show that at 500 mg/kg of protein, the majority of wheat allergic consumers and possibly celiac patients would be at risk when consuming less than the average portion of sauce. The risk calculations above do not include the part of the population with celiac disease that may in addition react to the sauce. The inclusion of celiac patients

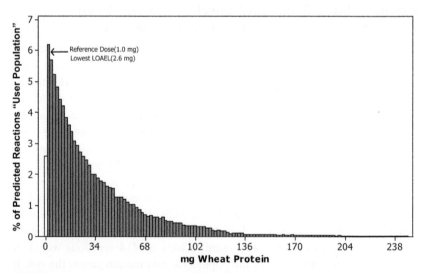

FIGURE 6.2 The histogram illustrates the predicted threshold values of wheat allergic individuals predicted to have a reaction in the user population. The histogram results from the repeated sampling of pairs of values from the wheat dose distribution and the distribution of wheat exposure based on consumption of the product (tomato sauce containing a flavor with a wheat ingredient) and illustrates the proportion of reactions attributable to a particular threshold level. The most sensitive LOAEL from DBPCFCs and the *Reference Dose* for wheat are indicated. Reactions occurring over the *Reference Dose* appear in red.

would increase the risk profile of the spice mix with wheat flour as a carrier and make a case to recall the product.

RESULT

The wheat example demonstrates a high level of risk, and the product would need to be relabeled to ensure the safety of the allergic population. In this instance, a probabilistic approach is not necessarily needed, nor does it always provide significant insight to a problem with high levels of risk. However, when the NOAEL/LOAEL and BMD approaches demonstrate a high level of risk, probabilistic modeling confirms the conclusions of the two approaches. As expected, additional information for wheat in spice mix confirms the high level of risk for the allergic user (41.4%) and finds nearly all predicted reaction thresholds are above the *Reference Dose* for wheat (Figure 6.2, Table 6.1).

Particulate Contamination

The approach described is applicable to food containing allergens in non-particulate distributions and cannot directly be used to assess the risk from sporadic contamination with particles such as whole seeds, pieces of nuts, or clots of dough. A slightly modified approach with additional data requirements will be necessary for the probabilistic approach risk assessment to properly characterize the risk when dealing with particulate contamination. Thus, in addition to the variables already included in the model previously described, particulate contamination requires consideration of the frequency with which particles of allergen are likely to find themselves in the product, as well as the size (or mass) distribution of the particles in question. The frequency of contamination will set an upper limit on the probability of any reaction. The particle size (mass) distribution will determine how likely a reaction will be if an allergic person consumes the contaminated product, since it will determine the dose. One of the reasons why particulate contamination often leads to an 'automatic' precautionary label is an assumption that any particle must be large enough to

Table 6.1 500 ppm Wheat Protein in Sauce Assessment Summary

Approach	Results	Conclusion	Suggested Action
NOAEL/ LOAEL	75 mg wheat protein in 150 g of sauce	500 mg/kg wheat protein may trigger an allergic reaction in a significant part of the wheat allergic population	Recall may be warranted, change product label to declare the presence of wheat
BMD	$MoE_{05} = 0.01$	(see above)	(see above)
Probabilistic	Risk of a reaction in the wheat allergic population is 1.2% (risk for the allergic user: 41.4%)	(see above)	(see above)

elicit a reaction. However this is demonstrably not the case. For instance, cashew pieces of approximately 5 mm in length on each side have a volume of approximately $0.125\,cm^3$ and, assuming a protein concentration of 25%, would deliver a dose of 31 mg per particle. This undoubtedly presents a risk to a proportion of cashew-allergic people. However, if we consider pieces of 2 mm in length, then the particle mass would be 8 mg and the protein content would be 2 mg, the proposed VITAL reference value for cashew. Unless there was a significant probability that more than one piece would contaminate a product, then a precautionary label would hardly be justifiable. Thus, probabilistic models can readily be extended and refined with additional information being available, and programs used for probabilistic modeling easily allow insertion of additional probability determining parameters.

EXAMPLES OF RISK ASSESSMENTS USING ALL THREE APPROACHES

The NOAEL/LOAEL approach and BMD approach could have been used exclusively in the wheat in spice mix example. These next two examples will examine lower levels of allergen contamination using all three risk assessment approaches to determine whether a risk exists to the allergic consumer.

Example: Lemonade Company Learns of Peanut Proteins in a Flavor Carrier Ingredient

An ingredient company finds peanut proteins due to cross contamination during routine testing of a flavor carrier product. Analyses revealed 50 ppm of peanut proteins in the highest sample, with other samples down to below the limit of quantitation (BLQ). A lemonade company uses the flavor carrier at 0.5% in its final product. Does 50 ppm peanut protein in a flavor carrier present a significant health risk to the peanut allergic population?

50 ppm peanut protein in flavor carrier	0.5% flavor carrier in final product	= 0.25 ppm peanut protein in lemonade

NOAEL/LOAEL APPROACH

Again, we must assume the 0.25 ppm peanut protein is homogeneously distributed through the lemonade. The average consumption of lemonade per eating occasion is 410 g. An average intake of 0.103 mg peanut protein would be expected, above the lowest objective LOAEL of 0.1 mg peanut protein [11a]. The most sensitive portion of the peanut allergic community could be at risk when consuming this product.

BMD APPROACH

As shown above, the average lemonade intake will yield a dose of 0.1 mg peanut protein. The Reference Dose (BMD$_1$) suggested for peanut is 0.2 mg and can be divided by the estimated average exposure to determine an MoE of 2.0 (0.2 mg/0.1 mg = 2.0). As previously stated, an MoE > 1 is ideal as it indicates the risk is lower than the risk posed by the Reference Dose. The larger the MoE, the less reason for concern, but a low MoE of 2 could still pose an unacceptable risk due to the uncertainty associated with the determination of the Reference Dose and the consumption estimates. Additionally, an allergic individual can purchase a 24 oz (710 ml) can of lemonade in the United States, which would decrease the MoE to 1.1 (0.2 mg/0.18 mg). The lemonade does not have peanut on the label, and the MoE indicates that there is a risk to sensitive peanut allergic persons if they drink the lemonade. However, the risk will not be higher than the 1% risk connected with the Reference Dose.

PROBABILISTIC APPROACH

As the NOAEL and BMD approaches are not conclusive in this instance, a probabilistic risk assessment of 0.25 ppm peanut protein in lemonade should be run. The model uses consumption data for lemonade from the 2003−2008 United States National Health and Nutrition Examination Survey. Intake data were conservatively based on a per eating occasion basis, and an estimated 5.7% of the US population consumes lemonade. The peanut allergic thresholds used in the probabilistic risk assessment were those used to derive the suggested Reference Dose for peanut in chapter 5. From the inputs above we can estimate that the risk of an objective reaction in the peanut allergic persons who drink lemonade is 0.3%. So if 1,000 representative peanut allergic individuals were to consume the contaminated lemonade, a total of three reactions would be predicted. But as only 5.7% of the US population drinks lemonade the risk of a reaction in the peanut allergic population as a whole is 0.018%. Further analysis of the predicted reactions in the simulations indicate that 5.8% of the predicted reactors had a threshold over the proposed Reference Dose and fell within the range of observed clinical thresholds (Figure 6.3). As shown in the figure, over 70% of the predicted reactors had thresholds below the most sensitive LOAEL recorded in clinical Double Blind Placebo Controlled Food Challenges (DBPCFCs). Again, due to the asymptotic nature of the distributions, the simulation is predicting a significant proportion of reactions in individuals who would have thresholds below the most sensitive peanut allergic individuals in the clinical threshold data set. In addition, the individuals who would be predicted to react in the simulation would need to consume larger than average amounts of lemonade in a single eating occasion. Thus, the three predicted reactions out of 1,000 allergic users would not be expected.

RESULT

Probabilistic modeling utilized the entire consumption pattern for lemonade in the United States in addition to the entire population distribution of peanut allergic individuals. The risk modeling demonstrates the need for an extremely sensitive individual to consume a larger than average serving in order for a reaction to occur. As a result, a corporate decision based on the quantitative

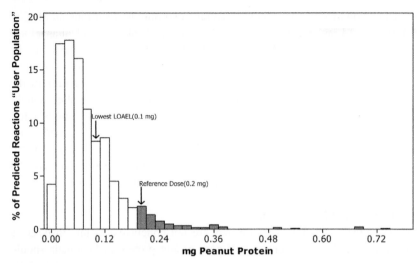

FIGURE 6.3 The histogram illustrates the predicted threshold values of peanut allergic individuals predicted to have a reaction in the user population. The histogram results from the repeated sampling of pairs of values from the peanut dose distribution and the distribution of peanut exposure based on consumption of the product (lemonade containing a flavor carrier contaminated with peanut) and illustrates the proportion of reactions attributable to a particular threshold level. The most sensitive LOAEL from DBPCFCs and the *Reference Dose* for peanut are indicated. Reactions occurring over the *Reference Dose* appear in red.

risk analysis could be to keep the product on the shelf (no market withdrawal) since the level of peanut protein would not pose a public health concern. From a risk management standpoint, the company would want to reassess its allergen control practices (and audit its suppliers if warranted) to isolate where the cross contamination could have occurred and minimize future peanut cross contact issues (Table 6.2).

Example: Egg in Bread

An international baking company finds egg proteins due to cross contamination during routine testing of finished bread. Analyses of brown wheat bread product in the United States revealed 1.2 ppm of egg proteins in the baked bread. Does 1.2 ppm egg protein in sliced bread present a significant health risk to the US egg allergic population?

NOAEL/LOAEL APPROACH

Again, we must assume the egg protein is homogeneously distributed through the bread. The average daily consumption of wheat bread is highest among teenage and adult males. In the United States, the average serving size is one slice (26 g) and the average consumption in males is two slices (52 g). If the intake calculations are made on serving size, intakes of 0.03 mg egg protein would be expected in the US. This value is above the lowest objective

Table 6.2 0.25 ppm Peanut Protein in Lemonade Assessment Summary

Approach	Results	Conclusion	Suggested Action
NOAEL/LOAEL	0.1 mg peanut protein in 410 g of lemonade	Could pose a risk to the most sensitive peanut allergic individuals	Not conclusive, risk of a reaction present
BMD	$MoE_{01} = 2.0$	(see above)	(see above)
Probabilistic	Risk of a reaction in the peanut allergic population is 0.018% (risk for the peanut allergic user: 0.3%)	Analysis shows nearly all predicted reactions are below the lowest reactive doses observed in clinical setting	No recall needed

LOAEL of 0.014 mg egg protein (unpublished). The most sensitive portion of the egg allergic community could be at risk when consuming a single slice of the contaminated wheat bread.

BMD APPROACH

As shown above, it is not necessary to consume more than one piece of bread for a sensitive egg allergic individual to be at a risk of a reaction. Intake of a single wheat bread slice will yield a dose of 0.03 mg egg protein in the US. The action level (BMD_1) suggested for egg is 0.03 mg, and this can be divided by the estimated exposure (0.03 mg/0.03 mg = 1.0). The MoE for a single slice of bread is therefore 1.0. If the average consumption of wheat bread is used, doses of 0.06 mg egg protein will be consumed. The MoE at an average consumption would be 0.5. The low MoE indicates that there is a risk for sensitive egg allergic persons when real consumption patterns are taken into account, so it is important to consider the realistic range in the quantity of product that could be consumed when conducting a risk assessment, rather than relying solely on a recommended serving size.

PROBABILISTIC APPROACH

The US National Health and Nutrition Survey (NHANES) database was used to conduct region-specific probabilistic risk assessments. In the US, 19% of men consume wheat bread. The egg allergen thresholds used in the probabilistic risk assessment were those used to derive the suggested Reference Dose for egg in chapter 5. The inputs above indicate that the risk of an objective allergic reaction in the egg allergic persons that eat bread is 3.0%. If 1,000 representative egg allergic individuals were to consume the contaminated bread, a total of 30 reactions would be expected. But as only 19% of the egg allergic population is expected to eat wheat bread, the risk in the whole egg allergic population is 0.6%. Further analysis of the predicted reactions in the risk simulations indicates that 45.4% of the predicted reactors had a threshold over 0.014 mg egg protein (LOAEL) and would therefore be in the observed clinical range

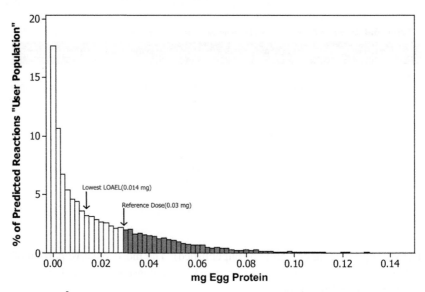

FIGURE 6.4 The histogram illustrates the predicted threshold values of egg allergic individuals predicted to have a reaction in the user population. The histogram results from the repeated sampling of pairs of values from the egg dose distribution and the distribution of egg exposure based on consumption of the product (bread contaminated with egg) and illustrates the proportion of reactions attributable to a particular threshold level. The lowest LOAEL found in food challenges (DBPCFCs) and the *Reference Dose* for egg are indicated. Reactions occurring over the *Reference Dose* appear in red.

(Figure 6.4). Egg has the most sensitive allergic individuals of any food allergen, and the simulation is predicting a number of reactions in the observed clinical range. Although some egg allergic individuals can tolerate baked forms of egg, there is still a possible risk to a sensitive egg allergic individual consuming this bread product.

RESULT

This scenario is not as clear cut as the spice mix or lemonade examples. Due to the low but appreciable level of risk, arguments could be made for or against a market recall of the bread product. The expert opinion of the authors would be to recommend a market recall of the bread as it presents a higher level of risk than the 1% allowed by the Reference Dose for advisory labeling as outlined in chapter 5. A number of allergic reactions could occur, as the bread does not have egg on the label in any form, and it would be wise to recall the product, but ultimately the final decision would need to be made by the food company.

Another option is to evaluate the results in light of the egg challenge data from EuroPrevall (chapter 5, Table 5.6b). In these data, the estimated dose giving objective reactions in 10% of the patients is 0.1 mg for children < 3.5 years and 5.3 mg (lognormal) for patients > 3.5 years of age. The challenge data set for eggs used in the probabilistic risk assessment is based on 206 individuals comprising 174 children, 20 adults, and 20 of undetermined age [11b].

Table 6.3 1.2 ppm Egg Protein in Bread Assessment Summary

Approach	Results	Conclusion	Suggested Action
NOAEL/ LOAEL	0.03 mg egg protein in 1 slice of bread	Above 0.014 mg egg protein and could pose a risk to the most sensitive egg allergic individuals	Not conclusive, risk of a reaction present
BMD	1 Slice $MoE_{01} = 1.0$ Avg. Consumption $MoE_{01} = 0.5$	Average consumption is higher than a single slice and the differences are illustrated by the MoE values. Could pose a risk to the most sensitive egg allergic individuals	(see above)
Probabilistic	Risk of a reaction in the egg allergic population is 0.6% (risk for the egg allergic user: 3.0%)	Analysis shows nearly all doses in predicted reactions are above the proposed VITAL reference dose and could be a health risk to the egg allergic community	Predicted to present a public health risk to egg allergic individuals; food company should consider a market recall of the product — see discussion above

This means that the data set used reflects the more sensitive child population, but the bread intake is based on adult males. If intake data for small children are used, the predicted risk of reaction will possibly be lower. This could, in this case, shift the conclusion from recall to no action (Table 6.3).

SUMMARY OF RISK ASSESSMENT EXAMPLES

Three food allergen risk assessment examples were presented to illustrate different scenarios and the information behind the risk analysis and decision process that may be needed to decide if a market recall (or no recall) is warranted. The wheat in spice mix example was a clear-cut allergen risk that could be assessed without the BMD or probabilistic approaches. However, it is advantageous to have the results of all three risk assessment forms to have references of what a high risk scenario output creates in terms of the MoE and risk to the allergic user. The peanut in lemonade example is not completely clear cut, but probabilistic modeling helps us to focus on where the risk actually occurs. The level of peanut in lemonade was not predicted to present a public health risk to peanut allergic individuals, as the probabilistic risk assessment predicted that high levels of consumption by very sensitive peanut allergic individuals (predicted to have individual thresholds well below the most sensitive individual in the threshold database of 750 individuals) were necessary for reactions to potentially occur. The egg in bread example is an example of an allergen cross contact issue that is a borderline safe/risky product. The use of a more quantitative risk analysis along with expert analysis of the data is important to make an informed risk assessment decision. The BMD and probabilistic models show that a risk does exist at a level that could be a public health risk to egg allergic individuals, and a corporate decision based on the guidance of the expert analysis would need to be made on whether to withdraw the product from the market.

Every situation is slightly different, and companies need to consider a multitude of factors when considering a recall. These factors include the levels of allergen present, the serving size and realistic consumption amounts of the product, the sensitivity of the allergic population, and how many contaminated products are on the shelf. Proper use of risk assessment methods and inputs available, along with expert analysis, will aid in ensuring that an informed decision is made in regard to product recall and consumer safety.

UNCERTAINTY FACTORS

On page 4 we discussed the use of uncertainty factors when using NOAELs or LOAELs. We argued that because the data used in the risk assessment comes from clinical challenge trials in food allergic humans, an uncertainty factor for interspecies extrapolation is not relevant. By conducting a risk assessment based solely on a single NOAEL or LOAEL, it is obvious that using one data point makes it very difficult to be confident that this estimated risk level is representative of the whole population in question.

The other approaches are based on larger data sets. The BMD approach uses a Reference Dose derived from the distribution of as much challenge data as possible, and the probabilistic risk assessment uses the whole distribution of challenge data. Both these methods rely on the assumption that the challenged population is representative of the allergic population for which the risk is going to be calculated. So if the Reference Dose is the dose estimated to give a reaction in 1% of the allergic population, 1% of the 'free living' allergic population will have a reaction at that dose. This is also true for the probabilistic method. If this method estimates that the probability of a reaction is 2%, it is based on the assumption that the challenge distribution used is a correct description of the allergic population.

When discussing if and how uncertainty factors are to be used in food allergy risk assessment it is worth bearing in mind how uncertainty factors are used in other areas of toxicology. Most toxicological data comes from experiments in small rodents. The default uncertainty factors are 10 to allow for interspecies differences and 10 for intraspecies differences. Small animals have a much higher metabolic rate than large animals such as humans. A large part of the interspecies uncertainty factor just covers this difference in metabolism. The intraspecies uncertainty factor is meant to take into account different susceptibilities within the human species, such as are caused by age and genetic differences, which affect toxicokinetics and toxicodynamics [12].

The food allergy challenge data used to estimate the dose-response distribution are a combination of data from humans with the (relevant) genetic makeup to develop an IgE-mediated food allergy, having different symptoms, from different geographical areas, and including both children and adults. As more and more data are included, the diversity of the human population, on which the data are based, will be even larger. This means that a lot of ground meant to be covered by the intraspecies uncertainty factors is already included in the way the data were derived.

This leaves two important questions about the need to use uncertainty factors, which are: i) whether the challenge population and ii) the conditions under which the challenge population were challenged are representative for the whole allergic population and the everyday lives they are living (outside of the clinical challenge setting).

There are also two possibilities. Either the Reference Dose/challenge distribution overestimates the risk or it underestimates it.

How could the risk be overestimated? As described in chapter 4, challenge studies require a high degree of food allergy expertise as well as a well-equipped hospital environment because challenges are potentially life threatening. Centers with this kind of expertise are often tertiary referral centers, seeing patients with the most severe symptoms and difficult conditions. This patient group may not be representative of the whole allergic population (for whom the risk assessment needs to be done) and hence the dose distribution may be shifted to the left and the Reference Dose is lower than relevant.

How could the risk be underestimated? As also described in chapter 4, patients are excluded from clinical challenges in certain circumstances where there is a possibility they are more vulnerable to challenge (e.g., having an uncontrolled asthma). In everyday life they may encounter an allergenic food by accident in situations where their asthma is unstable. Hence the challenge situation does not reflect everyday life. In addition, there are factors that may aggravate a challenge reaction that are not included in the challenge protocol. These aggravators could be exercise, pollen exposure, intake of drugs like aspirin (non-steroid anti-inflammatory drugs [NSAIDs]), or alcohol.

Whether the possibility exists for over- or underestimating the risk by using challenge data is an ongoing discussion. One of the unknowns in this discussion is how often and how severe are the reactions that food allergic patients, who successfully avoid their allergenic food as an ingredient, experience in everyday life. It is known that fatal food allergic reactions are almost exclusively caused by incidents where the allergenic food as an ingredient was accidentally ingested [13].

SEVERITY CONSIDERATIONS IN RISK ASSESSMENT

Allergic reactions can range in severity from barely perceptible subjective reactions to life threatening anaphylaxis. Clearly the implications of these different types of reactions differ greatly from both an individual as well as a public health perspective and need to be taken into account in risk assessments. Indeed, in the US, the concept of an effect being 'harmful to human health' forms part of legal texts, the converse possibly being that there are some effects that can be deemed not to be harmful to human health. The concept of risk itself, according to many definitions as well as in the public mind, is associated not only with the probability of an adverse effect but also with a consideration of its likely severity. One of the difficulties in applying this concept in food allergen risk assessment has been the perception that severity is unpredictable, such that a history of mild reactions is no guarantee that future reactions will not be more severe. There are clear indications that extrinsic factors, such as exercise, medication, etc., can modulate severity, and indeed might be predicted to do so because of their effects on physiology. However, in evaluating the conclusion that severity is unpredictable, due regard should be paid to the fact that this conclusion is often based on circumstances where no information exists on a very important modulator of effect, namely dose.

Current models for assessing the risk from allergenic foods focus solely on a calculation of the probability of any reaction (or any objective reaction) without any regard to its health impact. This reflects, rightly, that it has been possible to derive knowledge from food challenge studies about the relationship between dose and frequency of response. However, even a fairly cursory examination of the outputs of these models reveals that the number of predicted reactions far exceeds the numbers that are recorded. This may not be due solely to inevitable inaccuracies in the model but could reflect, among other factors, that a large proportion of reactions at low doses have such limited impact on health that they are, and will always remain, invisible to any monitoring system. For instance, in a small study on the relationship between dose of peanut and the occurrence of anaphylaxis, Wainstein et al. [14] showed that the minimum dose needed to progress to anaphylaxis was the equivalent of 5 mg of peanut protein, which is 25-fold greater than the proposed VITAL reference dose.

In addressing the issue of uncertainty factors, it is useful to remember that the model predictions for any given dose represent the total number of reactions, of which only a proportion will be more than mild. At the doses chosen as reference levels (e.g., in VITAL), this proportion is very likely to be quite low indeed, and the majority will be transient in nature. If the risk assessment goal is to minimize the number of reactions that could be considered harmful to human health, then one could thus argue that the models themselves incorporate their own inbuilt uncertainty factor. Notwithstanding this observation, more knowledge on the relationship between dose and severity would help to improve risk assessments and agreement on what could be considered an acceptable level of risk. Food challenge data provide relatively limited, but nevertheless useful, information in this regard, and analysis of the EuroPrevall data set has examined ways to use this in risk assessments, as discussed in chapter 5. Another, complementary, approach would be to use the Delphi technique to generate a consensus among clinical experts about the likelihood of a severe reaction for an appropriate range of low doses. This probability element could then be incorporated into probabilistic risk assessment models.

VITAL PROGRAM

VITAL is the Voluntary Incidental Trace Allergen Labeling program established by the Allergen Bureau of Australia beginning in 2007. The goal was to limit the use of precautionary labeling (also known as advisory labeling) related to the unintended presence of allergens, for instance through cross contact. It was based on the use of action levels derived through the application of risk assessment principles to potential exposure and standardized label messages. Initially, a VITAL grid was established with three action levels: Green (low risk; no precautionary labeling); Yellow (possible risk; precautionary 'may be present: ####' labeling recommended); Red (higher risk; definitive 'contains ####' labeling recommended).

With the creation of the VITAL grid, the initial action levels were established on the basis of the threshold doses of protein from allergenic foods for subjective and objective responses cited by the 2006 US FDA Threshold Working Group [4]. Because some uncertainty existed surrounding these initial FDA estimates and the general paucity of the available data, a 10-fold uncertainty factor was applied. Although it is generally recognized that food allergic people react to the dose of protein consumed, rather than its concentrations, action levels in the grid were expressed as concentrations (ppm), which corresponded to an amount of protein (mg) in a 5 g serving size (a teaspoon). Most of the green levels were set at < 2 ppm of protein from the allergenic food (exceptions: fish, milk, soy, gluten). While the VITAL program (version 1) expresses the action levels in terms of ppm of protein from the allergenic food, many of the analytical methods provide results in terms of ppm of the whole food; thus conversions are needed if one wishes to express the action levels in alternative units.

Recently the VITAL program (version 2) was updated with new Reference Doses. The basis for these is described in chapter 5. In addition, the three categories were reduced to two, and the intake data are recommended to be mean intake of the food. The consequences of these changes are to make the basis of precautionary labeling much less conservative, hence helping to limit it to situations where a significant risk exists. Furthermore, the transparent and quantitative basis for the reference values will help risk managers communicate the practical meaning and consequences of the new approach [15].

CONCLUDING REMARKS

As with all other risk assessments, the outcome of the assessment of risks of allergens relies on good quality input data. In particular, good hazard characterization enabling establishment of Reference Doses for allergenic foods based on a large number of challenge data, as described in chapter 5, is a big step forward, allowing high quality risk assessment. It also offers the possibility of more transparent safety/risk assessments that may even give similar results because they may be based on identical challenge distributions.

Allergen intake data are the combination of consumption data and contamination data. Consumption data may be based on portion size or rely on actual intake (e.g., mean intake or 95th percentile in each meal/eating occasion). For probabilistic risk assessment the consumption distribution will be used. It is therefore not necessary to decide if the mean or 95th percentile should be used. However, it must be decided if the analyses use consumers only or if the frequency of consumption is also included. This choice depends on the risk management question and the way the risk is to be expressed.

Concentration data may be an estimate of a concentration or an actual measured value. Probabilistic risk assessment also allows for the inclusion of the distribution of the contamination. Although the level of detail available for concentration distribution data is often limited, more data points are available in many situations (for instance in case more samples of a batch of food are analyzed). Such information can easily be incorporated in the risk assessment. A notable special case is where the contamination is particulate. In case of particulate contamination, the likelihood and size distribution of particulate contamination can be estimated as well as the dose distribution through such particles. The resulting overall contamination distribution can subsequently be used as an input parameter in the probabilistic risk assessment.

Consumption data from national dietary surveys is used, relying on the assumption that food allergic consumers of a food product on a population basis show comparable distributions in the amounts of foods consumed and have the same pattern of food choices as non-allergic consumers. The extent to which this assumption is valid under all circumstances is yet to be assessed, but it is likely that the risk assessment outcome based on this assumption will in most cases be of the right order of magnitude if the risk is expressed on a user basis (i.e., the percentage of responders among people using the food product under evaluation). It however is likely that on a population basis, the percentage of people using a food product may differ between allergic and non-allergic populations for certain products. Thus the risk assessment outcome may often be wrong if the risk is expressed on a total population basis (i.e., the percentage of responders among the total population, which includes users as well as non-users of the product). It is therefore crucial to always carefully discuss the possibilities and limitations of risk assessment between the risk assessor and the risk manager who is formulating the risk management questions. Based on an iterative interaction between these two disciplines, the risk assessment and the risk can best be expressed in a way avoiding unnecessary ambiguity.

The two safety assessment approaches described can be used where there are insufficient data or capabilities to make a probabilistic risk assessment or where a fast determination of 'is this product unsafe' is sufficient. The MoEs may be used to prioritize areas of concern for risk managers. For example, very large MoEs indicate that efforts do not need to be spent further improving cleaning procedures for some allergens. MoEs are thus complementary to the risk assessment process.

How and whether uncertainty factors should be used in food allergy risk assessment will probably depend on the goal of the risk assessment and the policy for which it is used. The way in which uncertainty factors may be used still needs to be resolved in the overall context of minimizing negative public health impacts. This discussion should be based on data on factors modulating the allergic response (severe, uncontrolled asthma, exercise, pollen exposure, intake of drugs like NSAIDs, or alcohol) and where and how these should be controlled or managed as well as on mathematical analyses of the appropriate Reference Dose to be used. In general, factors such as uncontrolled asthma, exercise, pollen exposure, intake of drugs like NSAIDs, or alcohol should be taken in consideration in the individual allergy management of allergic individuals in consultation with their doctors or dietician. In probabilistic risk assessment, the use of uncertainty factors is not necessarily needed, as uncertainties and variabilities can be part of the description of the output of the risk assessment.

As explained in the introduction to this chapter, a major area of application of risk assessment is the development of a harmonized approach for *may contain* labeling. A prerequisite for this is an agreement on acceptable residual risks, as a zero-risk approach would not be helpful. The latter would lead to impractically low limits that cannot be complied with in many food production environments and would thus lead to an increase in *may contain* labeling. It is not straightforward to define, on a population basis, the risk that may be tolerated. In relation to food allergy, a recent workshop concluded that risk tolerance depends, among other factors and unsurprisingly, on the nature of the effect. The more severe and widespread the effect, the lower the tolerance [16].

It was also recognized that at an individual level, perception of risk is important, and that if uncertainty is high then the tolerability of the risk decreases. A way to fight uncertainty is to be able to make transparent and reliable food labeling decisions in relation to allergenic foods. Thus, improvement in this respect depends on acceptability of a certain residual risk that at the same time will contribute to risk acceptance. The presence of allergenic food in normal food ingredients and the very large biological variation in the amount of food allergens that may trigger a reaction makes it impossible to set Reference Doses where the risk is zero. So a tolerable risk level needs to be decided upon. The increased number of challenge data makes it possible to develop dose response curves for a number of allergenic foods. This allows performance of an increasing number of probabilistic risk assessments calculating a specific risk relating to a specific level of contamination and vice versa. These data, in combination with new studies on food allergic reactions in the community in combination with one-dose challenges to validate Reference Doses, will inform the discussion on tolerable risk and hopefully make it possible to decide on tolerable risk on an informed basis.

The evolution of the VITAL system and the definition of new Reference Doses based on larger quantities of clinical data represent a major step forward, which can form the foundation for international harmonization making risk assessment and hence risk management decisions more transparent and hopefully easier to understand and rely on for the benefit of the food allergic consumer.

REFERENCES

[1] Larsen JC. Risk assessment of chemicals in European traditional foods. Trends Food Sci Technol 2006;17:471−81.

[2] Madsen CB, Hattersley S, Buck J, Gendel SM, Houben GF, Hourihane JO, et al. Approaches to risk assessment in food allergy: report from a workshop 'developing a framework for assessing the risk from allergenic foods'. Food Chem Toxicol 2009;47:480−9.

[3] EFSA. Opinion of the Scientific Panel on Dietetic Products, Nutrition and Allergies on a request from the Commission relating to the evaluation of allergenic foods for labelling purposes. EFSA J 2004;32:1−197.

[4] FDA Threshold Working Group. Approaches to establish thresholds for major food allergens and for gluten in food. J Food Protection 2008;71:1043−88.

[5] Food Standards Agency. Guidance on allergen management and consumer information − best practice guidance on managing food allergens with particular reference to avoiding cross-contamination and using appropriate advisory labeling (e.g., 'May Contain' labeling). Published by the UK Food Standards Agency. Crown Copyright; 2006. Available on-line at: http://www.food.gov.uk/multimedia/pdfs/maycontain guide.pdf.

[6] Ito K, Futamura M, Borres MP, Takaoka Y, Dahlstrom J, Sakamoto T, et al. IgE antibodies to omega-5 gliadin associate with immediate symptoms on oral wheat challenge in Japanese children. Allergy 2008 Nov;63(11):1536−42.

[7] EPA. The use of the benchmark dose approach in health risk assessment. EPA/630/R-94/007. Washington DC: Risk Assessment Forum; 1995.

[8] EFSA Guidance of the Scientific Committee on a request from EFSA on the use of the benchmark dose approach in risk assessment. EFSA J 2009;1150:1−72.

[9] Spanjersberg MQI, Kruizinga AG, Rennen MAJ, Houben GF. Risk assessment and food allergy: the probabilistic model applied to allergens. Food Chem Toxicol 2007;45:49−54.

[10] Kruizinga AG, Briggs D, Crevel RWR, Knulst AC, van den Borch LMC, Houben GF. Probabilistic risk assessment model for allergens in food: sensitivity analysis of the minimum eliciting dose and food consumption. Food Chem Toxicol 2008;46:1437−43.

[11] Rimbaud L, Heraud F, La Vielle S, Leblanc J-C, Crepet A. Quantitative risk assessment relating to adventitious presence of allergens in food: a probabilistic model applied to peanut in chocolate. Risk Anal 2010;30:7−19.

[11a] Taylor SL, Moneret-Vautrin DA, Crevel RWR, Sheffield D, Morisset M, Dumont P, et al. Threshold dose for peanut: risk characterization based upon diagnostic oral challenge of a series of 286 peanut-allergic individuals. Food Chem Toxicol 2010;48:814−9.

[11b] Allergen Bureau. Summary of the VITAL Scientific Expert Panel Recommendations. http://allergenbureau.net/downloads/vital/VSEP-Summary-Report-Oct-2011.pdf; 2011.

[12] Renwick AG. subdivision of uncertainty factors to allow for toxicokinetics and toxicodynamics. Human Ecol Risk Assess 1999;5:1035−50.

[13] Pumphrey RSH, Gowland MH. Further fatal allergic reactions to food in the United Kingdom 1999−2006. J Allergy Clin Immunol 2007;119:1018−9.

[14] Wainstein BK, Studdert J, Ziegler M, Ziegler JB. Prediction of anaphylaxis during peanut food challenge: usefulness of the peanut skin prick test (SPT) and specific IgE level. Pediatr Allergy Immunol 2010;21:603−11.

[15] www.allergenbureau.net/vital/vital, accessed February 2012, the Allergen Bureau Ltd.

[16] Madsen CB, Hattersley S, Allen KJ, Beyer K, Chan C-H, Godefroy SB, et al. Can we define a Tolerable Level of Risk in Food Allergy? Report from a EuroPrevall/UK Food Standards Agency workshop. Clin Exp Allergy 2012;42:30−7.

Risk Management
of Gluten

Chapter seven

Celiac Disease and Risk Management of Gluten

Steffen Husby[1], Cecilia Olsson[2], Anneli Ivarsson[3]

[1]Hans Christian Andersen Children's Hospital at Odense University Hospital, Denmark
[2]Department of Food and Nutrition, Umeå University, Sweden
[3]Department of Public Health and Clinical Medicine, Epidemiology and Global Health, Umeå University, Sweden

CHAPTER OUTLINE

Risk Management for Food Allergy. http://dx.doi.org/10.1016/B978-0-12-381988-8.00007-5

ABBREVIATIONS

CD celiac disease
EMA IgA endomysial antibodies
ESPGHAN European Society for Paediatric Gastroenterology, Hepatology and Nutrition
GFD gluten-free diet
HLA Human Leukocyte Antigen
IgE Immunoglobulin E
IgG Immunoglobulin G
ATIs amylase trypsin inhibitors
NICE National Institute for Clinical Excellence, United Kingdom
TG2 transglutaminase 2

INTRODUCTION

The purpose of this chapter is to describe the background of the adverse reactions to gluten, in particular in relation to celiac disease (CD). The recently described gluten sensitivity without markers of CD or of IgE-mediated allergy has primarily been studied in adults [1] and will be discussed in less detail. This chapter describes current concepts in CD in relation to clinical symptoms and diagnosis, CD being distinctly different from other food hypersensitivities. It will illustrate how CD has emerged over the last decades as a global public health problem, making preventive strategies and active case finding increasingly important. Lastly the chapter describes the possibilities for performing concrete disease risk assessment and management. The mainstay of the treatment is a lifelong gluten-free diet (GFD), which poses challenges for affected individuals and their families, and also for society at large.

DEFINITION OF CELIAC DISEASE

The definition of CD has changed over the last 20 years. The rather simple and straightforward definition by Anne Ferguson: 'a permanent, gluten-dependent inflammation of the gut (enteropathy)' [2] stresses the permanence of the condition, being quite different from food allergy. This definition was followed by a more sophisticated definition from Mäki [3]:

an autoimmune-like systemic disorder in genetically susceptible persons perpetuated by the daily-ingested gluten cereals wheat, rye, and barley with manifestations in the intestine and in organs outside the gut

Recently published guidelines on the diagnosis of CD in children and adolescents by an ESPGHAN working group on the diagnosis of CD [4,5] contain a more elaborate definition, as suggested by Riccardo Troncone [6], which encompasses the immunological and clinical advances during the last decade:

CD is an immune-mediated systemic disorder elicited by gluten and related prolamines in genetically susceptible individuals, characterized by the presence of a variable combination of gluten-dependent clinical manifestations, CD specific antibodies, Human Leukocyte Antigen (HLA)-DQ2 and DQ8 haplotypes and enteropathy.

This definition mentions CD specific antibodies and HLA haplotypes, utilizing the serological and HLA-related diagnostic measures. Furthermore, the definition states that the clinical manifestations may appear in a variable combination, touching on the variability of the disease. Demonstration of enteropathy is not obligatory.

CD may be classified into subgroups. Several classifications of CD have been used in the past, most importantly distinguishing between classical, atypical, asymptomatic, latent, and potential CD. As atypical symptoms ironically may be considerably more common than classic symptoms, the nomenclature given in Table 7.1 was suggested by the ESPGHAN working group.

Another set of definitions has been published at the same time; namely, the Oslo definitions, which clearly demarcate the definition of CD as 'a chronic small intestinal immune-mediated enteropathy precipitated by exposure to dietary gluten in genetically predisposed individuals', discarding typical/atypical and latent CD [7].

Table 7.1 Symptoms

Gastrointestinal symptoms and signs e.g., chronic diarrhea

Extra-intestinal symptoms and signs e.g., anemia, neuropathy, decreased bone density, and increased risk of fractures

Silent CD is defined as the presence of positive CD specific antibodies, HLA, and small bowel biopsy findings compatible with CD but without sufficient symptoms and signs to warrant clinical suspicion of CD.

Latent CD is defined by the presence of compatible HLA but without enteropathy in a patient who has had a gluten-dependent enteropathy at some other time in her/his life. The patient may or may not have symptoms and may or may not have CD specific antibodies.

Potential CD is defined by the presence of specific CD antibodies and compatible HLA but without histological abnormalities in duodenal biopsies. The patient may or may not have symptoms and signs and may or may not develop a gluten-dependent enteropathy at a later time.

GLUTEN SENSITIVITY

Recently, a gluten sensitivity different from CD has emerged, at least in adults [1]. This condition may not be diagnosed as CD (see below), is not related to autoimmunity, and has a clinical picture that is more diffuse than CD. The sensitivity may then be diagnosed as a firm clinical entity based on gluten avoidance and challenge [8]. The background for the condition is not known, but gluten components may be involved, and recently amylase trypsin inhibitors (ATIs) have been suggested to be related to gluten sensitivity.

WHEAT (FOOD) ALLERGY

Wheat allergy is a distinct entity with a clinical picture as an atopic disease corresponding to IgE-mediated allergy. It may be divided [9,10] into wheat food allergy, wheat-dependent exercise-induced anaphylaxis, and baker's asthma, which is a respiratory allergy that occurs in a significant proportion of bakers. Lastly, a condition named allergy to hydrolyzed wheat proteins has recently emerged. The allergens in wheat allergy also involve gluten but are not restricted to gluten components.

DIAGNOSIS OF CELIAC DISEASE

The mainstay of CD diagnosis has been the histological evaluation of small bowel biopsies (Figure 7.1) coined in the ESPGHAN criteria for CD from

FIGURE 7.1 Small intestinal mucosa with villous atrophy (to the left) and normal mucosa (to the right), shown as a histological specimen (at the top) and by a scanning electron microscope (bottom).

1990 [11]. In children above the age of two years, the diagnosis is made by appropriate symptoms accompanied by small bowel biopsy findings of villous atrophy and crypt hyperplasia, followed by symptom improvement and eventual disappearance of antibody markers after a period on a gluten-free diet. Below the age of two years and in unclear cases, the diagnostic process includes a biopsy after 1−2 years on a GFD showing normalized histology, followed by a gluten challenge for 3 months or after severe symptoms. Then a renewed intestinal biopsy should again show villous atrophy.

Several histological classification systems for the changes in CD have been suggested, but the most commonly used classification of the histological changes in CD is the Marsh classification [12]. The celiac lesions in the Marsh classification include infiltrative, hyperplastic, and atrophic patterns, graded from I−IV, modified by Oberhuber [13] and Corazza [14]. However, analysis of histological evaluations in several studies has only shown fair to moderate inter-observer correlations, with kappa values of 0.4−0.6 [15,16,17]. Transglutaminase 2 (TG2) is the main auto-antigen in CD. CD antibody determination mainly consists of IgA transglutaminase 2 (TG2) antibodies, which in the present reagent generation are based on genetically modified human reagents used in ELISA or radioimmunoassay techniques [18]. Serum IgA endomysial antibodies or EMA are detected by immunofluorescence and are directed towards a submucosal structure with abundant TG2 [19].

Possibly, serum IgA antibodies to deamidated gliadin [20] may be regarded as CD antibodies. Antibody determinations are now in general of high quality and based on standardized reagents. The ESPGHAN working group formulated a new set of guidelines for CD diagnosis in children and adolescents. The guidelines (4a) state that the presence of characteristic symptoms and high levels of IgA TG2 antibodies (corresponding to 10 times the upper normal limit) may be sufficient for a diagnosis of CD, thus omitting the intestinal biopsy in some cases. In order to strengthen the diagnostic determination, HLA-DQ2/DQ8 haplotypes are recommended, as more than 95% of subjects with CD have one of these HLA types. However, the specificity of the HLA determination is low, as approximately 25% of the population is HLA-DQ2 positive. In cases with equivocal symptoms, or with medium or low antibody levels, the diagnosis is also based on histological analysis as mentioned above. Similarly, for screening purposes the intestinal biopsy is necessary for diagnosis.

An evidence report made in conjunction with the ESPGHAN guidelines concluded that the accuracy of IgA anti-TG2 antibody assays as measured by ELISA was 90−95%, and of the IgA EMA antibodies as determined by immunofluorescence approached 100% [21]. These calculations were necessarily based on reference to histology as the reference standard.

A PUBLIC HEALTH PROBLEM

Over the last decades it has been increasingly recognized that CD is a widespread public health problem and not, as previously thought, a rare disease restricted to young European children [22,23]. CD is found worldwide, including in large populations such as India [24]. Furthermore, it has been shown that the disease can develop at any age, including among old people [25]. Added to this variability is the challenge of case recognition, as symptoms and signs often are non-specific and vary over time for each individual and therefore are often not thought of as CD related. Thus, most cases are still unrecognized and untreated, resulting in negative short- and long-term health consequences. Many accept a chronic state of vague ill health as normal, while others repeatedly approach the health care system to find an explanation. In a recent study from the US with long-term follow-up a four-fold increased mortality was reported for untreated CD adult cases, illustrating the ultimate drawback of the disease [26]. It is very likely that childhood CD also contributes to mortality, especially from diarrhea and, in the poorer parts of the world, largely due to an unawareness of the disease and the available treatment [27].

TIME, PLACE, AND PERSON

The CD prevalence often mentioned is 1%, which likely is a reasonable global estimate but also a crude simplification. The basic concepts in descriptive epidemiology — time, place, and person — can be used to illustrate the more complex pattern. Early studies based their estimates on CD cases diagnosed by symptoms, while most current studies include screening-detected cases, giving a more reliable estimate.

A change in CD occurrence over time was first described in the 1970s in Ireland, where the CD prevalence based on clinically diagnosed cases over a few decades decreased from 0.3% to 0.04% [28,29]. Shortly after, Sweden was surprisingly struck by an epidemic of symptomatic CD cases in children under two years of age (Figure 7.2) with an increased risk from 0.1% to 0.4% [30], which at nine years of age had increased to 0.6% [31]. Notably, additional cases were revealed when these and other populations were later screened for CD.

Lately, an increasing CD trend over time has been demonstrated in several Western populations, also taking into account screening-detected cases. In

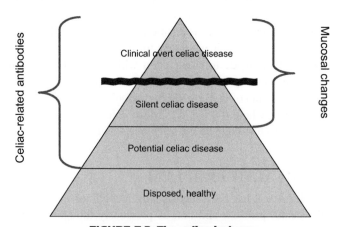

FIGURE 7.2 The celiac iceberg.

US adults, the CD prevalence increased from 0.2% in 1974 to about 1% in 2000 [32,26], and in Finnish adults its prevalence increased from 1.0% in 1978–1980 to 2.0% in 2000–2001 [33]. In Sweden an estimated prevalence of 2.9% in 2005–2006 occurred in 12-year-olds born during the CD epidemic [34], to be compared with the adult CD prevalence in the mid 1990s of 0.5% [35]. The variation in CD prevalence with place can be further illustrated by Germany, where it was as low as 0.3% in 1999–2001 [36]. The highest prevalence so far reported in a general population is among Saharawi children in Algeria, which was 5.6% in 1998 [37]. Over recent years, CD prevalence estimates have become available from an increasing number of countries but are still largely missing from Asia and Sub-Saharan Africa [27]. We discuss possible explanations for the variation with place and time later in this chapter.

CD risk also varies with personal characteristics, being more common in females than males [38] and more common with increasing age, likely due to the autoimmune features of the disease. Thus, in a certain population the CD prevalence is expected to increase with age. However, a decrease with age has also been seen recently [35,35,39], probably explained by the increasing trend over time firstly affecting the pediatric population.

CLINICAL PRESENTATION

A diverse presentation of CD has been appreciated [40,41], partly due to increased recognition of CD with minor or less intense symptoms and partly due to a change in the presentation from clear gastrointestinal symptoms (e.g., diarrhea) to general symptoms such as anemia and calcium deficiency leading to rickets in the child [24,25] and osteoporosis in the adult [42,43]. The British National Institute for Clinical Excellence (NICE) guidelines of 2009 compiled data for a series of symptoms and signs occurring in both adults and children (Table 7.2).

Gastrointestinal symptoms still occur frequently in clinically diagnosed childhood CD, including diarrhea [40,41] as well as chronic constipation

Table 7.2 Diseases Associated with Celiac Disease, Modified From the UK NICE Guidelines
(National Institute for Health and Clinical Excellence, CG86 NICE Guidelines. Diagnosis of Celiac Disease)

Type 1 diabetes mellitus	2–12 %
Down's syndrome	5–12 %
Autoimmune thyroid disease	up to 7 %
Williams' syndrome	up to 9 %
IgA deficiency	2–7 %
Autoimmune liver disease	12–13 %
First degree relatives with CD	10–20 %

[44]. Chronic abdominal pain may be indicative of CD, and this has been reported as a presenting symptom in 90% of Canadian children with CD [44]; however, chronic abdominal pain is very common in childhood. In a questionnaire study looking for subclinical symptoms (diarrhea, constipation, failure-to-thrive) in a cohort of 10,000 8–9-year-old children, symptoms were reported in approximately 1,800 children. Blood was drawn from these symptomatic children and 14 were found to have CD, in addition to five who had been previously diagnosed with CD, mainly due to abdominal pain or constipation [45]. The study suggests that CD cases are found with discrete symptoms in the general population. It is still unclear if the supposed shift from gastrointestinal symptoms to extra-intestinal symptoms in children with CD [41,46] reflects a true clinical variation or is merely the result of improved recognition of non-gastrointestinal forms of CD because of increased awareness of the disease. There is good evidence that failure-to-thrive and stunted growth may be caused by CD. Interestingly, the risk of CD in children with isolated stunted growth or short stature has been calculated to be as high as 10–40% [47], so a clinical work-up of such children should include consideration of CD. Notably, CD can also be found in obese persons [48]. Furthermore, CD is diagnosed in approximately 15% of children with iron deficiency anemia [21].

ASSOCIATED DISEASES

As part of the paradigm change for CD from a rather rare enteropathy to a common, strongly genetically-dependent disease with autoimmune manifestations [49], an increased focus on autoimmune diseases has taken place. As an example, the prevalence of CD in type 1 diabetes (T1DM) has been found to be 3–12% in Caucasian populations (Table 7.2) [50]. A study of the prevalence of CD in children and adolescents with T1DM in Denmark [51] found a prevalence of CD in this population of 12%, the highest reported in Europe. The majority of these patients actually had subclinical symptoms, which in most disappeared on a gluten-free diet. Remarkably, this data was seen in a population with an otherwise very low incidence of recognized CD. In this population an increase in the incidence and prevalence has been documented to occur during the last decade but with an unchanged proportion of children with associated diseases [5].

An overrepresentation of CD also has been observed in subjects with autoimmune thyroid disease [52]. Furthermore, CD occurs more frequently than expected by chance in conditions with chromosomal aberrations, as seen in both children and adults with Turner syndrome [53,54] and in children with Down's syndrome [55]. A 10–20-fold increase of CD prevalence has been reported in subjects with selective IgA deficiency [56], which is particularly noticeable as it may affect the diagnosis of CD (see below). Lastly, a number of conditions (e.g., epilepsy) have been suspected to be associated with CD, but the reported prevalences of 0.5–1% do not seem to differ significantly from the prevalence of CD in the respective background populations.

A MULTIFACTORIAL ETIOLOGY

Until a few decades ago CD etiology was considered simple, depending on the individual's genetic makeup and exposure to dietary gluten proteins. The thinking was that when a genetically predisposed child was exposed to gluten, CD was unavoidable. It was thus surprising when a large discrepancy in CD occurrence was reported from the neighboring countries Denmark and Sweden [57,58]. At about the same time, Sweden was struck by an epidemic of CD in children below two years of age (Figure 7.3), which was totally unexpected [30]. The epidemic pattern illustrated that there must be some hitherto unknown lifestyle or environmental factor contributing to CD development, changing over time on a national basis.

Two causal, interacting factors were later identified: the proportion of infants introduced to gluten while still being breast-fed and the amount of gluten given to the infants [59]. Both factors had changed almost simultaneously across Sweden due to a change in national infant feeding recommendations, with abrupt introduction of cereals and, independently, an increased gluten content of industrially produced infant foods [30,60]. Thus, our findings support recommendations of a gradual introduction of gluten to infants, preferably while still breast-feeding. In a later meta-analysis by Akobeng [61] the protective effect of breast-feeding was supported. The suitable age for introducing dietary gluten to infants has been much debated. Present recommendations support starting complementary feeding, including gluten, at four months at the earliest, and no later than six months, and the same for those with elevated

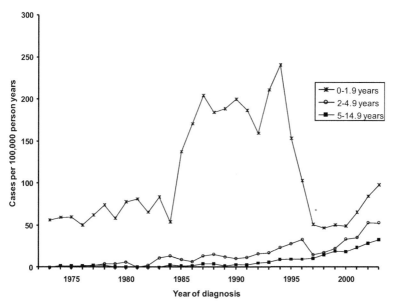

FIGURE 7.3 Incidence rates of CD in the Swedish childhood population from 1973 to 2003. Reproduced with permission [15].

CD risk [62]. This recommendation has support from both European and American expert organizations [63,64,65] but is in partial conflict with the WHO recommendation of exclusive breast-feeding during the infant's first half year. Studies are now in progress to further clarify the role of infant feeding in CD risk [39,62], and further revisions of infant feeding recommendations may be expected. Among other suggested causal factors, an interesting candidate seems to be infectious disease (especially rotavirus) [31,66]. However, changes in infant feeding or infectious panoramas can hardly explain the increasing CD trend noted in several Western countries, both among children and adults. So far, most research on CD and its risk factors has focused on infancy; however, causal lifestyle and environmental factors may also exhibit their effects during other periods of life, including adulthood. In our opinion, we consider it a priority to explore any contribution from the increasing consumption of wheat and fast foods in many countries; however, we are not aware of any such studies.

ETIOPATHOGENETIC CONSIDERATIONS

The understanding of CD has radically changed during the last 10—20 years. Today it is well established that CD has a multifactorial etiology. Genes and environment, and interactions between the two, influence immunological responses and may affect disease development and confer either increased or reduced CD risk [31]. CD is a strongly inherited chronic disease, as documented by twin studies [67] as well as family studies [43], which find high heritability levels. A major finding of these studies was the association of CD with certain HLA haplotypes, HLA-DQ2 and HLA-DQ8 [68], which are present in more than 95% of CD patients, underscoring the autoimmune nature of the disease. The demonstration in CD patients of gluten-reactive small bowel T cells that specifically recognize gliadin peptides in the context of HLA-DQ2 and -DQ8 combined the immunological elements of disease pathogenesis [69]. Furthermore, tissue transglutaminase 2 (TG2) has been recognized as the major auto-antigen in CD [70], and the detection of IgA TG2 antibodies has become the main serological diagnostic tool. The deamidation of gliadin by TG2 is recognized as a central process in CD pathogenesis, as it markedly increases the immunogenicity of the gliadin fragments [71,72], leading to augmented antigen presentation by the antigen presenting cells to the T cells. The antigen presenting cells are usually regarded to be dendritic cells or macrophages, but other cells may be involved.

Apart from the adaptive immune system as expressed by the action of T cells and auto-antibodies, the innate or constituent immune system seems to have an additive effect in CD pathogenesis. Epithelial cells in CD patients secrete the cytokine interleukin (IL)-15 after activation by gluten components of the major histocompatibility complex chain A (MICA), which binds the receptor NKG2D [73,1]. The importance of the innate immune system in CD is not known.

CASE IDENTIFICATION

Most CD cases are undiagnosed, and thereby also untreated, with a seemingly unnecessary negative impact on population health. The first step to improve the situation is to increase awareness of CD among both health professionals and the public. When CD is suspected, a blood sample for analysis of CD serological markers should be taken, followed by further diagnostic measures as required. Such an active case-finding strategy has been tried and proven effective among Finnish adults [74] and Danish children. In the latter study, a questionnaire was used as the screening first step [45]. In addition, the high risk groups such as T1DM and Down's syndrome patients described above should be offered CD screening. However, it is evident that the majority of CD cases will only be identified through mass screening efforts (i.e., approaching the general population).

Most of the World Health Organization's criteria for CD mass screening are fulfilled (i.e., CD is a fairly common disease with a known treatment and with available efficient screening tools) [75]. Acceptability to the population approached is also important, and recently we showed that this may be obtained in a Swedish school setting by approaching 12-year-olds [76]. Still, evidence to support a CD mass screening is not yet complete. It has been questioned whether the cases identified through screening have the same benefit from diagnosis and treatment as the clinically detected cases, assuming a difference in both disease severity and adherence to the dietary regime. However, in a recent one-year follow-up of screening-detected CD in Swedish 12-year-olds, 54% reported improved well-being. The dietary compliance was acceptable (91% always or often on a gluten-free diet). However, experiences of social sacrifices related to the treatment were also reported [77]. In a 10-year follow-up of screening-detected CD in Dutch 2−4-year-olds, 66% had improved health status and 81% adhered to the diet.

Another question is whether CD mass screening is an appropriate use of societies' resources from a health economy viewpoint. This aspect also requires further research. However, the few studies carried out so far favor CD screening in countries with comparatively high prevalence [78,79,80]. An additional question is at what age(s) the screening should be implemented to be most cost effective for both the individual and society. In conclusion, evidence favoring CD mass screening is being compiled but is still not sufficient even in high prevalence countries.

LIVING WITH CELIAC DISEASE

Lifelong Gluten-Free Diet − Cornerstone of the Treatment

The basis of the gluten-free diet is exclusion of specific storage proteins found in the botanically related cereals wheat, rye, and barley and their crossbred varieties, commonly referred to as 'gluten' (see Figure 7.4). The most immunogenic components reside in prolamines rich in proline and glutamine, the

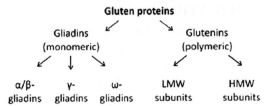

FIGURE 7.4 Gluten protein classification. Modified from Tatham and Shewry, 2008, and Howdle, 2006.

alcohol soluble fractions of the proteins. In wheat these prolamines are called gliadins, and in rye and barley, hordeins and secalins, respectively. Wheat, rye, and barley belong to the same tribe of the grass family (Figure 7.5) and have strong similarities in protein sequences but also differences, which most likely imply differences in CD immunogenicity. Despite these differences, life-long exclusion of all three grains (wheat, rye, and barley) is in practice a pre-requisite for complete recovery of the intestinal mucosa.

Today, oats are considered not to cause disease in most children and adults with CD, at least in daily amounts not exceeding 50–70 grams [81,82]. However, an immune response to oats has been reported in a few CD cases [83]. To enable the identification of possible adverse reactions, it is appropriate to introduce oats cautiously in the gluten-free diet and only after clinical recovery has been established. Moreover, contamination of oats by wheat, rye, and

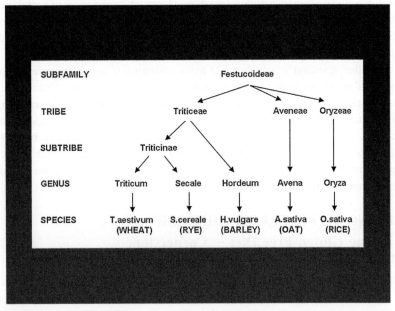

FIGURE 7.5 The family of grasses illustrated by the taxonomy tree and by the appearance in the field.

barley is a concern. Importantly, oats contained in foods manufactured for people with CD must be free from cross contamination at all stages: growth, harvest, transport, storage, processing, food preparation, and serving/catering.

Foods That Are Gluten-Free by Nature

Plain foods such as vegetables, fruits, potatoes, legumes, dairy products, meats, and fish are inherently gluten-free provided processing, preparation, cooking, and serving are gluten-free and free of contamination. Naturally gluten-free grains and flours such as buckwheat, corn, quinoa, and rice are available and may substitute for gluten-containing counterparts in bread making and cooking. More gluten-free grains and flours are presented in Table 7.3.

Where Is Gluten to Be Found?

Products, including flour, made from wheat, rye, and barley contain gluten. Bulgur and couscous are made from wheat. Kamut and spelt are wheat varieties and also contain gluten. Gluten-containing grains and flours are presented in Table 7.3.

Besides foods that obviously contain gluten (e.g., bread and pasta), a myriad of industrially processed foods contain wheat in varying amounts (e.g., soups, sauces, breakfast cereals, meat products, candies, and potato and tortilla chips). All gluten-containing ingredients should be indicated in an easily understandable way (e.g., bulgur (wheat) or modified starch from wheat). If a gluten-containing ingredient is included as an additive, it is possible that the gluten

Table 7.3 Common Gluten-Free and Gluten-Containing Grains and Flours

Gluten-Free Grains and Flours	Gluten-Containing Grains and Flours
Amaranth	Barley
Arrowroot	Bulgur
Buckwheat	Couscous
Corn	Dinkel
Millet	Durum
Oats (uncontaminated)	
Potato flour	Kamut
Quinoa	Rye
Rice	Semolina
Sorghum	Spelt
Tapioca	Triticale
Teff	Wheat

content is below the threshold of 20 ppm and thus safe to eat, but without an analysis the product is not allowed to be marketed as gluten-free. Foods with no gluten-containing ingredients in the list of ingredients may also be contaminated with gluten.

Special Dietary Foods for Persons with Celiac Disease

Special dietary foods are available for substituting for most gluten-containing counterparts such as flour, pasta, bread, pizza, and crackers. Ingredients in these foods can either be based on inherently gluten-free grains (e.g., maize, rice, potatoes, and buckwheat) or gluten-containing grains that have been specially processed to remove the gluten (e.g., purified wheat starch and oats). These foods are formulated, processed, or prepared to meet the Codex Alimentarius standard for gluten-free foods and should supply approximately the same amount of vitamins and minerals as the foods they replace [84].

According to the revised Codex Alimentarius standard from 2008 [84], gluten levels in foods labeled as *gluten-free* should not exceed 20 mg/kg (20 ppm). Foods labeled as *very low gluten* should have a gluten level above 20 and up to 100 mg/kg (100 ppm). Determination of gluten levels should be based on methods with a detection limit at or below 10 mg gluten/kg. The basis for the dual standard of foods is insufficient evidence for a single threshold in combination with the range of gluten sensitivity found in CD. The differential labeling enables individuals with CD to choose appropriate foods according to their individual sensitivity to gluten. The main adjustments made from previous regulation are the more stringent limits and regulation based on total gluten content rather than the ingredients of a product. The previous threshold level for 'gluten-free' of 200 ppm was tolerated by most individuals with CD, although evidence suggested a lower upper threshold for protection of all individuals with CD. The shift in focus from ingredients to gluten content allowed the inclusion of wheat starch and oats in products labeled as 'gluten-free' as long as the gluten level does not exceed 20 ppm. Marketing and labeling of foods with *very low gluten* should be determined on a national level because of the different routine uses of wheat starch, for example between Scandinavia and the US. In Scandinavia, special dietary foods containing wheat starch are widely used and accepted, whereas in the US, the residual content of gluten in wheat starch is still a concern.

RISK ASSESSMENT IN CELIAC DISEASE
Consequences of Gluten Ingestion

In CD, gluten triggers inflammation of the intestinal mucosa resulting in the impaired absorption of nutrients [85]. Exclusion of gluten contributes to the reduction of inflammation, healing of the intestinal mucosa, and improvement of symptoms. Active and untreated CD is associated with a variety of negative health consequences related to immunological processes and the impaired absorption of nutrients (e.g., diarrhea, abdominal pain, anemia, fatigue, and osteoporosis) as mentioned above. Whether adherence to the gluten-free diet reduces the risk of conditions associated with CD, such as diabetes type I and malignancy, needs to be further explored. Also, health risks for those CD patients who have only minor intestinal damage need further assessment.

How Much Is Too Much? A Safe Threshold of Gluten

Contamination with gluten may occur, even in products labeled as *gluten-free* or *very low gluten* [86,87], and total exclusion of gluten from the diet may be impossible. Moreover, the exact tolerable daily amount of gluten is not yet established, and further efforts to establish a safe total amount of daily gluten intake need to be made.

Defining a single threshold of daily gluten intake that is protective for the celiac disease population is complicated due to individual differences in gluten sensitivity [80] and methodological difficulties in both the assessment and the control of human dietary intake. A limited number of randomized controlled trials have been conducted to test different gluten exposures, both with regard to amount and time [81–83]. At this time, it is suggested that a total daily intake of gluten below 10 mg is harmless for all individuals with CD and that a level of 50 mg should not been exceeded [88,89]. Tolerability in the range of 10 to 50 mg daily is probably due to individual sensitivity. Ensuring that the daily intake of gluten does not exceed 10 mg corresponds to an upper limit of approximately 20 slices of bread baked with *gluten-free* flour (≤ 20 ppm) or four slices of bread baked with *very low gluten* flour (≤ 100 ppm), provided that no other foods containing gluten are added. It is important to note that the bread with *very low gluten* flour might contain less gluten than stated but that the celiac consumer cannot know this with any certainty.

The issue of safe thresholds for gluten intake is further complicated by differences in individual energy requirements, which depend on body weight, muscle mass, gender, growth, and physical activity level. A higher energy requirement means larger quantities of food are needed, giving the risk that individuals with high energy requirements in combination with high gluten sensitivity exceed their tolerable gluten intake. Most likely, the more stringent limits of the revised Codex Alimentarius standard for gluten-free foods [78] compensate for this possibility in most cases, but the fact remains that a totally safe concentration of gluten in gluten-free foods is still unknown.

Nutritional Aspects of a Gluten-Free Diet

There is a need to consider both short- and long-term aspects of nutrition in CD. In newly diagnosed cases, short-term aspects focusing on recovery from malnutrition are important together with the implementation of the gluten-free diet. For long-term health, it is important to compose a gluten-free diet providing both macro- and micronutrients in accordance with national dietary guidelines.

Cereals containing gluten are important sources of dietary fiber, different B vitamins, and iron. Consequently, elimination of those cereals limits the range of foods providing these essential nutrients. Some gluten-free substitutes are nutrient-enriched and might compensate for this deficiency, but substitutes are not always nutritionally equivalent [90]. Lower intakes of dietary fiber, B vitamins, and iron together with a more unbalanced composition of macronutrients in the diet compared to that of controls have been reported in both children and in adults with CD [91–94]. In a longer perspective, an unbalanced diet may increase the risk of other diet-related diseases such as obesity, coronary heart disease, and type 2 diabetes. Still, the picture of nutrient intake and food habits among individuals with CD is not complete and needs to be studied further.

ADHERENCE PROBLEMS

Cereals are basic foods in most populations, and gluten is found in a number of manufactured products. Hence, complete, lifelong exclusion of gluten from the diet is often a challenge. Adherence problems with the gluten-free diet have been demonstrated to exist in both children and adults with CD, and only 42% to 91% of individuals with CD are reported to be strictly compliant with the gluten-free diet [95].

Factors affecting compliance are complex and multifactorial [95]. Younger children have higher compliance rates than older children [96], and young adults diagnosed before four years of age comply better than those diagnosed at an older age [97]. A regular CD diagnosis confirmed by intestinal biopsy and based on more than clinical suspicion increases the compliance rate [98,99]. Disease-related knowledge in CD adolescents [96] and in parents of children with CD [100,101] has also been shown to increase compliance with a gluten-free diet. Moreover, decreased compliance over time has been reported in asymptomatic children diagnosed through mass screening imposing specific problems [102].

HOW TO KEEP CELIAC CUSTOMERS HAPPY AND SAFE?

Difficulties when eating out is a well-described reality for individuals with CD [103]. Gluten-free meals are not easily available, and when asking for information, significant lack of gluten-free assurances is often obvious. Provision of safe and easy access to gluten-free catering requires adequate knowledge, service, and professionalism concerning different aspects of gluten-free eating.

Availability of Gluten-Free Alternatives

When eating at public restaurants, clear statements on menu cards that meals are gluten-free or can be easily adjusted to become gluten-free substantially facilitate life with CD. When this clear information is lacking, convincing answers from waiting staff and chefs are a necessity when asking questions about the food in order to be served a guaranteed gluten-free meal.

Palatability of Gluten-Free Foods and Meals

The physical properties of gluten give structure to bread and other gluten-containing foods [24]. The formulation of palatable gluten-free cereal-based products is challenging, and different ingredients such as starches, hydrocolloids, and dairy products are used to improve sensory properties by mimicking the properties of gluten. The quality of these products has improved remarkably during the last decades, but the structure, taste, and appearance still differ slightly from their gluten-containing counterparts. As an example, the deterioration of sensory properties due to heat retention is much more pronounced in gluten-free pasta in comparison with gluten-containing pasta.

Flexibility of Social Norms

Social norms around food and meals are strong [25]. Adherence to a gluten-free diet implies that you often end up in situations where you need to ask questions about the food served, and it might be necessary to refuse to eat at all if no gluten-free option is available. Such behavior may be perceived as impolite if not following expected social 'rules', and to avoid disapproval or expected diapproval eating the gluten-containing meal might be an option, even though it may imply serious health consequences to the subject with CD [104]. A greater understanding in the general public of the importance of special dietary needs will most likely reduce the number of unpleasant experiences for the CD subject.

SUPPORT STRATEGIES

Adequate social support and easily available gluten-free alternatives facilitate life with CD. Awareness of CD and the importance of a gluten-free diet within society at large increases the likelihood that individuals with CD will have easy access to gluten-free alternatives even outside the home and when traveling. Clear gluten-free labeling on menu cards gives access to a safe gluten-free meal with a minimum of comments or questions, and tasty, gluten-free cooking can be achieved by skilled chefs. For this to happen, it is evident that celiac patient organizations play an important role through the dissemination of information, promotion, and communication of research, improvement of the availability of gluten-free alternatives, promotion of accurate food labeling, and financial compensation for excess costs.

PERSPECTIVES FOR THE FUTURE

The increasing public health burden of CD over the last decade calls for action, which should be evidence based and requires knowledge and expertise within several fields. Increased knowledge of the pathogenesis of CD with elements of adaptive as well as innate immunity may lead to improved and earlier diagnosis. The option for primary prevention, reducing the proportion of people developing the disease, needs to be further explored [31]. Infant feeding practices have been suggested to play a role in CD development [59]. As a consequence, current European and American guidelines recommend complementary feeding, including all foods (also gluten), not be initiated before four months of age and not later than around six months [76]. Epidemiological research drives a paradigm shift in complementary feeding — the celiac disease story and lessons learned [105,64,65]. Also it appears urgent to explore any contribution to disease risk from the increase in wheat consumption, which is occurring all over the world and is typically related to fast food products. This has considerable implications for both food production and dietary advice. The institution of a diagnosis of CD as early as possible is preferred, and treatments linked to support measures also need to be developed. More CD cases will be diagnosed as awareness of the disease increases among health professionals and the public and as active case-finding strategies and high risk group screening measures are implemented. However, it is probable that the majority of CD cases will only be identified if mass screening programs are introduced, which for several reasons still is a controversial issue [75]. A gluten-free diet is known to be an effective treatment for CD. Novel treatment strategies under investigation for CD include vaccination and enzyme treatment as well as probiotic adjuncts to the daily diet.

CASE STORIES
Complicated Diagnostic Efforts in a Child

A 5-year-old girl with Down's syndrome was diagnosed one year ago with type 1 diabetes mellitus. She is on daily insulin treatment and a diabetic's diet, and the diabetes is well-regulated with hemoglobin A1c (HBA1c) levels of 7—8 mmol/l. The girl grows well and develops psychomotorically as expected in a child with Down's syndrome. During the last six months she has seemed tired, and a lab work-up disclosed iron deficiency anemia with a hemoglobin (Hb) of 5.9 mmol/l, mean corbuscular volume (MCV) 73 fl, and mean corpuscular hemoglobin content (MCHC) H 17.0. The parents resist further testing, but after another six months a blistering rash develops. Celiac antibodies in blood are determined with a transglutaminase 2 (TG2) antibody of 56 U/l. A gastroscopy including duodenal biopsies is offered, but the parents are still reluctant to have further diagnosis. After another six months the parents agree to a gastroscopy in full anesthesia. The procedure is uneventful and the duodenal histology shows partial villous atrophy corresponding to Marsh IIIa. A diagnosis of celiac disease is made and a gluten-free diet instituted. After another six months hematological parameters are normal.

Teenage Problems: Lack of Adherence

Anne, 17 years old, was diagnosed with CD three years ago. At the time of diagnosis, she had mild symptoms of diarrhea, abdominal distention, and iron deficiency anemia. Her symptoms improved after diagnosis, but TG2 antibody elevation is still present and a visit to the dietician was initiated to evaluate her adherence to the gluten-free diet.

At the visit, it became clear that Anne has adequate knowledge and understanding about CD and the gluten-free diet. However, embarrassment and fear of social disapproval put her in situations where she sometimes chooses to eat normal food, especially when she is together with peers. During the visit, Anne discussed with the dietician how to manage her adherence problem. She expressed a need to meet others with CD and therefore decided to become a member of a celiac association. At follow-up three months later, Anne was more comfortable with keeping to the gluten-free diet and had found sharing common experiences with other teens with CD to be empowering.

Celiac Disease in Old Age: A 75-Year-Old Man

Carl is a 75-year-old man, previously a carpenter, who 20 years ago had a heart attack and at that time was moderately overweight with a BMI of 26. Until recently he weighed 75 kg with a body mass index (BMI) of 21 but had bone pains and a low-impact fracture of his hip two years ago. He has had a 5 kg weight loss and diarrhea has started and worsened during the last six months. Blood serology testing showed positive IgA transglutaminase (TG2) antibodies and a subsequent duodenal biopsy disclosed partial avillous mucosa (Marsh IIIb). Symptoms recovered after six months on a gluten-free diet.

REFERENCES

[1] Schuppan D, Junker Y, Barisani D. Celiac disease: from pathogenesis to novel therapies. Gastroenterology 2009;137(6):1912−33.
[2] Ferguson A. New perspectives of the pathogenesis of celiac disease: evolution of a working clinical definition. J Intern Med 1996;240(6):315−8.
[3] Lohi O, Maki M. [Wheat to celiac disease]. Duodecim 2004;120(6):645−6.
[4] Husby S, Koletzko S, Korponay-Szabo IR, et al. European Society for Pediatric Gastroenterology, Hepatology, and Nutrition guidelines for the diagnosis of celiac disease. J Pediatr Gastroenterol Nutr 2012;54(1):136−60.
[5] Dydensborg S, Toftedal P, Biaggi M, et al. Increasing prevalence of celiac disease in Denmark: a linkage study combining national registries. Acta Paediatr 2011.
[6] Troncone R, Jabri B. Celiac disease and gluten sensitivity. J Intern Med 2011;269(6):582−90.
[7] Ludvigsson JF, Leffler DA, Bai JC, et al. The Oslo definitions for celiac disease and related terms. Gut 2012.
[8] Biesiekierski JR, Newnham ED, Irving PM, et al. Gluten causes gastrointestinal symptoms in subjects without celiac disease: a double-blind randomized placebo-controlled trial. Am J Gastroenterol 2011;106(3):508−14.
[9] Mittag D, Niggemann B, Sander I, et al. Immunoglobulin E-reactivity of wheat-allergic subjects (baker's asthma, food allergy, wheat-dependent, exercise-induced anaphylaxis) to wheat protein fractions with different solubility and digestibility. Mol Nutr Food Res 2004;48(5):380−9.
[10] Pastorello EA, Farioli L, Conti A, et al. Wheat IgE-mediated food allergy in European patients: alpha-amylase inhibitors, lipid transfer proteins and low-molecular-weight glutenins. Allergenic molecules recognized by double-blind, placebo-controlled food challenge. Int Arch Allergy Immunol 2007;144(1):10−22.
[11] Walker-Smith JA. Management of infantile gastroenteritis. Arch Dis Child 1990;65(9):917−8.

[12] Dhesi I, Marsh MN, Kelly C, et al. Morphometric analysis of small intestinal mucosa. II. Determination of lamina propria volumes; plasma cell and neutrophil populations within control and celiac disease mucosae. Virchows Arch A Pathol Anat Histopathol 1984;403(2):173−80.

[13] Oberhuber G, Granditsch G, Vogelsang H. The histopathology of celiac disease: time for a standardized report scheme for pathologists. Eur J Gastroenterol Hepatol 1999; 11(10):1185−94.

[14] Bottaro G, Cataldo F, Rotolo N, et al. The clinical pattern of subclinical/silent celiac disease: an analysis on 1026 consecutive cases. Am J Gastroenterol 1999;94(3): 691−6.

[15] Weile B, Hansen BF, Hagerstrand I, et al. Interobserver variation in diagnosing celiac disease. A joint study by Danish and Swedish pathologists. APMIS 2000;108(5): 380−4.

[16] Ravelli A, Bolognini S, Gambarotti M, et al. Variability of histologic lesions in relation to biopsy site in gluten-sensitive enteropathy. Am J Gastroenterol 2005;100(1): 177−85.

[17] Weir DC, Glickman JN, Roiff T, et al. Variability of histopathological changes in childhood celiac disease. Am J Gastroenterol 2010;105(1):207−12.

[18] Design of a family-based lifestyle intervention for youth with type 2 diabetes: the TODAY study. Int J Obes (Lond) 2010;34(2):217−26.

[19] Leonard JN, Chorzelski TP, Beutner EH, et al. IgA anti-endomysial antibody detection in the serum of patients with dermatitis herpetiformis following gluten challenge. Arch Dermatol Res 1985;277(5):349−51.

[20] Prause C, Ritter M, Probst C, et al. Antibodies against deamidated gliadin as new and accurate biomarkers of childhood celiac disease. J Pediatr Gastroenterol Nutr 2009; 49(1):52−8.

[21] Breidert M. Carpal tunnel syndrome as an occupational disease by Dr. med. Klaus Giersiepen und Dr. med. Michael Spallek in volume 14/2011. Recognize acromegaly. Dtsch Arztebl Int 2011;108(24):424.

[22] Dube C, Rostom A, Sy R, et al. The prevalence of celiac disease in average-risk and at-risk Western European populations: a systematic review. Gastroenterology 2005; 128(4 Suppl. 1):S57−67.

[23] Cataldo F, Montalto G. Celiac disease in the developing countries: a new and challenging public health problem. World J Gastroenterol 2007;13(15):2153−9.

[24] Makharia GK, Verma AK, Amarchand R, et al. Prevalence of irritable bowel syndrome: a community based study from northern India. J Neurogastroenterol Motil 2011;17(1):82−7.

[25] Vilppula A, Collin P, Maki M, et al. Undetected celiac disease in the elderly: a biopsy-proven population-based study. Dig Liver Dis 2008;40(10):809−13.

[26] Rubio-Tapia A, Kyle RA, Kaplan EL, et al. Increased prevalence and mortality in undiagnosed celiac disease. Gastroenterology 2009;137(1):88−93.

[27] Byass P, Kahn K, Ivarsson A. The global burden of childhood celiac disease: a neglected component of diarrhoeal mortality? PLoS One 2011;6(7):e22774.

[28] Mylotte M, Egan-Mitchell B, McCarthy CF, et al. Celiac disease in the West of Ireland. Br Med J 1973;3(5878):498−9.

[29] Gumaa SN, McNicholl B, Egan-Mitchell B, et al. Celiac disease in Galway, Ireland 1971−1990. Ir Med J 1997;90(2):60−1.

[30] Ivarsson A, Persson LA, Nystrom L, et al. Epidemic of celiac disease in Swedish children. Acta Paediatr 2000;89(2):165−71.

[31] Olsson C, Hernell O, Hornell A, et al. Difference in celiac disease risk between Swedish birth cohorts suggests an opportunity for primary prevention. Pediatrics 2008;122(3):528−34.

[32] Catassi C, Kryszak D, Bhatti B, et al. Natural history of celiac disease autoimmunity in a USA cohort followed since 1974. Ann Med 2010;42(7):530−8.

[33] Lohi S, Mustalahti K, Kaukinen K, et al. Increasing prevalence of celiac disease over time. Aliment Pharmacol Ther 2007;26(9):1217−25.

[34] Myleus A, Ivarsson A, Webb C, et al. Celiac disease revealed in 3% of Swedish 12-year-olds born during an epidemic. J Pediatr Gastroenterol Nutr 2009;49(2): 170−6.

[35] Ivarsson A, Persson LA, Juto P, et al. High prevalence of undiagnosed celiac disease in adults: a Swedish population-based study. J Intern Med 1999;245(1):63−8.

[36] Mustalahti K, Catassi C, Reunanen A, et al. The prevalence of celiac disease in Europe: results of a centralized, international mass screening project. Ann Med 2010;42(8):587−95.

[37] Catassi C, Ratsch IM, Gandolfi L, et al. Why is celiac disease endemic in the people of the Sahara? Lancet 1999;354(9179):647−8.

[38] Ivarsson A, Persson LA, Nystrom L, et al. The Swedish celiac disease epidemic with a prevailing twofold higher risk in girls compared to boys may reflect gender specific risk factors. Eur J Epidemiol 2003;18(7):677−84.

[39] Myleus A, Ivarsson A, Webb C, et al. Celiac disease revealed in 3% of Swedish 12-year-olds born during an epidemic. J Pediatr Gastroenterol Nutr 2009;49(2): 170−6.

[40] Bottaro G, Volta U, Spina M, et al. Antibody pattern in childhood celiac disease. J Pediatr Gastroenterol Nutr 1997;24(5):559−62.

[41] Garampazzi A, Rapa A, Mura S, et al. Clinical pattern of celiac disease is still changing. J Pediatr Gastroenterol Nutr 2007;45(5):611−4.

[42] Mustalahti K, Lohiniemi S, Collin P, et al. Gluten-free diet and quality of life in patients with screen-detected celiac disease. Eff Clin Pract 2002;5(3):105−13.

[43] Mustalahti K, Sulkanen S, Holopainen P, et al. Celiac disease among healthy members of multiple case celiac disease families. Scand J Gastroenterol 2002;37(2):161−5.

[44] Rashid M, Cranney A, Zarkadas M, et al. Celiac disease: evaluation of the diagnosis and dietary compliance in Canadian children. Pediatrics 2005;116(6):e754−9.

[45] Toftedal P, Hansen DG, Nielsen C, et al. Questionnaire-based case finding of celiac disease in a population of 8 to 9 year-old children. Pediatrics 2010; 125(3):e518−24.

[46] Bottaro G, Failla P, Rotolo N, et al. Changes in celiac disease behavior over the years. Acta Paediatr 1993;82(6−7):566−8.

[47] van Rijn JC, Grote FK, Oostdijk W, et al. Short stature and the probability of celiac disease, in the absence of gastrointestinal symptoms. Arch Dis Child 2004;89(9): 882−3.

[48] Venkatasubramani N, Telega G, Werlin SL. Obesity in pediatric celiac disease. J Pediatr Gastroenterol Nutr 2010;51(3):295−7.

[49] Rewers M. Epidemiology of celiac disease: what are the prevalence, incidence, and progression of celiac disease? Gastroenterology 2005;128(4 Suppl. 1):S47−51.

[50] Rostom A, Dube C, Cranney A, et al. Celiac disease. Evid Rep Technol Assess (Summ) 2004;104:1−6.

[51] Hansen D, Brock-Jacobsen B, Lund E, et al. Clinical benefit of a gluten-free diet in type 1 diabetic children with screening-detected celiac disease: a population-based screening study with 2 years follow-up. Diabetes Care 2006;29(11):2452−6.

[52] Savastano S, Tommaselli AP, Valentino R, et al. A quick method to detect circulating anti-thyroid hormone autoantibodies. J Endocrinol Invest 1995;18(1):9−16.

[53] Bonamico M, Pasquino AM, Mariani P, et al. Prevalence and clinical picture of celiac disease in Turner syndrome. J Clin Endocrinol Metab 2002;87(12):5495−8.

[54] Ekelund CK, Jorgensen FS, Petersen OB, et al. Impact of a new national screening policy for Down's syndrome in Denmark: population based cohort study. BMJ 2008;337:a2547.

[55] Goldacre MJ, Wotton CJ, Seagroatt V, et al. Cancers and immune related diseases associated with Down's syndrome: a record linkage study. Arch Dis Child 2004; 89(11):1014−7.

[56] Korponay-Szabo IR, Dahlbom I, Laurila K, et al. Elevation of IgG antibodies against tissue transglutaminase as a diagnostic tool for celiac disease in selective IgA deficiency. Gut 2003;52(11):1567−71.

[57] Weile B, Cavell B, Nivenius K, et al. Striking differences in the incidence of childhood celiac disease between Denmark and Sweden: a plausible explanation. J Pediatr Gastroenterol Nutr 1995;21(1):64–8.

[58] Bode S, Gudmand-Hoyer E. Incidence and prevalence of adult celiac disease within a defined geographic area in Denmark. Scand J Gastroenterol 1996;31(7):694–9.

[59] Ivarsson A, Hernell O, Stenlund H, et al. Breast-feeding protects against celiac disease. Am J Clin Nutr 2002;75(5):914–21.

[60] Ivarsson A. The Swedish epidemic of celiac disease explored using an epidemiological approach — some lessons to be learnt. Best Pract Res Clin Gastroenterol 2005; 19(3):425–40.

[61] Akobeng AK, Ramanan AV, Buchan I, et al. Effect of breast feeding on risk of celiac disease: a systematic review and meta-analysis of observational studies. Arch Dis Child 2006;91(1):39–43.

[62] Hogen Esch CE, Rosen A, Auricchio R, et al. The PreventCD Study design: towards new strategies for the prevention of celiac disease. Eur J Gastroenterol Hepatol 2010; 22(12):1424–30.

[63] Agostoni C, Decsi T, Fewtrell M, et al. Complementary feeding: a commentary by the ESPGHAN Committee on Nutrition. J Pediatr Gastroenterol Nutr 2008;46(1): 99–110.

[64] Agostoni C, Braegger C, Decsi T, et al. Breast-feeding: a commentary by the ESPGHAN Committee on Nutrition. J Pediatr Gastroenterol Nutr 2009;49(1): 112–25.

[65] Greer FR, Sicherer SH, Burks AW. Effects of early nutritional interventions on the development of atopic disease in infants and children: the role of maternal dietary restriction, breastfeeding, timing of introduction of complementary foods, and hydrolyzed formulas. Pediatrics 2008;121(1):183–91.

[66] Kondrashova A, Mustalahti K, Kaukinen K, et al. Lower economic status and inferior hygienic environment may protect against celiac disease. Ann Med 2008;40(3): 223–31.

[67] Greco L, Romino R, Coto I, et al. The first large population based twin study of celiac disease. Gut 2002;50(5):624–8.

[68] Lundin KE, Sollid LM, Qvigstad E, et al. T lymphocyte recognition of a celiac disease-associated cis- or trans-encoded HLA-DQ alpha/beta-heterodimer. J Immunol 1990;145(1):136–9.

[69] van de Wal Y, Kooy YM, van Veelen PA, et al. Small intestinal T cells of celiac disease patients recognize a natural pepsin fragment of gliadin. Proc Natl Acad Sci U S A 1998;95(17):10050–4.

[70] Dieterich W, Ehnis T, Bauer M, et al. Identification of tissue transglutaminase as the autoantigen of celiac disease. Nat Med 1997;3(7):797–801.

[71] Sjostrom H, Lundin KE, Molberg O, et al. Identification of a gliadin T-cell epitope in celiac disease: general importance of gliadin deamidation for intestinal T-cell recognition. Scand J Immunol 1998;48(2):111–5.

[72] Tollefsen S, Arentz-Hansen H, Fleckenstein B, et al. HLA-DQ2 and -DQ8 signatures of gluten T cell epitopes in celiac disease. J Clin Invest 2006;116(8):2226–36.

[73] Meresse B, Chen Z, Ciszewski C, et al. Coordinated induction by IL15 of a TCR-independent NKG2D signaling pathway converts CTL into lymphokine-activated killer cells in celiac disease. Immunity 2004;21(3):357–66.

[74] Virta LJ, Kaukinen K, Collin P. Incidence and prevalence of diagnosed celiac disease in Finland: results of effective case finding in adults. Scand J Gastroenterol 2009; 44(8):933–8.

[75] Mearin ML, Ivarsson A, Dickey W. Celiac disease: is it time for mass screening? Best Pract Res Clin Gastroenterol 2005;19(3):441–52.

[76] Nordyke K, Olsson C, Hernell O, et al. Epidemiological research drives a paradigm shift in complementary feeding - the celiac disease story and lessons learnt. Nestle Nutr Workshop Ser Pediatr Program 2010;66:65–79.

[77] Rosen A, Ivarsson A, Nordyke K, et al. Balancing health benefits and social sacrifices: a qualitative study of how screening-detected celiac disease impacts adolescents' quality of life. BMC Pediatr 2011;11:32.

[78] Hershcovici T, Leshno M, Goldin E, et al. Cost effectiveness of mass screening for celiac disease is determined by time-delay to diagnosis and quality of life on a gluten-free diet. Aliment Pharmacol Ther 2010;31(8):901−10.

[79] Shamir R, Hernell O, Leshno M. Cost-effectiveness analysis of screening for celiac disease in the adult population. Med Decis Making 2006;26(3):282−93.

[80] Norstrom F, Ivarsson A, Lindholm L, et al. Parents' willingness to pay for celiac disease screening of their child. J Pediatr Gastroenterol Nutr 2011;52(4):452−9.

[81] Storsrud S, Olsson M, Arvidsson LR, et al. Adult celiac patients do tolerate large amounts of oats. Eur J Clin Nutr 2003;57(1):163−9.

[82] Hogberg L, Laurin P, Falth-Magnusson K, et al. Oats to children with newly diagnosed celiac disease: a randomised double blind study. Gut 2004;53(5):649−54.

[83] Lundin KE, Nilsen EM, Scott HG, et al. Oats induced villous atrophy in celiac disease. Gut 2003;52(11):1649−52.

[84] Codex Alimentarius. Codex Standard for gluten-free foods. Codex Stan 118−1979, revised 2008:2008.

[85] Troncone R, Auricchio R, Granata V. Issues related to gluten-free diet in celiac disease. Curr Opin Clin Nutr Metab Care 2008;11(3):329−33.

[86] Hischenhuber C, Crevel R, Jarry B, et al. Review article: safe amounts of gluten for patients with wheat allergy or celiac disease. Aliment Pharmacol Ther 2006;23(5):559−75.

[87] Collin P, Thorell L, Kaukinen K, et al. The safe threshold for gluten contamination in gluten-free products. Can trace amounts be accepted in the treatment of celiac disease? Aliment Pharmacol Ther 2004;19(12):1277−83.

[88] Collin P, Maki M, Kaukinen K. Safe gluten threshold for patients with celiac disease: some patients are more tolerant than others. Am J Clin Nutr 2007;86(1):260−1.

[89] Akobeng AK, Thomas AG. Systematic review: tolerable amount of gluten for people with celiac disease. Aliment Pharmacol Ther 2008;27(11):1044−52.

[90] Niewinski MM. Advances in celiac disease and gluten-free diet. J Am Diet Assoc 2008;108(4):661−72.

[91] Kinsey L, Burden ST, Bannerman E. A dietary survey to determine if patients with celiac disease are meeting current healthy eating guidelines and how their diet compares to that of the British general population. Eur J Clin Nutr 2008;62(11):1333−42.

[92] Hopman EG, le CS, von Blomberg BM, et al. Nutritional management of the gluten-free diet in young people with celiac disease in The Netherlands. J Pediatr Gastroenterol Nutr 2006;43(1):102−8.

[93] Ciacci C, Cirillo M, Cavallaro R, et al. Long-term follow-up of celiac adults on gluten-free diet: prevalence and correlates of intestinal damage. Digestion 2002;66(3):178−85.

[94] Mariani P, Viti MG, Montuori M, et al. The gluten-free diet: a nutritional risk factor for adolescents with celiac disease? J Pediatr Gastroenterol Nutr 1998;27(5):519−23.

[95] Hall NJ, Rubin G, Charnock A. Systematic review: adherence to a gluten-free diet in adult patients with celiac disease. Aliment Pharmacol Ther 2009;30(4):315−30.

[96] Ljungman G, Myrdal U. Compliance in teenagers with celiac disease−a Swedish follow-up study. Acta Paediatr 1993;82(3):235−8.

[97] Hogberg L, Grodzinsky E, Stenhammar L. Better dietary compliance in patients with celiac disease diagnosed in early childhood. Scand J Gastroenterol 2003;38(7):751−4.

[98] Mayer M, Greco L, Troncone R, et al. Compliance of adolescents with celiac disease with a gluten free diet. Gut 1991;32(8):881−5.

[99] Bardella MT, Molteni N, Prampolini L, et al. Need for follow up in celiac disease. Arch Dis Child 1994;70(3):211−3.

[100] Jackson PT, Glasgow JF, Thom R. Parents' understanding of celiac disease and diet. Arch Dis Child 1985;60(7):672–4.

[101] Anson O, Weizman Z, Zeevi N. Celiac disease: parental knowledge and attitudes of dietary compliance. Pediatrics 1990;85(1):98–103.

[102] Fabiani E, Taccari LM, Ratsch IM, et al. Compliance with gluten-free diet in adolescents with screening-detected celiac disease: a 5-year follow-up study. J Pediatr 2000; 136(6):841–3.

[103] Leffler DA, Edwards-George J, Dennis M, et al. Factors that influence adherence to a gluten-free diet in adults with celiac disease. Dig Dis Sci 2008;53(6):1573–81.

[104] Olsson C, Lyon P, Hornell A, et al. Food that makes you different: the stigma experienced by adolescents with celiac disease. Qual Health Res 2009;19(7):976–84.

[105] Koletzko B, Koletzko S, Ruemmele F. Drivers of innovation in pediatric nutrition. Preface. Nestle Nutr Workshop Ser Pediatr Program 2010;66:VII–VIII.

Practical Food Allergen Risk Management

Chapter eight

Food Allergen Risk Management in the Factory — From Ingredients to Products

Stella Cochrane[1], Dan Skrypec[2]

[1]*Unilever Safety and Environmental Assurance Center, Sharnbrook, Bedfordshire, UK*
[2]*Kraft Foods, Glenview, IL, US*

CHAPTER OUTLINE

INTRODUCTION

Allergens are a special food safety issue. They are in principle harmless, natural substances; however, in allergic individuals they can cause reactions including some that may be life-threatening.

155

Risk Management for Food Allergy. http://dx.doi.org/10.1016/B978-0-12-381988-8.00008-7

In order for food manufacturers to protect their food-allergic customers, there must be systems in place in production facilities to assess and manage the risks posed by food allergens. The accurate declaration of allergens is key to risk management through labeling both those allergens used as ingredients and those that may be present unintentionally, as a result of cross contact, in amounts judged to pose a non-negligible risk. Indeed several countries and regions including the European Union, United States of America, Australia, and New Zealand have legislation for the labeling of important allergens used as food ingredients [1,2,3].

It is recognized that in some manufacturing operations, total avoidance of cross contact, and therefore absence of specific allergens from products where they are not part of the formulation, is not always practicable. It is therefore very important to ensure that unintentionally present allergens only occur at levels that will not harm allergic individuals or that precautionary ('may contain') labeling is used when this is not possible. An analysis of the risk arising from the residual allergen is required to ensure that precautionary ('may contain') labeling is only applied when absolutely necessary in order to maintain the value of such warnings. As well as being a risk communication tool, precautionary labeling is an important risk management measure, and correct application is essential to maintain its effectiveness and thereby minimize risk to allergic consumers. This requires an integrated approach to managing the allergen risk, taking into account factors from raw material sourcing and product formulation, through processing and retailers to the ultimate end user − the consumer.

This chapter describes the key areas of consideration in allergen risk management in food factories.

ALLERGEN RISK MANAGEMENT: PRINCIPLES AND SYSTEMS

Allergen risk management in any establishment needs to be a fully integrated approach within an overall food safety management system, requiring consideration within the wider context of such food safety management systems including those for chemical and microbiological risks. Such systems include elements such as GMP (Good Manufacturing Practice) and HACCP (Hazard Analysis and Critical Control Point), and the requirement for such integration is recognized under ISO 22000:2005 [4,5,6].

Within such a framework, allergen risk management can then be embedded and supported, and the impacts of one aspect of food safety on another and beyond can be considered. For example, sustainability drivers may impact on allergen management and through an integrated approach the needs of both can be addressed, such as balancing a need to save water with maintaining effective cleaning to prevent allergen cross contact.

Each manufacturer should have a defined allergen risk management policy, in which the aims and objectives are clearly described. Alongside this policy, guidelines or guidance documents should also be drawn up to cover the details and procedures required to deliver the aims of the policy. Any allergen risk management program should include all levels of employees, including senior management, and within the allergen management policy and associated documentation, the roles and responsibilities of all relevant personnel should be clearly defined.

Depending upon on the manufacturing setting and product portfolio, the control of common allergens within a manufacturing environment may not be easily achievable, and allergen management should be considered at all stages in a product life cycle, from development through to delivery to the consumer. The principles of HACCP can be used to effectively identify where the cross contact hazards arise from manufacturing operations and to provide a rationale for reduction of the hazard to acceptable levels.

The following sections of this chapter cover allergen management considerations at each product life cycle stage in further detail.

EMPLOYEE TRAINING AND AWARENESS

Successful and effective allergen management requires understanding and commitment from everyone involved in the life cycle of a food product, from senior management to those working on the production lines. Therefore staff training and awareness is a critical component of allergen management.

All relevant personnel (including all temporary staff) should be trained so that they are aware of the hazard posed by common allergens, the site allergen management policy, and the site procedures for allergen management. Steps should also be taken to ensure the safety in the workplace of individuals with specific allergies.

Training may need to be differentiated according to people's roles and needs to be maintained through regular updates. Processes should be put in place to ensure all new employees are trained in a timely manner and have an acceptable level of understanding. Training records should be kept for all personnel, and untrained personnel should not be permitted to handle common allergens.

Use of signs and pictures throughout the facility, for example highlighting where allergens are being handled, can also act as a daily reminder of the need for allergen awareness and management.

If outside contractors are brought onto a site, any impact that they could have upon allergen management should be considered and relevant training provided if required.

INGREDIENTS/RAW MATERIALS — SUPPLIERS, STORAGE, AND HANDLING

The first step in effective allergen management at a site is gaining a detailed knowledge of the allergenic materials being handled there. All allergens that require management (including derivatives) that are used on site should be identified, along with all ingredients and formulations that contain these allergens.

Once identified, if there is the possibility to explore whether or not the materials containing specific allergens can be formulated out of products this should be investigated. Where this is not possible, these materials should be clearly identified upon receipt at the site and then stored and handled so as to minimize the risk of cross-contaminating other materials.

Suppliers should be required to have allergen management systems in place that minimize the unintentional presence of an allergen in raw materials supplied and are of an acceptable quality. Therefore a system of approving and regularly reviewing suppliers should be in place; for example, a supplier questionnaire could be used to assess the scope and quality of the supplier's allergen controls.

Accurate specifications form the basis of accurate risk assessments and accurate product labels. Therefore it is vitally important that suppliers provide detailed, preferably quantitative information on allergens. For example, rather than just indicating that an ingredient, for example a flavoring, contains milk protein, if the amount of milk protein present is provided in a specification then the risk that the ingredient may pose with respect to cross contact can be more accurately assessed. This also applies to any allergens unintentionally present in supplied ingredients. It should also be clear that any changes the supplier makes to material specifications, which could impact allergen management, are agreed to in a timely manner with, and acceptable to, the receiving site.

All allergen-containing materials received on site should be assessed to ensure they have not been compromised during transport and then clearly labeled (either by using the information printed on packaging or via labels placed on the material) and stored in such a way as to prevent or minimize possible contamination of non-allergenic materials (e.g., in dedicated areas within warehouses/storage areas if possible). For example allergen-containing materials should be stored on the bottom of shelving systems so they cannot fall onto non-allergen-containing materials. An allergen cleanup procedure should also be available for use in the event of a spillage.

Dedicated, color-coded utensils, containers, or other equipment should be used to separate the handling of different allergens and non-allergenic materials and also to clearly indicate where allergenic materials are being handled within a facility.

Open bags or containers of materials returned from production should be tightly sealed again and checked to ensure they are still clearly labeled.

PRODUCTION — MATERIAL FLOW, FACTORY AND EQUIPMENT DESIGN, PRODUCT SCHEDULING, AND REWORK

Once there is a detailed understanding of the allergens being handled at a site, a 'map' of the flow of materials through a facility can be made and steps taken to minimize any points of high risk for cross contact that are identified.

In the rare but privileged situation of building a new factory, allergen management should be considered during the site's design to minimize areas of 'hang up' and ensure the maximum ease of cleaning and inspection to minimize cross contact. For existing establishments, understanding the site equipment and flow of materials to identify areas of high risk may mean that equipment or design changes can be implemented to reduce said risk; however where this is not feasible the challenge is to implement processes within the existing infrastructure to manage allergens as well as possible. When new equipment is required, any potential to improve allergen management should also be considered (e.g., an alternative, new design of mixer may be easier to clean than the mixer that needs to be replaced).

One of the most effective means of minimizing cross contact in a facility is by separating the production of products containing specific allergens from products that do not contain the same allergen. This can be achieved by separation in space by physically separating products containing allergen from those not containing the same allergen (e.g., through the use of dedicated production areas, dedicated lines, dedicated equipment, or even separate factories). It can also be achieved by separation in time, by using a combination of production scheduling and effective cleaning (e.g., it may be possible to produce all allergen-containing products at the end of a week and then follow this by a deep clean over a weekend before returning to production on non-allergen-containing products at the start of the following week). Good control may also be achievable while minimizing downtime by scheduling the production of recipes containing low or medium levels of allergens after a recipe containing high levels, rather than switching straight to a product that does not contain any allergen at all.

During production, systems should be in place to manage formulation controls (e.g., the use of checklists for weighing and dosing can reduce the risk of formulation mistakes), and during changeover from one product to another, special care must be taken and robust measures should be in place to ensure that any potential for cross contact is adequately managed. There is an additional risk during changeover of cross packing (i.e., a product being put in the wrong packaging), and steps should also be taken to ensure this risk is managed. This is discussed further in the labeling section of this chapter.

Rework, the reincorporation of finished or semi-finished products into the manufacturing process, is a particular risk with regard to allergens. It is therefore essential that a factory has a rework handling system in place that specifically addresses the management of allergens. Allergen-containing rework should be clearly labeled and should be stored separately from non-allergen-containing rework (e.g., where possible the rework should be covered so that cross contact is minimized). Segregation of rework containers (e.g., by color coding) provides a clear, visible method for managing cross contact of different rework types in production. Ideally, allergen-containing rework should only be used in products containing the same allergen and if possible exact-into-exact rework is recommended. Rework matrices can help in determining which rework can be used in which finished product and ensuring accurate ingredient labeling of such products.

PRODUCTION – CLEANING AND CLEANING VALIDATION

Where dedicated production facilities are not available, cleaning provides the break between allergen- and non-allergen-containing production runs. Cleaning should therefore effectively remove the allergens from the line to a level that results in an acceptable level of risk and thereby minimizes the requirement for products to carry precautionary labels. The effectiveness of cleaning should be assessed to enable a judgment to be made as to whether current practice is sufficient to meet the above objectives or needs to be reviewed. Such a review should determine whether cleaning systems need to be improved or whether current limitations due to cleaning practices or line design cannot reduce the risk to an acceptable level and precautionary labeling is required.

Where cross contact is likely to be homogeneous, a risk assessment approach may be used to determine if levels of allergen remaining after cleaning are such that precautionary labeling is required or not. Where contamination is likely to be non-homogeneous (e.g., with particulates such as nut pieces), the risk posed is very different and such an approach may not be suitable.

All cleaning procedures should be validated to confirm their adequacy, and the allergen management program should include verification of these procedures (i.e., processes to ensure the validated cleaning procedures are being correctly implemented). Validation should be repeated periodically (e.g., yearly) to check that the assumptions that have been made are still valid and also repeated if the process changes significantly.

Starting with a qualitative risk assessment and then moving on to a semi-quantitative assessment is recommended in order to determine whether or not an analytically based validation study is required or applicable. For example, it is sometimes possible to estimate levels of allergen carryover from one production run to another by 'worst case scenario calculations'; that is measuring how much material is left behind in a process (e.g., based on film thickness on equipment or weighing brushed out residual), what the levels of such material would be after dilution with the next product (or in the next process step), what amount of the material is allergen, and therefore allergen levels in the final product that could be consumed.

If an analytically based study is required, accurate and robust analytical results are only useful if the samples analyzed have been taken as part of a correctly designed study. The aim of any validation study should be clearly defined and understood so that the sampling procedures and subsequent analyses are correctly designed/selected and implemented.

For a food product, development of a scientifically sound sampling plan should include a statistical analysis of the probability that all allergens are detected and ensure that any allergens present are accurately measured. Important sampling questions that should be considered include whether the allergen is likely to be evenly distributed within the batch; the number of samples per batch that should be tested; which batches should be tested; which portion of a run should be tested; and how to obtain a specific degree of confidence (e.g., 95% confidence) that no allergen is present.

In brief, the main steps in the design of an analytical validation study are as follows:

1) Define and document the procedure to be validated. Always validate the worst case scenario that could occur.

2) Define and understand what the potential 'contaminating' material is to ensure analytical methods are available and correctly chosen.

A variety of allergen detection methods are used for allergen cleaning validation and each has its own advantages and disadvantages. There are three main types of methods in current use, which are:

A. Non-specific methods including visual checks, adenosine triphosphate (ATP) measurements, and total protein measurements.

Visual inspection is an important first step in assessing cleanliness, but it only applies to accessible areas, and the relationship between visually clean and allergen levels is unknown and will depend upon the surface, allergen(s), and matrix. ATP and total protein measurements are rapid and cheap but they are non-specific. Negative results do not confirm a lack of allergen and positive results are difficult to interpret.

B. PCR (polymerase chain reaction).

This method is very specific; DNA is stable and less affected by processing. However measuring DNA is not measuring the allergen (i.e., the protein) and therefore this approach only provides an indication of the potential presence/absence of an allergen. This can be a useful confirmatory technique but should only be used with caution where other methods are unavailable.

C. ELISA (Enzyme-Linked Immunosorbent Assay) and Lateral Flow/ Dipstick devices.

These are antibody-based detection methods that measure the allergen by detecting allergenic/antigenic protein. These methods are sensitive and relatively fast. ELISAs are quantitative though affected by a number of factors and not available for all allergens. Lateral flow devices are simple to use though only qualitative (semi-quantitative at best).

At present ELISAs are the best and most widely used method for measuring levels of allergens. However, as these tests rely on an antibody reaction with a protein(s) they do suffer from a number of important limitations that require consideration.

Proteins exist in different forms and relative abundances in different foodstuffs. Thus the antibodies in an ELISA may have been raised against a different mixture of proteins to those present in the potential contaminating material. Target protein(s) can also differ between ELISA kits that have the same purpose (e.g., ELISAs for milk may detect beta-lactoglobulin **or** casein **or** a mixture of milk proteins).

Therefore, knowledge of the protein composition of the allergen source is required in order to ensure the correct ELISA kit is chosen to detect it and for the interpretation of the data. For example, if the source of potential cross contact is a whey concentrate, then beta-lactoglobulin would be a suitable choice of target protein; however if the source is skimmed milk powder, then the dominant protein present would be casein. It is also important to understand the reporting units of the chosen ELISA (e.g., in the case of whey as a potential source of cross contact a reporting unit of ppm beta-lactoglobulin would be required, but in the latter results could be in ppm casein or ppm skimmed milk powder).

Food processing can also alter the ability of an ELISA to detect an allergen, for example due to changes in the protein such that the antibody can no longer recognize the target. ELISAs thus may have difficulty recognizing, and produce a false-negative result/reduced quantification, for allergenic materials that are heated, fermented, hydrolyzed, or otherwise processed.

ELISAs require the extraction of the protein into an aqueous environment prior to analysis, and this extraction depends on the solubility of the protein(s) of interest and the formulation of the food from which they are to be extracted (e.g., high fat matrices or recipes rich in polyphenols can affect extraction). To check extraction efficiency for a given sample matrix, a 'spike and recovery test' is recommended as a minimum; for example if the aim is to detect skimmed milk powder in a milk-free product, then a known milk-free sample of product (e.g., prepared in the Quality Assurance kitchen) can be 'spiked' with a known amount of skimmed milk powder and the level of extraction quantified. The food matrix can also affect ELISAs directly (e.g., some ingredients could cross-react with the antibodies in the ELISA to give a false positive reading and others may produce colored backgrounds that need to be controlled for). Thus provision of a known, allergen-free sample as a control has further value.

Whatever analytical method is chosen, the ability to provide a reliable service will depend on the experience and expertise of the analytical laboratory. It is therefore strongly recommended that a dialog is established with the selected analytical laboratory to assess this. Furthermore a good laboratory should offer a confidential service and welcome early discussion of the validation study providing advice on correct test selection and study design.

D. Mass spectrometry.

Until recently a fourth method for detecting allergens was only available in a limited capacity: mass spectrometry. Mass spectrometry is highly specific, can detect multiple allergens in a sample, and is much less affected by processing and food matrices. The technique has now been further developed and has become commercially available for some allergens. As such, mass spectrometry should be considered when selecting an analytical method for cleaning validation.

3) Define what to sample. There are three main types of sampling that can be carried out to assess the presence of allergen after cleaning:

A. Direct Surface Sampling — swab sampling can be used to identify contaminated surfaces. However not all areas can usually be swabbed and interpretation of a positive result can be very difficult as it is impossible to relate swab results to allergen levels in products.

B. Sampling of Rinse/Push Materials — two advantages of using rinse samples are that a larger surface area is assessed, and inaccessible systems or ones that cannot be routinely disassembled can be evaluated. This method assumes that any allergen residue is uniformly removed from the equipment by the rinse material and that if the rinse material is clean then the equipment is clean. If the allergenic material is not soluble in the rinse material, or is physically trapped in the equipment, it will be missed in the validation and may lead to unexpected spot contamination. ELISAs may be affected by high levels of alkali and acid, so when sampling rinse water it is important to make sure that this is neutral (i.e., at the end of the flushing cycle). All parts of production processes should be considered, including loops and bulk dead ends and whether or not there are filler heads or nozzles that require special cleaning. To enable some relationship to be drawn between rinse material and final product, it is important to know how the volume of rinse material used compares with the volume of product that would normally pass through the system being assessed.

C. Final Products — this approach has the great advantage of being directly relevant to consumer exposure and therefore assessment of consumer risk. However potential issues associated with using ELISAs to measure allergens in different product types must be considered and controlled for, and there may also be legal or regulatory considerations to take into account.

4) Ensure control samples are considered. In addition to the controls supplied with an ELISA kit, which must be run to ensure the test kits are working within expected parameters, extra positive and negative controls should be included in a validation study to correct for possible matrix effects. These include a sample known to be free from the allergenic ingredient (a negative control) and a negative control sample to which a known amount of allergenic ingredient has been added (a 'spike and recovery' sample) to enable assessment of extraction efficiency.

Standards supplied with a kit may not be the best for the target ingredient in the formulation (e.g., an allergen kit for casein, supplied with a pure casein standard, may not give a true representation of the level of contamination with cheese powder, which will contain a mixture of milk proteins including casein). Therefore standards based on the target ingredient may be considered.

5) Clearly define in the study protocol how to take, label, and store samples to avoid contamination, sample leakage, confusion over results, and microbial spoilage.

6) Finally the results of the sampling need to be evaluated and a risk assessment needs to be performed to establish whether levels are acceptable. When the validation has been finalized the work should be documented in the Quality Management System.

LABELING

General Considerations

Once a product is packed, opportunities for cross contact are limited but do still exist. Packed products should be stored and handled in such a way as to prevent damage and potential cross contact with allergens. In addition, warehouse and transport personnel should be made aware of the need for cleanup procedures if stock is damaged and leaking.

The most important way to protect allergic consumers from intentionally used allergens is by labeling, and therefore it is essential that a robust process is in place for ensuring the accuracy of allergen labeling on products. Furthermore incorrect labeling is the main cause of allergen-related product recalls, and it is therefore essential that procedures and checks are in place to ensure artwork is correct and the correct artwork is on the correct product.

Accurate artwork requires accurate specifications to ensure ingredient labels are correct, and any precautionary labeling should reflect the results of risk assessment. There should therefore be clear roles and responsibilities regarding ensuring accuracy of specifications and carrying across the information to artwork. Artwork should also be checked when received back from printers to ensure accuracy in the final product.

Even the most accurate and best designed product labels will have no value in allergen management if they are on the wrong product, and therefore processes and checks must be in place to minimize any potential for cross packing. Obsolete labels should be immediately discarded, and simple checklists are often effective in ensuring labels are changed during product change over. There are also a number of more 'high tech' options available for checking packaging, such as bar code scanners, should they be suitable.

Ingredient Labeling

Although ingredient labeling is legislated in some countries the format is generally not; although within Europe the Food Information Regulation (FIR), which has just been passed (July 2011), will result in more prescriptive labeling requirements. As a result, currently a variety of labeling layouts are used, each with different advantages and disadvantages. For example, some manufacturers list allergenic ingredients both in the ingredient list and a separate panel, and while this approach has improved visibility to allergic consumers, more care is required to ensure that the increased risk of potential labeling errors is managed. All labeling should be accurate, clear, and unambiguous, and a simple but effective approach is to list allergenic ingredients in bold text within a product's ingredient list [7].

Legal requirements should be considered as just a starting point for labels, and manufacturers should think beyond legal compliance when designing labels to communicate with their allergic consumers. For example, in the case of adding a allergen to a product that previously did not contain it, simply adding the allergen to the ingredient list may not sufficiently protect the allergic consumer. In such cases, to ensure allergic consumers who have safely eaten the product before are made aware, the addition of the allergen should be flagged on the front of the pack to alert consumers to the change. Another example would be a product where an allergen might not be expected. Indeed, bottled water developed as a 'bridge for the hunger gap' between meals and formulated such that it contained a higher concentration of b-lactoglobulin than cow's milk induced anaphylaxis in two cow's milk allergic children despite correct ingredient labeling [8]. Despite the inclusion of whey protein, the product was still a clear transparent liquid.

Additional product information can be provided through care lines and websites, but this information must be correct and kept up to date. For those manufacturers with product care lines, care line staff should have sufficient, accurate, and up-to-date information to be able to answer questions about allergens. If product information is provided via websites, there should be processes in place to ensure the accuracy of such information and timely updating to reflect any formulation changes or other needs for amendments.

Precautionary Labeling

Food allergen labeling legislation exists in a number of countries across the globe, with each country having defined requirements for the labeling of specified allergens and derivatives that are part of the formulation of the product. While this legislation is very prescriptive for allergens used as ingredients in a product, the unintentional presence of allergens through cross contact is still not directly legislated, although in some countries such as Japan and Argentina the use of precautionary labels is restricted. As discussed, through a site's allergen management program, the risk posed by any cross contact should be evaluated and only when deemed necessary to minimize risk should a precautionary label be applied.

A wide variety of formats have been used for precautionary labels (e.g., made in a factory that handles X or produced on a line that is also used for Y). The range of such labels can be confusing, and many consumers do not know how to interpret such information. It is therefore recommended that simple and clear wording be used for any required precautionary labels such that any potential consumer confusion or misunderstanding is minimized [7].

Currently there is no consensus on safe limits for unintended food allergens, although there have been some attempts in different areas of the world to provide precautionary labeling guidance. Furthermore in the European

Union, some derivatives of allergens have been permanently exempted from allergen labeling based on opinions from the European Food Safety Authority (EFSA). These exempt products have been assessed as not likely to cause severe allergic reactions as they only contain trace amounts of protein. In Switzerland, an action limit for labeling of 1 g/kg (one part per thousand) was defined in 2001 [9], and in Australia and New Zealand the Allergen Bureau (an initiative of the Australian Food and Grocery Council) developed the Voluntary Incidental Trace Allergen Labeling (VITAL) system, which includes a set of action levels that specify whether or not a precautionary label is required based on the level of cross contact identified [10]. Also, in Japan any food containing allergen proteins at greater than 10 mg/kg must be labeled under law. It is clear that there is a need for agreed upon, acceptable limits for the labeling of non-deliberately added allergens in foods, and indeed there is a great deal of time and effort currently being invested in addressing this challenge, as discussed in other chapters of this book.

REFERENCES

[1] European Parliament and Council. Directive 2003/89/EC. Official J Eur Union 2003; L308:15—8.
[2] US Food and Drug Administration. Food Allergen Labeling and Consumer Protection Act of 2004. (Title II of Public Law 108—282); 2004. http://www.fda.gov/Food/GuidanceRegulation/GuidanceDocumentsRegulatoryInformation/Allergens/ucm10687.htm.
[3] Food Standards Australia New Zealand (FSANZ) 2005. Australia and New Zealand Food Standards Code Std 1.2.3
[4] Institute of Food Science and Technology (IFST). Food & Drink — Good Manufacturing Practice A Guide to its Responsible Management. 5th ed.; 2007.
[5] ISO 22000:2005 International Standard 22000. Food safety management systems — requirements for any organization in the food chain. First edition. Geneva, Switzerland: International Standards Organization; 2005. 2005-09-01.
[6] European Commission: Directorate — General Health and Consumer Protection. Guidance document on the implementation of procedures based on the HACCP principles and facilitation of the implementation of the HACCP principles in certain food businesses; 2005.
[7] Barnett J, Leftwich J, Muncer K, Grimshaw K, Shepherd R, Raats MM, et al. How do peanut and nut-allergic consumers use information on the packaging to avoid allergens? Allergy 2011;66(7):969—78.
[8] Ameratunga R, Woon S- T. Anaphylaxis to hyperallergenic functional foods. Allergy. Asthma Clin Immunol 2010;6:33.
[9] Lebensmittelkennzeichnungsverordnung. Verordnung des Eidgenössischen Departements des Inneren über die Kennzeichnung und Anpreisung von Lebensmitteln 817.022.21. Art 2005;8:7—8.
[10] Australian Food and Grocery Council. Food Industry Guide to Allergen Management and Labeling 2007 revised edition. www.allergenbureau.net/downloads/allergen-guide/Allergen_Guide_2007.pdf.

Chapter nine

Managing Food Allergens: Case Histories and How They Were Handled

René W.R. Crevel[1], Steven L. Taylor[2], Sylvia Pfaff[3], Anton Alldrick[4]

[1]*Safety and Environmental Assurance Center, Unilever, Sharnbrook, Bedfordshire, UK*
[2]*Food Allergy Research & Resource Program, University of Nebraska, Lincoln, NE, US*
[3]*Food Information Service (FIS) Europe, Bad Bentheim, Germany*
[4]*Campden BRI, Chipping Campden, UK*

CHAPTER OUTLINE

Risk Management for Food Allergy. http://dx.doi.org/10.1016/B978-0-12-381988-8.00009-9

INTRODUCTION

The principles of allergen management are described in some detail in other chapters in this book, and the body of knowledge that makes them up can often be reasonably derived from first principles. For instance, the principle of separating allergens from other food ingredients and from each other as a means of managing them is almost self-evident, even if implementation is rather more complex. Similarly, few, if any, could argue against the need for a thorough knowledge of the allergen status of supplied materials or the need for staff training. However, presented as a series of practices on their own, without direct reference to context, they cannot convey the full complexity of allergen management in operational circumstances. This chapter aims to overcome this issue by describing a series of case studies that have led directly to lessons being learned. The authors hope that these case studies will prove informative to readers seeking to put an allergen management system in place, as well as those who may wish to check that they have taken into consideration all necessary elements in their allergen management plans. The case studies have been selected to cover specific components of the supply chain and manufacturing and are based on the authors' combined experiences and knowledge. To protect the companies involved, some details may have been changed, but the essence of the issues have been retained.

CASE STUDY 1 – SUPPLY CHAIN

Background

The company received a report that a schoolgirl had had a reaction, which was severe enough to require hospital treatment. The girl had a severe allergy to egg and milk and suffered the allergic reaction after eating a meal at school that consisted of breaded fish and a white sauce provided by a caterer and supplied by the company in question. The ingredients of the white sauce were listed by the manufacturer and did not include milk or egg. Analysis of a sample of white sauce powder by the authorities indicated that it contained 553 mg casein/kg (ppm). None of the other foods eaten contained casein or egg protein, thereby firmly implicating the white sauce as the cause of the reaction.

Incident Investigation

The white sauce contained no milk by formulation, nor did any of the ingredients used in its preparation. The milk was ultimately traced to a creamer, which constituted 23.7% of the sauce. Analyses by Enzyme-Linked Immunosorbent Assay (ELISA) (Neogen whole milk kit) of the implicated batch of non-dairy creamer showed that it contained 6650 mg/kg milk protein. Further analyses of retained samples of all the batches of creamer received by the factory that made the implicated product revealed milk protein contents ranging from 90 mg/kg to 1155 mg/kg (see table). These analyses also showed that the lower milk protein content was found in the earliest batches received, following which the milk protein content not only increased but fluctuated more. The supplier of the creamer undertook their own analyses using the Kjeldahl total nitrogen assay (lower limit of detection 1000 mg/kg). Results broadly correlated with the ELISA results. They showed an association between the protein content of the preceding product and the milk protein content of the following creamer batch, a result consistent with observations made during studies to validate allergen management protocols. The conclusions drawn from these results were:

- A change in some aspect(s) of the production process for the creamer took place between the early batches and the later ones.
- Insufficient consideration was given to allergen management as part of the process for creamer production.
- No or inadequate methods were used to monitor possible changes in the extent of allergen cross contact during creamer production.
- The supplier of the creamer showed inadequate understanding of the factors affecting allergen cross contact and consequently made no attempt to mitigate them.

Batch	Goods Receipt Date	ppm Milk (ELISA)	ppm Milk Protein (ELISA)	Results of Kjeldahl Analysis (% Protein)	Protein Content of Previous Product	Partial Wet Cleaning?
1	15/01/2008	~250	~90	0.33%, <0.1%		
2	12/02/2008	~250	~90	0.33%, <0.1%	8%	No
3	17/04/2008	~300	~105	<0.1%, <0.1%	5%	No
4	27/05/2008	~3000	~1050	0.29%,0.12%	2%	No
5	10/06/2008	ND		0.29%, 0.12%		
6	26/08/2008	~18000	~6300	0.4%, 0.37%	15%	No
7	23/10/2008	~33000	~11550	0.42%, 0.15%	23%	No
8	18/11/2008	~19000	~6650	0.17%, <0.1%	9%	Yes

Supplier's Response

The supplier agreed that the analytical results showed that measurable amounts of milk protein were present in the creamer. However, it also asserted that the specification for the ingredient included the possibility of traces of milk protein and that the sauce manufacturer had implicitly accepted that specification. It claimed that the ingredient at all times complied with the specification. It also asserted that an external audit of its facilities did not highlight any allergen issues. From these discussions three questions therefore arose:

- Could the amounts of milk protein found in the product legitimately be described as 'traces'?
- Did the supplier use adequate methods to monitor the possible changes in allergen cross contact?

- In view of the increasing and fluctuating milk protein content in the creamer, could the supplier maintain that its allergen control processes could be considered to be adequate?

Further information obtained from the supplier revealed that scheduling of the spray-drying of the creamer, which took place in the same spray-drying tower as milk powder, had changed. Instead of being confined to the start of the week, following a thorough cleaning of the tower over the weekend, it took place during the week with no special cleaning.

Traces

According to the Oxford English Dictionary a trace is *'a very small quantity, especially one too small to be accurately measured'*. The definition implies that what a trace is depends on the analytical method used. Analytical measurements provided by the supplier imply a limit of detection of 0.1% (1000 mg/kg) using the Kjeldahl method, since its cites the lowest values as being < 0.1%. Using the conservative assumption that the lower limit of quantification is twice the limit of detection, five measurements on batches containing high amounts of milk protein (by ELISA) were above that limit. Since those quantities could be measured accurately by the supplier's method, they cannot be described as 'traces', and the relevant batches fail the specification on that measure.

Analytical Methodology

The analytical method used should be able to detect the milk protein at relevant levels, that is, at levels close to those known to provoke reactions (allowing for any dilution effects in the use of the ingredient, etc.). Knowledge of the incident available at the time indicated that the amount of milk protein capable of provoking reactions was of the order of low milligrams [1]. A milk content in the white sauce of 1 mg per portion corresponds to a milk content in the creamer of 100 mg/kg. Any assay used to manage milk cross contact in the creamer should therefore be able to measure 100 mg/kg milk protein. The Kjeldahl method, as applied by the supplier, does not meet this requirement. Very sensitive protein assays exist that would be adequate, but the preferred methodology is an ELISA-based immunoassay because of its specificity.

Allergen Management

Allergen management principles and practices have been well defined and accepted over recent years. Guidelines have been produced by a number of regulatory authorities and other organizations and are available publicly, often at no cost. Some of the key principles, which should be incorporated into an allergen management plan, include:
- Risk assessment:
    - Determine allergen(s) of concern, which need to be managed

- Determine how and where they can get into products of which they are not ingredients
- Estimate the extent and limits of cross contact at different stages of production
- A Hazard Analysis Critical Control Point (HACCP) study is a key part of the mechanism for the above.
- Risk management
  - Establish comprehensive measures to reduce cross contact to a minimum
  - Validate the measures, using appropriate quantitative methods
  - Re-examine the measures if anything changes (supplier, process, etc.)
  - Periodically verify that the measures are still operating as intended
  - Communicate any residual risk to users of product
  - Document the above.

Implicit in the approach to risk management is the concept that the extent of cross contact can be confined within limits applicable to the process (e.g., wet processes will differ from dry ones).

The available evidence showed that the supplier of the creamer did not have an effective allergen management plan in place, based on the following observations:

- The risk posed by the presence of milk in creamer was inadequately assessed, without sufficient consideration of the amount that could be harmful.
- Measures to reduce cross contact to a minimum were lacking. For instance, cleaning following a product with a high milk content was only introduced before the last batch supplied and there was no evidence that scheduling to reduce the amount carried over was used.
- Although a change was probably introduced into the process of manufacturing the creamer, there is no evidence that allergen control measures were re-validated.
- There is no evidence for the validation of any allergen control measures.
- The extent of cross contact was not accurately communicated to the manufacturer of the white sauce in such a way that it could safeguard its consumers.

Actions

Following on from this incident, the company whose product had produced the reaction instituted more frequent auditing of this supplier and reviewed procedures for other suppliers. It also required external auditors to demonstrate appropriate competencies for auditing allergen management processes. The company also reminded the supplier of the contractual requirement to notify and seek approval for, as well as document, any change in process, however apparently minor. It also sought evidence from the supplier that comprehensive allergen management training would be instituted to address the issues highlighted by the incident and thereby avoid repetition.

Concluding Comments

The above episode illustrates how allergen management is only as good as the weakest link in the supply chain. It also shows the importance of good communication between suppliers and their customers, as well as a common understanding of allergens and allergen management. Failures occurred at several points in this chain. Among the most glaring, the supplier was unaware of the implications of scheduling production differently and did not deem it necessary to communicate any change to the customer. Audit of the supplier did not highlight any allergen issues, which could be a reflection on the frequency of audits or the training and knowledge of the auditor, or indeed both simultaneously.

CASE STUDY 2 – SUPPLY CHAIN, MANUFACTURING – 'ALL NUTS ARE EQUAL'

Background

A premium chocolate confectioner out-sourced production of its almond praline filling to a third party. The product specification required compliance with the company's food allergen policy, and in particular that the filling should be free of other specified food allergens (including peanut). The third party manufacturer accepted this requirement, and supplier quality assurance inspection of the manufacturer confirmed that appropriate onsite systems for allergen control were in place. Initial validation and verification analyses of the praline paste for peanut showed for some time that the paste was compliant with the specification. However, a subsequent surveillance exercise by the national enforcement agency revealed the presence of significant amounts of peanut in the finished product. This was confirmed by additional analyses commissioned by the manufacturer and a product recall was effected. The cost of this exercise was in the region of one million Euros.

Investigation and Actions

Subsequent investigations revealed that the company that supplied kibbled almonds to the paste manufacturer had decided to extend its business and had commenced kibbling peanuts on the equipment that was used for processing the almonds. On further questioning the supplier advised that it had decided not to inform its customers of this fact since it considered that, despite their obvious botanical differences, almonds and peanuts were allergenically equivalent and that there was therefore no need to provide additional information. The confectioner reiterated to its third party manufacturer the importance of establishing the understanding and application of appropriate allergen management procedures among its own suppliers and ensuring that any change is agreed with prior to implementation.

Concluding Comments

This case highlights a number of points. Firstly, the more complex the supply chain, the greater the risk of an untoward event occurring. In this case the issue was not with the third party manufacturer but with one of its suppliers. Secondly, although laboratory analysis plays an important role in the management of the food allergen hazard, it has limitations due to sampling issues. These are particularly acute in the case of particulate contamination associated with both peanuts and tree nuts, where distribution can be very heterogeneous. Furthermore, analyses can only provide a snapshot of the process as it is when the sampling is undertaken. Any change in the process, such as occurred at the supplier of almonds, negates any previous validation exercise. Finally, despite their botanical differences there is still a widespread misunderstanding in some quarters that peanuts and listed tree nuts are equivalent in terms of the food allergy risk they present and the people they affect.

CASE STUDY 3 − LABELING
Background

Errors with artwork can result in allergenic ingredients being undeclared and, as a result, a product posing a risk to people with allergies to the implicated allergen. However, in some cases, while the allergen is correctly declared in the ingredient list, the label bears contradictory information. In this case, a popsicle provoked an allergic reaction in a child. Investigation showed that while the ingredient list was correct and complete, the packaging was also accidentally (mis)labeled as 'non-dairy'. That label was more prominent than the ingredient list, and the product was bought and given to the milk-allergic child on that basis. According to the recipe, the amount of milk protein in the ice lolly was 1600 mg, mostly in the form of whey proteins, an amount sufficient to cause a severe reaction in a large proportion of milk-allergic individuals. Once notified of the reaction, the company immediately initiated a public recall of the product, supplemented by a direct notification to the national allergic patients' association so that they could warn their members.

Investigation

The company immediately instigated an investigation to find out how a contradictory label indication had been firstly inserted into the artwork and then remained undetected. This revealed that the mistake was made on the draft artwork by the graphic designer at the design agency and remained undetected, although all the stages of the artwork approval process were followed. No quality assurance process was in place at the graphic design agency, and the company noted that draft artwork was frequently presented for approval while bearing many mistakes, making it more likely that any one mistake would remain undetected. The non-dairy designation used was not considered as an integral part of the safety information in the way that the ingredients list is and was therefore not subject to the same stringent checks.

Actions

The company implemented a system of preliminary artwork review with the design agency prior to delivery of the initial draft artwork, with checks against the initial design brief. The company reinforced its own systems for formal artwork review and compliance to ensure that they were fully understood both with respect to requirements and to their importance in assuring consumer safety.

Concluding Comments

The presentation and labeling of pre-packaged foods is critical to the safety of allergic consumers. While the ingredient list is a vital element of this presentation, it is by no means the only one, and additional information that contradicts the ingredient list or otherwise detracts from it can lead to consumers being exposed to the allergens that they must avoid, as in this case. Thus a part of the final label check for the manufacturer is to ensure that nothing could possibly detract from the allergen information. From a wider perspective, this case illustrates one way in which a pack may carry contradictory or incomplete allergen messages. Another one relatively frequent in the United Kingdom involves a mismatch between the ingredient list and the 'allergen box', where this is used. Of course, incomplete or misleading allergen information can occur in many other ways too, ranging from simple omission of the allergenic ingredient to mispacking and even sale of products labeled in the wrong language.

CASE STUDY 4 – MANUFACTURING (LARGE COMPANY)

Background

A consumer complaint was received from the local food safety authority reporting that a 9-year-old child with a hazelnut allergy had suffered an allergic reaction after consumption of a chocolate-coated vanilla ice cream on a stick. Hazelnut was not declared in the ingredient list and a check against the ice cream recipe confirmed that it was not a component of the product, nor were any other nuts. The product also had no precautionary labeling for nuts on the pack.

Investigation

A reference sample was sent for analysis to an external contract laboratory, which reported the presence of 150 ppm hazelnut (protein) in the chocolate coating. The manufacturing plant produced stick ice cream on two lines. Line 1 produced products without any nut ingredients, including the product implicated in the incident, while line 2 made products that contained nuts.

During manufacture, the chocolate mix used to coat the ice cream is stored before use in a tank. At the plant there is normally no link or piping between the tanks for lines 1 and 2. However investigation revealed that, due to production demands, an operator had connected a flexible hose between the tanks serving lines 1 and 2. Thus, chocolate that should only have been used on line 2 (because it could possibly be cross contaminated with nuts, due to their presence in previous batches) was used on the non-nut products from line 1. Thus 'cross contamination via tank and or piping on line 2' was identified as the main root cause of the incident. As the chocolate coating affected is a 'homogeneous puree (chocolate with hazelnut paste)', there was no visible indication of the presence of hazelnuts.

Actions

RISK MITIGATION

As an immediate response to the event, the company undertook a risk assessment. The amount of hazelnut protein per ice cream stick was calculated to be 1.3 mg, based on the analytical results. Based on the reaction already observed, as well as published data from a Dutch [2] and a Danish population [3], that amount of hazelnut protein was judged to pose an unacceptable risk to public health considering the number of product units still on the market. As a result, a public recall was undertaken.

CORRECTIVE ACTIONS

An immediate review of allergen management at the manufacturing plant was undertaken, involving meetings with warehouse, planning, purchasing, quality assurance, and manufacturing staff to ensure that processes around the use of chocolate coating were robust. Additionally the chocolate supplier was checked over the risk of contamination of the ingredient.

The nut-free line and associated equipment were cleaned to eliminate any remaining hazelnut contamination. Production records spanning the preceding two years were reviewed, and these revealed that the wrong HACCP protocol was executed on two occasions, resulting in the chocolate systems of line 1 and line 2 being linked. All supervisors were briefed about the incident and then cascaded the learnings to all other personnel. An immediate ban was implemented to preclude any possibility (e.g., flexible hose) of connecting the line 1 system to the line 2 system.

In order to embed good practice over the longer term, external HACCP training for the manufacturing site was instigated. Engineering measures were also put in place to make it physically impossible to repeat the error. Lock nuts were placed on pipes in the tank storage area, which could only be undone for cleaning. Couplings of different sizes were installed for the transfer of nut-free/nut-containing chocolate mix. A further review of chocolate mix handling systems led to a rebuilding of the chocolate coating system to include dedicated tanks for the nut-free mix.

An immediate allergen self-assessment for the implicated manufacturing plant and all the company's other sites was implemented and the results shared, while HACCP plans were refreshed for all operational teams with a focus on key roles at each site.

Concluding Comments

This case study illustrates two points. The first and perhaps most critical is the importance of keeping allergen awareness high among those working on the lines in manufacturing facilities. This should be largely achievable through embedding regular and focused training, the challenge being to keep it interesting for longer-serving staff. Regular reviews of allergen management measures, including HACCP plans, would aid this purpose, too. The second lesson is to look at the feasibility of designing out potential failure points (e.g., the connections between lines in this instance).

CASE STUDY 5 – MANUFACTURING (REWORK)
Background

An ice cream manufacturer received a complaint from a peanut-allergic consumer alleging that he had suffered an allergic reaction after eating chocolate ice cream. The company arranged for the remainder of the ice cream in the consumer's container to be tested for peanut residues by an independent laboratory. The test revealed that the chocolate ice cream contained undeclared peanuts at a level of 250 ppm. Of course, since the container of ice cream had been opened, the source of the peanut residues was in some question. Nevertheless, simultaneously, the company arranged for the same laboratory to test retained samples of chocolate ice cream from the same lot code. These samples also revealed peanut residues at similar concentrations. The analytical data thus strongly indicated that the manufacturer was responsible for the introduction of undeclared peanut. As the levels of peanut residue were reasonably consistent across the different samples analyzed, the results also implied that a consistent source of peanut had been added to the ice cream.

Investigation and Actions

An inspection of manufacturing records revealed that the chocolate ice cream had been made on equipment also used to produce peanut butter swirl ice cream. Good manufacturing practices would dictate that the peanut butter swirl ice cream should be scheduled as the final run before a full cleaning-in-place (CIP) sanitation of this equipment. Records indicated that the CIP had been done as scheduled. Further analyses revealed that several batches of different ice cream flavors produced immediately after the CIP contained no detectable residues of peanut. Furthermore the implicated chocolate ice cream was made following these other batches. These observations demonstrate that the CIP process had neither been skipped inadvertently nor was it inadequate, since either of those scenarios would have meant that the first batches of ice cream made following the peanut butter swirl flavor would have likely contained detectable peanut residues. It would also have been the case that the concentrations in each successive batch would have been lower due to the nature of ice cream and the equipment used to make it.

The investigation revealed that an additional quantity of unpackaged peanut butter swirl ice cream was produced during the shift in question due to a shortage of properly labeled containers. This peanut butter swirl ice cream had been set aside for later addition into a subsequent batch of peanut butter swirl ice cream, in accordance with the manufacturer's like-into-like (or exact-into-exact) rework policy. However, records detailing the use of rework were incomplete and a strong suspicion developed that the peanut butter swirl ice cream had been mistakenly reworked into chocolate ice cream. Of course, the manufacturer instituted a market withdrawal of the product immediately after obtaining analytical confirmation that the consumer's complaint was justified. The company also reiterated the importance of keeping complete records of rework and ensuring clear and prominent labeling of such materials.

Concluding Comments

Rework almost always constitutes a vulnerability in manufacturing systems, where it is held over until the manufacture of a new batch of product. Its nature also means that the amount of undeclared allergen introduced and its distribution within the batch pose a serious risk to allergic consumers. The manufacturer in question obviously appreciated this and had instituted good practices for the use of rework, for instance exact-into-exact. Nevertheless, failures were revealed in record-keeping and in the labeling of the rework to avoid incorporation into the wrong product.

CASE STUDY 6 – MANUFACTURING AND DESIGN – 'NUT SNOW'

Background

The following incident illustrates the importance of good factory design in managing the hazard presented by food allergens. A biscuit manufacturer produced a range of products in the same factory, some of which contained tree nuts (hazelnut-based) and some of which did not. Primary production was segregated, and the packaging of nut-containing products was undertaken on dedicated lines. Risk assessments indicated that operations were such that precautionary labeling ('may contain') was unnecessary. However, the company received reports on two separate occasions that an almond-allergic individual had reacted after eating a non-nut-containing product, and this became the subject of an investigation.

Investigation and Actions

Laboratory analysis failed to demonstrate the presence of any tree nut residues in the remaining biscuits within the pack, and the company advised that almonds were not used in its factory. Despite these observations, the company decided to review its risk assessments and undertook further investigations. A review of operations revealed that at one point a conveyor carrying nut-containing biscuits passed over another that carried non-nut biscuits. Examination of the rollers and bearings used to drive the conveyors revealed a slow but steady accumulation of crumb which, as a result, was only occasionally removed. It was therefore possible that accumulating crumb could be dislodged in small quantities ('nut snow') at random intervals, and could therefore contaminate biscuits traveling below. The resulting risk was removed by changing sanitation regimes and the introduction of screens to catch any crumbs falling off the upper conveyor.

Concluding Comments

In this case the company's expertise in allergen management enabled it to recognize the phenomenon of cross reactivity (that persons sensitized to one particular food may also exhibit adverse reactions to another). The scope of its investigation was therefore widened after its initial negative analytical results, leading to a thorough review of the design of its plant. This case illustrates that no matter how 'purpose built' a food factory is, in many cases changes in product portfolio and/or advances in food safety knowledge will present or identify new cross contamination risks. This can frequently be the case when a new plant is 'retro-fitted', which can, if improperly managed, compound the risk of food allergen contamination. It also shows that in identifying such risks, there is no substitute for a thorough knowledge of the facility and attention to detail. Additionally, this case highlights the limitations of analytical testing already discussed in earlier cases.

CASE STUDY 7 – MANUFACTURING (SMALL COMPANY)

Background

Food safety enforcement authorities analyzed cookies produced by a small bakery for the presence of almonds and detected 2600 mg almonds per kg in one sample and 20–25 mg/kg in another. Based on those figures, they assumed that almonds were an ingredient of the cookies and that they had inadvertently not been declared. They confronted the bakery with their findings and demanded that the labels be corrected to include the listing of almonds as ingredients. However, the almonds were actually not ingredients.

Investigation and Actions

The bakery conducted a thorough root cause study in order to determine the reason for the allergen contamination in both samples. Using the lot number of the products, the contamination was traced back to its source. In the case of the higher level contamination the causative mistake was quickly established: one member of staff had added the rework from a dough that contained almonds to the batch of cookies, which was recorded in the production sheet. The staff member was questioned and it emerged that he could not identify the almond-containing dough clearly.

The management decided to implement several corrective measures to address the high level contamination issue. Firstly, all staff received formal allergen management training to raise awareness of this problem. The training stressed that the cookies contaminated with the larger amount of almonds had the potential to trigger severe reactions within the allergic population. The individual responsibility of each member of staff for good manufacturing practices and the avoidance of any unintentional contamination was also emphasized. Secondly, new rules were applied to rework, including clearer labeling and identification and 'exact-into-exact' use.

Identifying the cause of the low level contamination proved more challenging. The small amount of almonds detected indicated unintentional contamination during the production process. A check on production records showed that the contaminated cookies had been made on the same day as production of an almond-containing bakery product. The manufacturing order showed that the almond-containing cookies were produced first because of the volume ordered. After these cookies were finished, the preparation tables were cleaned with hot water containing cleaning agents, and the non-almond cookies were then produced. Since detectable almond was found, the management concluded that the cleaning step was not sufficient to remove the almond protein from the surface.

In order to improve the processes and reduce contamination, the management introduced two measures: firstly, products that did not contain the specific allergen would always be scheduled first in the production order. Secondly, cleaning protocols would be revised and monitored in order to improve understanding and ensure that they were adequate for removing the almond protein to the required extent.

Concluding Comments

The action of the food safety enforcement authorities in bringing to the attention of the management the instances of contamination with almonds acted as a 'wake-up call' to the management of the bakery. Again, lack of allergen awareness played a significant role in the development of the incident. The company acted commendably in recognizing the gaps in its systems, leading to the problems observed. Putting systems in place can be particularly challenging in small- and medium-sized companies (SMEs) because they will often not have the relevant expertise in place and will therefore need to be able to access it externally. In addition, training and other activities can prove more onerous than in larger companies, since it may require interruption to production. Nevertheless, such actions are non-negotiable.

CASE STUDY 8 — ALLERGEN AWARENESS — 'NO PROTEIN = NO FOOD ALLERGEN: TRUE OR FALSE?'

Background

At the end of the 1990s, when the reality of managing allergens as a food safety issue within a food business was gaining wide acceptance, an EU cake manufacturer was producing a product containing very small amounts of cold-pressed almond oil for flavoring purposes. In line with labeling legislation at the time, the almond oil was referred to as 'flavoring' in the ingredients declaration. Being a responsible company and being aware of current developments in food safety, it sought and received assurances from the supplier that cold-pressed almond oil did not present a risk to people with almond allergy. The company was therefore somewhat surprised to receive a consumer complaint from an almond-allergic individual who had experienced an adverse reaction from eating the product.

Investigation and Actions

A review of the supplier quality assurance system revealed that the only potential source of almonds was the cold-pressed oil. When challenged, the supplier justified its assurance on the basis that on analysis, protein was never detected in the oil. What the supplier had failed to understand was that protein determination in foods is based on an indirect method of analysis (Kjeldahl Nitrogen), and given that the amounts of allergenic protein needed to elicit an adverse reaction are extremely low, such a method of analysis simply does not have the sensitivity to detect relevant levels of allergenic protein. Parenthetically, it should also be noted that preceding the incident there had been reports in the literature of food allergy incidents relating to cold-pressed oils from other sources (e.g., sesame). Concern about the potential of cold-pressed oils, as well as debate about the safety of highly refined oils, also led to studies with peanut oil that unequivocally demonstrated that cold-pressed oils presented a significant risk to people with an allergy to the specific source [4]. A later review of the evidence documents significant amounts of residual protein in unrefined vegetable oils, albeit well below the lower limit of detection of the Kjeldahl technique [5].

Concluding Comments

In many jurisdictions (e.g., the EU), food labeling legislation now requires that cold-pressed oils from regulated allergenic foods be accurately described so that the food-allergic consumer can make an informed choice. However, even now, follow-up inquiries to supplier quality assurance questionnaires occasionally find that the 'no protein = no food allergen' argument is deployed to justify an answer to the effect that a supplied raw material is free from a particular food allergen. The risk posed by allergenic residues is now well-accepted to be a function of dose, although it may be modulated by other factors [6]. A valid assessment of that risk based on analytical results can only be made if the analytical techniques used are appropriate to the task.

CASE STUDY 9 – PRODUCT DEVELOPMENT, TRAINING, ALLERGEN AWARENESS: – 'FOOD ALLERGENS: NEVER HEARD OF THEM!'

Background

This case study highlights internal failures in training and communication within an individual company in this case a business manufacturing ready meals on behalf of other brands (e.g., supermarket private label). The company concerned operated from a factory that had declared itself a 'nut-free' (no tree nuts or peanuts permitted in any way, shape, or form) site. However, this fact was either not recognized or ignored by the marketing department and the new product development team, although both were based on-site.

Development and Actions

In response to an inquiry from a potential customer, the marketing department instructed its colleagues in new product development to produce specimen products, one of which contained almonds and another peanuts. These products were submitted to the potential customer, who eventually placed a large order for them. Both almonds and peanuts were subject to legislative controls in terms of labeling within the jurisdiction where the factory was located and were also listed as allergenic foods for which additional controls were required for compliance with the company's third party food safety certification. The production and technical departments only became aware of the situation when they were advised of the launch date agreed upon with the client. As a consequence, the company had to advise its other customers at short notice that the factory would lose its 'nut-free' status. It also had to undertake substantial risk assessments and changes to the production facilities to mitigate the risks arising from the introduction of these new allergens. In the end, the changes effected were considered by some of the company's existing customers to be insufficient and resulted in them transferring production to the firm's competitors.

Concluding Comments

This case emphasizes the need for all operations within a food business to have an appropriate level of training in general food safety and food allergen awareness in particular. A particular cause for concern is that, contrary to established best practices, the new product developers failed to undertake a basic risk assessment of the proposed raw materials before even starting to make the specimen products. If they had done so, they would have realized the issue, and more informed decisions could have been taken before commitments were made. This case therefore also emphasizes the necessity for food businesses to have the necessary commercial and safety cultures in place to ensure that potential conflicts in requirements are minimized.

CASE STUDY 10 — TRAINING, ALLERGEN AWARENESS, AND SUPPLIER VERIFICATION
Background
Allergen awareness training of all manufacturing employees is an excellent practice. Often such employees can alert management to the presence of possible allergen hazards, which might otherwise go unnoticed. An element of best practices in this area involves encouraging an open discussion with employees following such training to determine if they are aware of any possible, existing allergen hazards within the operations for which they are responsible. During such a session at a baking company, an employee who was in charge of inspecting raw material shipments revealed that he occasionally observed peanuts in totes of tree nuts (pecans in particular). Peanut-allergic individuals can often safely consume tree nuts and would not therefore necessarily actively avoid them. The presence of peanuts in pecans could therefore pose a significant allergen hazard to such people.

Investigation and Actions
The company launched an immediate inspection of the existing totes of all tree nuts. Initially the suspicion was that the presence of peanuts in pecans was mostly likely due to the fact that both agricultural commodities are grown on farms in the same geographic region of the south-eastern US. Because of the episodic occurrence of peanuts in pecan shipments, the company decided against taking samples of the tree nut totes for laboratory analysis because of the low probability of detecting peanut by such a method. Instead, since whole peanuts had been reportedly observed, the inspection was done visually. Totes of pecan, almond, and macadamia nut were opened and sampled using large scoops. Occasional peanuts were observed in some samples from totes of all three tree nuts. These tree nuts had been obtained from the same supplier, which sorted, inspected, and repackaged multiple tree nuts and peanuts. Auditing of that supplier revealed that it used the same equipment to handle all tree nuts and peanuts at the location at which they were packed. This equipment included a bucket conveyor that could occasionally harbor an unwanted nut or peanut if not cleaned properly. At this point, the baking company had several choices:
1) Insist that the supplier do an effective job of allergen control; but this would require the baking company to do frequent audits and inspections of incoming tree nuts,
2) Apply an advisory label to the products indicating the possible presence of peanuts and other tree nuts in bakery items, or
3) Switch suppliers of pecans, almonds, and macadamia nuts to a company that either handled only one of those tree nuts or had a more effective allergen control plan.

Concluding Comments

This case study clearly illustrates the importance of employee awareness and responsibility in the effective operation of an allergen control plan, and shows how critical attention to even small details can be to successful management. The case also covers an area where risk assessment can be particularly challenging, as it involves particulate contamination, where the amount of allergen presented in a single particle can be sufficient to provoke a severe reaction. In this instance, a peanut cotyledon (approximately half a peanut) would weigh about 500 mg and deliver a dose of 125 mg of peanut protein, enough to cause reactions in 60% of people with a peanut allergy [7,8], a significant number of which would be severe and potentially life-threatening [9]. Against that, the manufacturer needs to balance the detrimental effect resulting from over-use of precautionary labeling, where the event that is cautioned about occurs very rarely. The discovery of the sporadic presence of peanuts among the tree nuts presents an opportunity for a discussion on allergen management with the supplier. However the outcome of that discussion and the actions that result will also depend to some extent on the relationship between supplier and manufacturer. Where the latter is a large company, more options will usually be open since the supplier could be reluctant to lose an important customer. On the other hand, a small company would likely have little leverage in the same situation and might therefore need to resort to precautionary labeling.

CASE STUDY 11 – AUDITOR'S ALLERGEN AWARENESS AND UNDERSTANDING

Background

External auditing by various organizations is becoming more commonplace. An evaluation of the allergen control plan often forms part of the audit. A salad dressing manufacturer underwent one such external audit. One of the outcomes was an adverse finding concerning allergen management. Specifically, the auditor criticized the company for failure to do a full allergen clean-up after manufacturing a salad dressing formulation that contained a small quantity of soy lecithin as one of the ingredients. The soy lecithin was the only ingredient in this particular formulation that was derived from a commonly allergenic source. The presence of soy lecithin was correctly declared on the label of the salad dressing in full compliance with the law. However, a full clean-up was not conducted before manufacturing a salad dressing formulation that did not include soy lecithin. That subsequent product did not declare the presence of soy lecithin in the ingredient statement, again correctly since it was not present as an ingredient, but it also did not have any form of precautionary (advisory) allergen statement. The affected company challenged the auditor's recommendation of the need for full clean-up and sought external expert advice about the risk posed by any carry-over of soy protein.

Risk Analysis

Food-grade soy lecithin can contain legitimately up to 3000 ppm hexane-insoluble solids (HIS) according to specifications based on the Codex Alimentarius. If all of the HIS was soy protein, then soy protein (allergen) levels could be up to 3000 ppm in the lecithin ingredient. In commercial practice, however, soy lecithin generally contains 50–200 ppm soy protein. If the soy lecithin-containing batch of salad dressing contained 2% soy lecithin as an emulsifier, then soy protein would be present at a maximum level of 4 ppm (200 ppm × 0.02), or 60 ppm in the very unlikely situation that the ingredient only met the Codex specification. The lower detection limits of most soy ELISAs fall in the range of 2.5 to 10 ppm soy. Thus, the original salad dressing would almost certainly fall into the marginally detectable range. Furthermore, even under worst case scenarios, the extent of carryover from a previous batch in a liquid handling system such as that used for dressings is likely to be very low, indeed much less than 1%. Even at 1% carryover, the unlikely maximum concentration of soy protein in the non-soy dressing would be unlikely to exceed 0.6 ppm. To put this in perspective, assuming that a consumer ate a rather large helping of 100 g of the dressing, they would still only be exposed to 60 μg of soy protein (and realistically considerably less). This represents approximately 1/17th of the reference dose established by the Voluntary Incidental Trace Allergen Labeling (VITAL) Scientific Expert Panel as the threshold for application of a precautionary label [10]. The level of soy protein would also be well below the limit of detection of the assay methods. In the absence of detectable soy protein, no allergenic hazard is known to exist. In this case with such a low allergen load in the initial formulation, expert opinion supported the company's view that a full allergen clean-up was not necessary to manage the allergen risk.

Concluding Comments

This case illustrates the situations that can arise in the absence of agreement over the risk posed by particular amounts of allergen, which lies at the heart of the issue of when it is appropriate to apply risk mitigation measures, including precautionary labeling. Any risk mitigation measure, be it additional cleaning or a precautionary label, comes with its own downside, which must be considered when deciding on whether to use it. In this case, an allergen clean would in the first instance entail costs for the manufacturer, some of which would be in the materials used, but also in the opportunity costs, whereby the plant is not available for production while being cleaned. The impact is wider however, since the cleaning cycle would use resources, such as water, which then would need to be disposed of. Both disposal and use of water have potential environmental consequences. Precautionary labeling also does not come without its disadvantages, which are the subject of considerable current debate and are discussed in detail in chapters 5, 6, and 15.

CASE STUDY 12 – AN ETHICAL DILEMMA
Background

The company concerned in this case manufactures oatmeal, packages it into large totes, and sells it to other companies to make oatmeal cookies or granola bars. During a walk-through of the facility, the manager noticed a mangled wrapper of some sort lying on top of a full tote of oatmeal. Close inspection revealed that the wrapper was from M&M® Peanut Candy. Due to the mangled nature of the wrapper, the manager could not determine whether the wrapper had been through the oatmeal production equipment. No candies were visible in the tote. Because the candies are hard-shelled, the possibility that candy pieces could be present in the large tote could not be excluded. Additionally the candies could have been crushed if they had been put through the oatmeal production equipment. Faced with this situation, the manager could have merely removed the wrapper and shipped the tote to the customer. However such action would clearly be unethical because of the possibility of contamination and the ensuing risk to the consumers of the products made by the manufacturer to whom the oatmeal was supplied.

Actions

The company immediately placed the suspect tote on hold. Then it took multiple samples from the tote and sent them to an external laboratory to test for peanut residues by ELISA. No peanut residues were detected. However, especially if the candies were intact within the tote, any sample would need to contain a candy piece, a possibility with low likelihood. Visual inspection of additional samples from the tote also did not reveal the presence of any peanut candies. However in an abundance of caution, the oatmeal company decided to divert this tote to animal feed. Clearly the presence of an intact peanut candy would have constituted a significant allergenic hazard and would have damaged this supplier's reputation with its customer, a chance that it chose not to take. As an additional measure to reduce the risk of recurrence, the company also removed peanut-containing candy from the vending machine in the company cafeteria.

Concluding Comments

This case demonstrates an interesting situation where the manager was called upon to make a judgment of the potential risk based only on circumstantial evidence of potential contamination. Clearly the simplest course of action would have been to assume that only the discarded wrapper was involved, the contents of the packet having been consumed prior to its appearance on top of the tote, thereby allowing the ingredient to be delivered without further consideration. However, he correctly realized that in pursuing such a course, he could be putting consumers of his customers' products at risk for the sake of expediency and that this could be damaging to his long term relationship as a supplier. It is interesting to note that, despite the testing and inspection failing to show the presence of any contaminating material, the company chose to sell the contents of the affected tote as animal feed. It could be argued that, given that the results of the tests and inspection were not sufficient to dispel the uncertainty about the risk in the company's mind, it would have been more appropriate to take that further action from the outset.

CONCLUSION

This chapter describes a number of case studies covering different stages in the manufacture of products, spanning the range from raw material and ingredients to the information provided to the final consumer. Clearly it would be possible to expand the number of such examples several-fold to highlight specific points, but the selected case studies are, in the opinion of the authors, of sufficiently broad interest to illustrate where and how allergen management systems may fail even though the facilities or situations where they occurred differ in detail from the readers' experiences. Perhaps one over-riding message is that failure may occur at any point in the system and must therefore be planned for, with systems ideally designed to be fail safe. This observation reinforces the message that allergen management must form an integral part of food safety management systems. As in other complex systems, incidents in which a failure becomes apparent, for instance through an allergic reaction in a member of the public, rarely if ever result from a single omission or mistake. Rather, they are the accumulation of a series of undetected errors and multiple missed opportunities to correct them. It follows that improvements could be fostered by the adoption of systems that encourage the reporting of such occurrences, if necessary on an anonymous basis, as has been successfully done in other fields, such as commercial aviation most notably.

REFERENCES

[1] Malmheden YI, Eriksson A, Everitt G, Yman L, Karlsson T. Analysis of food proteins for verification of contamination or mislabelling. Food Agric Immunol 1994;6:167−72.
[2] Wensing M, Penninks AH, Hefle SL, Akkerdaas JH, van RR, Koppelman SJ, et al. The range of minimum provoking doses in hazelnut-allergic patients as determined by double-blind, placebo-controlled food challenges. Clin Exp Allergy 2002;32(12):1757−62.
[3] Eller E, Hansen TK, Bindslev-Jensen C. Clinical thresholds to egg, hazelnut, milk and peanut: results from a single-center study using standardized challenges. Ann Allergy Asthma Immunol 2012;108(5):332−6.
[4] Hourihane JO, Bedwani SJ, Dean TP, Warner JO. Randomised, double blind, crossover challenge study of allergenicity of peanut oils in subjects allergic to peanuts. BMJ 1997;314(7087):1084−8.
[5] Crevel RW, Kerkhoff MA, Koning MM. Allergenicity of refined vegetable oils. Food Chem Toxicol 2000;38(4):385−93.
[6] Madsen CB, Hattersley S, Buck J, Gendel SM, Houben GF, Hourihane JO, et al. Approaches to risk assessment in food allergy: report from a workshop 'developing a framework for assessing the risk from allergenic foods'. Food Chem Toxicol 2009;47(2):480−9.

[7] Taylor SL, Crevel RW, Sheffield D, Kabourek J, Baumert J. Threshold dose for peanut: risk characterization based upon published results from challenges of peanut-allergic individuals. Food Chem Toxicol 2009;47(6):1198–204.

[8] Taylor SL, Moneret-Vautrin DA, Crevel RW, Sheffield D, Morisset M, Dumont P, et al. Threshold dose for peanut: Risk characterization based upon diagnostic oral challenge of a series of 286 peanut-allergic individuals. Food Chem Toxicol 2010;48(3):814–9.

[9] Perry TT, Matsui EC, Conover-Walker MK, Wood RA. Risk of oral food challenges. J Allergy Clin Immunol 2004;114(5):1164–8.

[10] Allergen Bureau Voluntary Incidental Trace Allergen Labeling (VITAL) august 2013 Available from http://www.allergenbureau.net/downloads/vital/VITAL-Guidance-document-15-May-2012.pdf; Accessed 27 August 2013.

Chapter ten

Catering — How to Keep Allergic Consumers Happy and Safe

Sue Hattersley[1], Rita King[2]
[1]*Food Standards Agency, London, UK*
[2]*British Beer and Pub Association, London, UK*

CHAPTER OUTLINE

INTRODUCTION

There have been significant improvements in recent years in the labeling of allergenic ingredients used in pre-packaged foods across Europe, following the implementation of Directive 2003/89/EC [1] and its subsequent amendments. This legislation has resulted in clearer declarations of the specified

189

Risk Management for Food Allergy. http://dx.doi.org/10.1016/B978-0-12-381988-8.00010-5

allergenic ingredients. For example, there has to be a clear reference to the common name of the food, such that casein has to be labeled with reference to milk, which makes it easier for allergic consumers to recognize foods that they should not eat.

However this legislation only covers intended ingredients in the food and foods that are pre-packaged. Inadvertent presence of the allergenic food as a result of cross contamination at some point during the growing, harvesting, transport, and manufacture of the finished food product does not have to be declared, although many food manufacturers do use advisory labeling to warn of this possibility. The food service sector consists of a broad range of food businesses, from market stalls selling loose food, through sandwich bars to cafes and restaurants, and it also includes staff canteens, schools and nurseries, hospitals, and prisons. Information about intended allergenic ingredients is currently not legally required to be provided for any unpackaged food sold in any European Union country, including that sold in catering businesses.

As noted by the Anaphylaxis Campaign in its submission of evidence to the UK House of Lords Science and Technology Select Committee inquiry on allergy, published in 2007:

> *The risks increase significantly when people eat out, largely because consumers do not have the benefit of comprehensive food labeling and must often rely on verbal assurances of catering staff.* [2]

Pumphrey [3] and Pumphrey and Gowland [4] have reported that three quarters of recorded deaths linked to food allergy in the UK occurred when food was bought in catering establishments. Pumphrey also noted that some people had fatal reactions to a type of nut that had not previously caused a reaction, and that substitution of one nut with another was common in catering. Advice to consumers allergic to at least one type of nut is normally to avoid all nuts, including peanuts. In some cases the person involved had asked for a meal that did not contain nuts and the caterers themselves did not know that the allergenic food was present.

WHY DO CONSUMERS CURRENTLY NOT HAVE THE INFORMATION THEY NEED TO MAKE SAFE CHOICES WHEN EATING OUT?

Given that there is currently no legal requirement for businesses selling unpackaged foods to provide information about allergenic foods, consumers with food allergies adopt a number of strategies when eating out to make decisions on which foods to eat. Some consumers use information provided on menus or elsewhere, including descriptions of particular dishes, to make judgments about whether or not a particular allergenic food is likely to be present. This can be influenced by previous experience of consuming foods described in the same way in that establishment or elsewhere. In some cases, consumers may be reluctant to ask, or to persist in asking, when receiving unsatisfactory or incomplete answers, as they do not wish to draw attention to themselves. There may also be language barriers that impede communication between the customer and the member of staff, particularly when eating out abroad, but also because many staff working in catering businesses may not speak the language of the country as their first language.

Consumers may seek information from staff in the catering business who may or may not be able to supply accurate information. Leitch reports the results of a survey [5] carried out in takeout establishments in Northern Ireland, where attempts were made to purchase meals that did not contain peanuts in premises where peanuts were included as an ingredient in some dishes. Twenty percent of the meals sold as suitable for someone with a peanut allergy were found to contain detectable amounts of peanut. In some cases the levels of peanut detected indicated possible cross contamination, but in about half, the levels were such that it was likely that peanut was a deliberate ingredient.

Although caterers are increasingly being asked about possible allergenic ingredients in the dishes they offer, their knowledge about food allergy and the training that they receive on this subject can be limited. A study by Pratten et al. [6] in the UK investigated the level of knowledge about food allergy in a range of catering businesses, as well as how the proprietors dealt with requests for special meals. The study showed that many businesses were accustomed to providing special dietary requirements, particularly relating to avoidance of gluten and nuts, and that better quality restaurants using fresh ingredients were the most confident about their ability to meet such requests as they could control the ingredients going into a particular dish. However, those businesses that brought in prepared meals for heating and serving were reliant on the information provided by their suppliers when trying to meet a special dietary request. It was pointed out that some products supplied to caterers did not have ingredient information on the packaging. Although this information is supplied separately to caterers, for example with invoices, it may not be supplied with every delivery for regular customers.

A study conducted in Ireland during 2009 [7] found that approximately 10% of meals sold in catering establishments as 'gluten-free' contained some gluten, with nearly 8% containing more than 100 mg/kg and over 5% having gluten levels in excess of 1000 mg/kg, demonstrating that staff confidence, 'gluten-free' notices, or menu choices were no guarantees of risk-free dining for people with celiac disease.

A more recent study has shown that there are still misunderstandings and a lack of knowledge about food allergy among catering staff. Bailey et al. [8] conducted a survey in the UK of staff working in catering (owners, managers, waiters, and chefs) using a structured telephone questionnaire. This study reports that while the vast majority of those working in catering are aware of food allergy, and most (90%) had undergone food hygiene training, only 33% reported having had food allergy training. Bailey also reports that, despite over 80% of those interviewed saying that they were confident (very or somewhat) in providing a safe meal for a customer with a food allergy, there were considerable misunderstandings. Sixteen percent thought that cooking the food prevented it causing allergy and 12% were unaware that food allergy was potentially fatal. More worryingly, nearly 40% though that a customer having an allergic reaction should drink water to dilute the allergen, nearly one quarter thought that it was safe to consume a small amount of the allergen, and one fifth thought that removing the allergen from a finished meal would make it safe. Although about half of those interviewed were interested in further training on food allergy, those whose knowledge was poor or those working in catering businesses where there was no food allergen separation and control were no more likely to identify that food allergy training would be important than those whose knowledge was better.

WHAT IS THE CURRENT LEGAL POSITION AND ARE ANY CHANGES EXPECTED?

Currently food labeling legislation in Europe does not require allergenic ingredients used in unpackaged foods to be specifically declared, nor does it require possible allergen cross contamination to be highlighted. However there is an expectation that, if asked, businesses supplying unpackaged foods should be able to provide information to their customers or, if they do not have the information, to say that they do not know.

Although there are currently no specific requirements in these areas, the provisions of the General Food Law Regulation 178/2002 [9] prohibit unsafe food from being placed on the market. Paragraphs 3(b) and 4(c), respectively, of article 14 state that:

> In determining whether any food is unsafe, regard shall be had to the information provided to the consumer, including information on the label, or other information generally available to the consumer concerning the avoidance of specific adverse health effects from a particular food or category of foods.
> In determining whether any food is injurious to health, regard shall be had to the particular health sensitivities of a specific category of consumers where the food is intended for that category of consumers.

In practice, this means that any information that a business chooses to supply to a consumer regarding the presence or absence of allergenic ingredients must be accurate and complete, and that the provision of incorrect or misleading information is potentially illegal. In particular, if a food is marketed specifically to consumers with a food allergy, by using claims that the food does not contain a particular allergenic food ingredient, extra care needs to be taken to ensure that such claims can be justified.

In 2008, the European Commission announced that it would be reviewing food labeling legislation, and a range of separate pieces of legislation have been consolidated into a single Food Information for Consumers Regulation.

While provisions relating to the declaration of allergenic ingredients in pre-packed foods are essentially unchanged (although the allergenic ingredients now have to be highlighted in the ingredients list), the new Regulation places further emphasis on the need to protect the health of the allergic consumer. Regulation 1169/2011 [10], which was published in November 2011, has introduced a new requirement to **provide** information on allergenic food ingredients for foods sold non-pre-packed in the European Union, covering foods sold loose in retail situations, in catering, and foods pre-packed for direct sale (packed on the premises from which they are sold). However, this Regulation does not specify **how** this information should be provided, although there is the option for individual Member States to adopt national rules concerning the means through which the information is to be made available and their form and presentation. These new provisions will come into force and businesses will need to comply from 13 December 2014.

In addition to the legislation described above, there is new European legislation that came into effect at the beginning of 2012 relating to foods for people who are intolerant to gluten. Commission Regulation (EC) No. 41/2009 [11] sets out specific requirements relating to the composition and labeling of foods for people with gluten intolerance, such that foods that are described as 'gluten-free' must not contain more than 20 ppm of gluten in the food as sold to the consumer. This Regulation applies to all foods, and therefore foods sold unpackaged, including in catering businesses, will need to meet these strict compositional requirements if they are making claims about gluten. The foods could be specifically prepared for this particular section of the population by using substitute ingredients or they could be everyday foods that normally do not contain gluten-containing ingredients and where cross contamination is controlled such that they can meet the 20 ppm limit. To assist food businesses, the UK Food Standards Agency (FSA) has provided information explaining the new legislation and has produced a fact sheet for caterers helping them decide how to label their products if they want to provide information about gluten to their customers [12]. The Agency has also produced a fact sheet for celiac consumers to explain the new rules [12].

BEST PRACTICE GUIDANCE PUBLISHED BY THE UK FOOD STANDARDS AGENCY IN 2008 ON THE PROVISION OF ALLERGEN INFORMATION FOR NON-PRE-PACKED FOODS

Given that allergic consumers were more likely to have a reaction following consumption of food that was unpackaged, and the lack of specific legislative controls on the provision of allergen information for unpackaged foods in Europe, the FSA decided to produce best practice guidance in this area. All the interested stakeholders (caterers, catering suppliers, retailers, enforcement officers, and allergic consumers) were asked to participate in the development of this guidance so that the resulting documents would balance the needs for information of those consumers that are allergic or intolerant to certain foods against the practical problems faced by businesses in the food service sector who are being asked to provide such information. All stakeholders agreed that the guidance should concentrate on the provision of information about allergenic foods used as ingredients in a particular food product, although possible cross contamination risks also needed to be considered. It was also agreed that the guidance should cover all those allergenic foods that are covered by European legislation relating to pre-packed foods. The guidance needed to cover the wide variety of businesses within the food service sector, ranging from unpackaged foods sold retail in bakeries and delicatessens through sandwich bars and coffee shops to fast food outlets and fine dining restaurants. It was also recognized that foods provided in schools, nurseries, hospital, and prisons also came into the category of unpackaged foods.

The guidance was published in 2008 [13] and is accompanied by a simple leaflet aimed at small businesses and a training poster for use by businesses [14,15]. As well as providing background information on food allergy and why it is an important food safety issue, the guidance includes a number of annexes that address the types of issues that are relevant for different types of food service sector business. The following sections of this chapter set out the approach adopted in the guidance document and the key messages that it delivers.

The Guidance and Its Key Messages

In order for a catering business to meet the needs of a food allergic customer and supply food that is safe for that person, there are three key areas that need to be considered. Firstly, the caterer needs to understand exactly which foods the customer needs to avoid and ensure that appropriate information is communicated to others in the business. Secondly, it is critical that staff receive training in food allergy so that they understand the seriousness of an inquiry about food allergy from a customer and know how such inquiries should be dealt with in that particular business. Thirdly, the catering business needs to have procedures in place that will allow staff to find out about the ingredients used in particular dishes, whether these are made on the premises from fresh ingredients or brought in prepared or part-prepared. In addition, information needs to be communicated back to the customer so that he or she can make safe, informed choices about what to eat. If staff cannot be sure whether or not a particular allergenic food is used as an ingredient in a certain dish, they should explain this to the customer and they should never guess or make assumptions.

Communication

Communication starts with the allergic consumer, who should make his or her dietary needs known to the catering business. However, catering businesses will get many requests from people who do not want to eat certain ingredients for reasons other than food allergy or food intolerance, such as for religious, moral, or dietary reasons or because they simply do not like a particular food. If the consumer has a severe food allergy, then he or she should inform the business about this, as cross contamination issues as well as the ingredients used in the dish then become relevant.

Some businesses will choose to make detailed allergy information available on their menus, notice boards, or websites. If they do so, allergic consumers are likely to use this and not engage directly with staff in the business. It is therefore essential that, if this approach is adopted, information is complete, accurate, and up to date. This may be an effective approach for catering businesses that are part of large chains, where the menus are standardized and the specifications of the different meal items are fixed, although possible allergen cross contamination risks associated with the individual catering business would not be included within this information. Some businesses may also include generic advice about allergen cross contamination risks on their menus, notice boards, or websites, but it is not helpful for allergic consumers to use warnings such as 'all the products on our menu may contain nuts', without assessing that such risks really exist.

Many large catering suppliers provide detailed information on the allergenic ingredients used in all the products that they supply, and catering businesses can refer to this information when responding to inquiries from customers. However other businesses may source a particular menu item, such as a dessert option, from more than one supplier and the allergen content of the dish may vary. In such cases, it may be more effective for the allergen information from the labeling for the particular product to be referred to and, if necessary, shown to the allergic consumer to allow him or her to make the decision whether or not to eat that item.

Other businesses may opt not to provide detailed information for each menu item but to have a general statement on menus, notice boards, or websites inviting customers to ask for information about allergenic ingredients. This may make customers who are less self-assured feel more comfortable asking for allergen information and can initiate a dialog between the caterer and the customer. During such a dialog it may become apparent that none of the regular items on the menu are suitable for the particular allergic individual. However a business may be able to prepare a special meal for that person, especially if it has been given advance notice. Allergic consumers should consider contacting businesses in advance, if possible, to make their needs known. In addition, businesses catering for conferences or special events, such as weddings, should actively encourage those with food allergies or other dietary needs to communicate these to the business so that separate meal options can be provided. For example, where food is supplied as a self-service buffet, then suitable food can be prepared and held separately.

Communication between different parts of the food business is just as important as communication between staff and the customer. If chefs change the ingredients used in a particular dish, for example due to supply issues, this should be communicated to the staff waiting tables. Equally, staff waiting tables should always check the ingredients in a particular meal item with the chef or manager, even if they have asked the same question previously, as the recipe being used or the supplier may have changed.

While most allergic reactions are caused by allergenic ingredients that were not recognized, either by the customer or the caterer, it is also important that cross contamination risks are taken into account when information is provided. The way in which food is displayed may increase the risk of cross contamination, for example in salad bars, dessert trolleys, or self-service buffets. In such situations, it may be possible for food for the allergic consumer to be served from items not yet put on display to minimize cross contamination risks. It is also important to make sure that utensils and preparation areas are properly cleaned before preparing and serving the food for the allergenic customer.

Staff Training

Staff working in catering businesses have basic food hygiene training when they start work, and it is also important that all staff receive training about food allergy at the same time. The training that is appropriate in different types of catering businesses will vary considerably but, as a minimum, all staff need to be told about the importance of dealing responsibly with food allergy questions and what they should do if a customer asks for food allergy information. It may be that a business could designate specific people to deal with all food allergy inquiries, such as the head chef or duty manager. In the UK, training materials for catering businesses are available from Local Councils or other providers, and there are also free training materials available from the FSA, such as the 'Safer Food, Better Business' for caterers [16]. There is also an e-learning training module on food allergy that is freely available from the FSA's website that includes catering as well as food manufacturing situations [17]. This can be particularly helpful for small catering businesses that do not have the resources or time to attend specific external training courses.

The guidance produced by the FSA on the provision of allergen information for non-pre-packed foods also includes advice on what staff in a catering business should do if they think customers may be having an allergic reaction, particularly if they are finding it hard to breathe, their lips or mouth become swollen, or they collapse.

Ingredient Information

When a person working in a catering business is asked whether or not a particular dish contains a certain allergenic food, he or she needs to be able to find that information or to tell the customer that he or she does not know. When a dish is made from fresh ingredients on the premises this information can be obtained from the chef, but if it is brought in prepared or part-prepared, information about allergenic ingredients and possible cross contamination risks needs to be ascertained from labeling or other documentation accompanying the delivery of the product or from the supplier. In the EU, foods sold to mass caterers are required to have the same allergen ingredient labeling as foods sold directly to the consumer. However this information may be on the outer packaging of large containers, which are often discarded after delivery, or on accompanying documentation. Catering businesses need to consider how they can retain this information and make it accessible to staff.

Large catering chains with central distribution chains may store ingredient and allergen information centrally in electronic form, which can be accessed at individual premises. Many large catering suppliers provide their customers with breakdowns of the allergenic ingredients used across their product range, and this can be made available to staff when they are dealing with a request from a food allergic customer. For other brought-in products typically used by smaller catering businesses, the ingredient and allergen information will be on the labeling of the individual product, although if the product is delivered in large containers and decanted into smaller ones for daily use, businesses need to have procedures that ensure allergen information is also transferred or retained.

The procedures set out above help a catering business to provide information that relates mainly to allergenic ingredients intentionally used in prepared or part-prepared dishes that are brought in, but it is also important for a caterer to be able to provide information on the allergen cross contamination risks on their own premises. For example, it can be difficult to prevent any cross contamination with nuts in many Asian cuisine restaurants, where nuts are widely used in many of the dishes offered and cross contamination risks with wheat flour can also be high.

Caterers' Perspective

For caterers, food allergy is an important safety issue. All catering staff, regardless of the type of venue in which they work, be it fast food, pub fare, or fine dining, need to be aware that food allergy exists and how serious it can be for the individuals affected. Those in the front line of serving customers need to understand the issues, what they can do to help, and what to do in the event of a customer having an allergic reaction.

The dialog between the customer and the catering staff is extremely important. Caterers should certainly encourage such dialog, but it is also important for customers to highlight their food allergy and ask questions where they judge that there may be a risk that food could contain their particular allergen and to ensure that they do not put themselves unnecessarily at risk. Customers can be allergic to a wide variety of different food ingredients and it is not practical for caterers to have different procedures to deal with different types of allergy. They do, however, have procedures to manage requests from customers with food allergy, regardless of the allergen concerned. Such procedures are triggered once a customer has alerted staff to his or her particular allergy.

Cross contamination can be a real risk in catering premises, particularly where food is being prepared fresh on the premises as opposed to being provided pre-prepared. Where a customer has requested a meal that doesn't contain a certain food, caterers will take certain precautions to avoid cross contamination, for example, by ensuring that worktops and all the equipment staff use is thoroughly cleaned with hot water and soap before they use them, including chopping boards, knives, food mixers, bowls, pans, and utensils used for stirring and serving. This is to prevent small amounts of the food that the person is allergic to from getting into his or her meal. It is also important to ensure that oil that has already been used to cook other foods is not used. For example, if food is cooked in oil that has already been used to cook prawns, this could cause a reaction in someone who is allergic to shellfish. Staff should also wash their hands thoroughly with soap and water before they prepare the meal and avoid touching other foods until they have finished preparing it. In self-service areas, such as salad bars or serve-yourself ice cream counters, it is good practice to put up signs warning of possible cross contamination hazards. For example, if nuts are used in one particular dish, they could be transferred to another by customers using the same spoon.

The vast majority of catering establishments are very accommodating and will prepare dishes that are not on the menu for a customer with a specific food allergy. They will also provide information on particular dishes to the best of their knowledge. Most restaurants and pubs are small businesses, and the retention and management of ingredient information can be challenging as they will tend to have a more ad hoc supply chain. Larger companies that operate chain businesses are better able to centrally manage ingredients information and data and are more likely to be able to provide more detailed, and often web-based, menu information.

In the event that it is not possible to confirm that a particular allergen is not present in a particular dish, then it is crucial that catering staff advise the customer of this so that the customer may assess the risk and make an informed choice.

THE RESPONSIBILITIES OF THE FOOD ALLERGIC CONSUMER

As mentioned previously, communication is one of the key factors in ensuring that food allergic consumers have the information they need to make safe food choices when eating out. This chapter has set out the importance of communication between the different members of staff within a catering business and between the business and the customer, but the allergic consumers themselves have an active role to play in ensuring that they receive sufficient information. Such consumers should actively seek information when they are eating out and make clear why they need to avoid the particular food(s). There are tools that allergic consumers can use to help them communicate their needs to a catering business. The FSA provides blank 'chef cards', which can be filled in on-line and then printed out, that food allergic people can then hand to catering staff setting out the foods they need to avoid [18], and similar services are offered by some of the allergy support organizations. Mandabach [19] recommends that 'food allergy buddy' cards developed in the US for food allergic customers to present to caterers should be integrated with Point of Sale systems used by caterers. Such systems could incorporate a feature allowing the servers to access allergen ingredient information and then block the server from ordering any item containing that allergen as an ingredient.

FUTURE DEVELOPMENTS AND RECOMMENDATIONS

Currently most food safety training for people working in the catering industry concentrates on food hygiene issues, with Bailey reporting that only one third of restaurant staff surveyed in a UK study had undergone formal food allergy training [8]. While training courses for those working in the catering and hospitality area should be expanded to include food allergy issues, there is also a need for simple training tools for those people to use when training all other staff working in their businesses.

In the European Union, the legislative changes introducing a legal requirement to provide allergen information for foods sold unpackaged, which were published in October 2011, are likely to drive the development of training for people working in this sector so that they can meet their legal obligations. The FSA already provides best practice guidance to help catering businesses provide accurate and helpful allergen information to their customers and offers a freely available e-learning module on food allergy that includes both manufacturing and catering scenarios. While catering businesses in the EU are likely to have a transition period of three years before they would need to comply with these new requirements, much work remains to be done if appropriate training courses and materials are to be developed in time.

REFERENCES

[1] EC (European Commission). Directive 2003/89/EC of the European Parliament and Council of 10 November 2003 amending Directive 2000/13/EC as regards the indication of ingredients present in foodstuffs. Official J Eur Union 2003;L.308:15−8.
[2] House of Lords. Science and Technology Committee 6th Report of session 2006−07-Allergy, Volume II. Evidence, HL paper 166-II, http://www.publications.parliament.uk/pa/ld200607/ldselect/ldsctech/166/166ii.pdf; 2007. Accessed June 13, 2011.
[3] Pumphrey RSH. Lessons for management of anaphylaxis from a study of fatal reactions. Clin Exp Allergy 2000;30:1144−50.
[4] Pumphrey RSH, Gowland MH. Further fatal reactions to food in the United Kingdom. J Allergy Clin Immunol 2007;119:1018−9.
[5] Leitch IS, Walker MJ, Davey R. Food allergy: gambling your life on a takeaway meal. Int J Environ Health Res 2005;15:79−87.
[6] Pratten JD, Towers N. Food allergies: a problem for the catering industry. Br Food J 2003;105(4/5):279−87.
[7] McIntosh J, Flanagan A, Maden N, Mulcahy M, Dargan L, Walker M, et al. Awareness of coeliac disease and the gluten status of 'gluten-free' food obtained on request in catering outlets in Ireland. Int J Food Sci Tech 2011;46(8):1569−74.

[8] Bailey S, Albardiaz R, Frew AJ, Smith H. Restaurant staff's knowledge of anaphylaxis and dietary care of people with allergies. Clin Exp Allergy 2011;41:713−7.

[9] EC (European Commission). European Community, 2002, Regulation (EC) No 178/2002 of the European Parliament and Council of 28 January 2002 laying down the general principles and requirements of food law, establishing the European Food Safety Authority. Official J Eur Communities 2002;L31:1−24.

[10] Regulation (EU) No 1169/2011 of the European Parliament and of the Council of 25 October 2011 on the provision of food information to consumers, amending Regulations (EC) No 1924/2006 and (EC) No 1925/2006 of the European Parliament and of the Council, and repealing Commission Directive 87/250/EEC, Council Directive 90/496/EEC, Commission Directive 1999/10/EC, Directive 2000/13/EC of the European Parliament and of the Council, Commission Directives 2002/67/EC and 2008/5/EC and Commission Regulation (EC) No 608/2004. Official Journal of the European Union. L304, 18−63.

[11] EC (European Commission). Commission Regulation EC No 41/2009 of 20 January 2009 concerning the composition and labelling of foodstuffs suitable for people intolerant to gluten. Official J Eur Union 2009;L16:3−5.

[12] Labeling of 'gluten free' foods. http://www.food.gov.uk/business-industry/guidancenotes/allergy-guide/gluten.

[13] The provision of allergen information for non-pre-packed foods — Voluntary Best Practice Guidance. http://www.food.gov.uk/multimedia/pdfs/loosefoodsguidance.pdf, accessed March 13, 2011.

[14] Food Allergy: What you need to know. http://www.food.gov.uk/multimedia/pdfs/publication/loosefoodsleaflet.pdf.

[15] Think Allergy! http://www.food.gov.uk/multimedia/pdfs/publication/thinkallergy.pdf

[16] Safer food, better business for caterers. http://www.food.gov.uk/foodindustry/regulation/hygleg/hyglegresources/sfbb/sfbbcaterers/, accessed June 13, 2011.

[17] Food Standards Agency's food allergy training online. http://allergytraining.food.gov.uk/, accessed June 13, 2011.

[18] Think Allergy! Chef cards. http://www.food.gov.uk/multimedia/pdfs/chefcard.pdf, accessed June 13, 2011.

[19] Mandabach KH, Ellsworth A, VanLeeuwen DM, Blanch G, Waters HL. Restaurant managers' knowledge of food allergies: a comparison of differences by chain or independent affiliation, type of service and size. J Culinary Sci Tech 2006;4:63−77.

Chapter eleven

Food Allergen Risk Management in the United States and Canada

Steven M. Gendel

Food and Drug Administration, Center for Food Safety and Applied Nutrition, US

CHAPTER OUTLINE

INTRODUCTION

Food allergen risk management is a complex process, involving public health officials at the national, state or provincial, and local levels in the United States and Canada. Although many of the issues and concerns are the same at each level, the regulatory authorities and distributions of responsibility differ. At the national level, allergen management is focused on processors producing packaged food products for wide distribution. The risk management tools available consist primarily of labeling laws and regulations and good manufacturing practice (GMP) regulations and guidelines. At the local level, risk management focuses on allergen control in the retail and food service industries. The tools of risk management differ from state to state or province to province depending on the laws and administrative structures of the public health authorities in each. Active communication among risk managers at all levels, and among government, academia, and industry, ensures that the overall approach to allergen risk control is scientific, effective, and consistent.

201

Risk Management for Food Allergy. http://dx.doi.org/10.1016/B978-0-12-381988-8.00011-7

US NATIONAL FOOD ALLERGEN RISK MANAGEMENT

The Food and Drug Administration (FDA), a component of the Department of Health and Human Services, is the federal agency with the most responsibility for food allergen risk management in the US. FDA is responsible for ensuring the safety of most foods, with the exception of meat, poultry, and some egg products, which are the responsibility of the US Department of Agriculture (USDA). FDA's most important risk management tools for food allergens are labeling regulations, particularly as articulated in the Food Allergen Labeling and Consumer Protection Act of 2004 (FALCPA) and GMP regulations. In addition, the newly enacted Food Safety Modernization Act (FSMA) identifies food allergen control as a component of preventive controls.

FALCPA defines the term 'major food allergen' as one of eight foods or food groups (milk, egg, peanut, soy, fish, crustacean shellfish, tree nuts, and wheat) or an ingredient derived from one of these. Foods regulated by FDA must declare the presence of any of the major food allergens using the name of the food source (in plain language). For example, a food containing whey as an ingredient must use the term 'milk' as the food source. For fish, crustacean shellfish, and tree nuts the specific type or species must be identified. Unlike the allergen labeling regulations in other countries, FALCPA does not limit the scope of these food groups to a few specific examples. FDA provides guidance on the labeling of foods that contain ingredients from these groups in an on-line question and answer document available through the agency home page. FALCPA explicitly states that the allergen labeling requirement also applies to flavorings, colors, and incidental additives.

There are two acceptable label formats for declaring allergens, either as part of the standard ingredient list or in a separate 'contains' statement. When a 'contains' statement is used, all of the major food allergens (and only the major food allergens) that are present in the food must be listed in that statement. FALCPA does not address advisory labeling (such as 'may contain'). FDA has stated that advisory labeling should not be used as a substitute for GMP and that it must be truthful and not misleading.

FALCPA includes an exemption for highly refined oil derived from a major food allergen. Also, Congress recognized that in some cases the manufacture of an ingredient derived from a major food allergen might degrade or reduce the allergenic proteins such that the ingredient is no longer a health risk for allergic consumers. Therefore, FALCPA includes mechanisms for exempting ingredients from the labeling requirement through either a petition or notification process. In the petition process, an exemption can be obtained by providing scientific evidence showing that an ingredient does not cause an allergic response that poses a risk to human health. In the notification process, an exemption can be obtained by providing scientific evidence showing that an ingredient does not contain allergenic protein.

FALCPA plays a critical role in food allergen risk management in several ways. First and foremost, the law ensures that sensitive consumers are better able to practice avoidance by providing them with the information that they need in a clear, consistent manner. Second, by focusing regulatory and control resources on those food allergens that are considered to be of greatest public health concern, the law ensures that allergen control efforts will have the greatest possible benefit for the sensitive population.

The chief limitation of FALCPA as a risk management tool is that the labeling requirement only applies to ingredients; that is, to components that are intended to be part of a food product. Allergens may also be present in a food (and not declared on the label) through cross contact or as a result of labeling or packaging errors. Cross contact is controlled or eliminated through the use of GMPs, which are defined by regulation in the US. In FSMA, allergen control has been recognized as part of an overall preventive control approach to food safety. GMPs and preventive controls address practices as diverse as equipment design, cleaning and sanitation, process layout, and supplier controls. In a broad sense, the goal of an allergen control program is to provide some form of separation between products containing different food allergens and between products with and without food allergens. This separation can be physical (e.g., dust control, dedicated equipment), procedural (e.g., personnel controls), or temporal (e.g., cleaning and sanitation between product runs, product sequencing).

Labeling controls, which are also part of a preventive control program, are risk management tools used to ensure that the ingredient declaration (including any allergen declaration) on a food package accurately represents the composition of the food in that package. This includes formulation review, checks on label design, verification that printed labels or packaging are as intended, and controls to ensure that the correct label is used with each food product. Labeling controls can also include supplier and materials controls to ensure that any allergen information related to ingredients used in a product is carried through to the finished product label. It is important to recognize that product and ingredient formulations can change over time and that label review needs to be a recurring process.

Allergen risk management also relies on systems that monitor and track problems and trends to identify emerging issues. The FDA monitors food allergen problems through the analysis of consumer complaints, recalls, and entries in the Reportable Food Registry. Consumers who experience an adverse reaction to a food can report it to the agency through a consumer complaint coordinator in each geographic district. The coordinator interviews the consumer to collect all available information on the food and the nature of the adverse event and forwards that information to agency medical officers (MOs) and subject matter experts (SMEs). The MOs and SMEs evaluate each complaint and recommend an appropriate follow-up action, such as sample collection or label review.

Companies can become aware of problems related to allergen content or labeling in several ways including consumer complaints, internal audits, external inspections, notifications from suppliers or customers, or process reviews.

Depending on the nature of the problem, a company might notify the agency through the Reportable Food Registry, initiate a recall, or do both. In many cases, a root cause analysis can identify the gaps or failures in the control program that led to the problem. In addition to directing the efforts needed to correct the immediate problem, this root cause knowledge helps in understanding which problems are widespread and which control procedures are and are not consistently effective. Ongoing analysis and monitoring of these problems and their causes is a critical tool for identifying issues that can be addressed on an industry or sector basis. This analysis has resulted in updating guidance and training for both agency inspectors and industry on issues such as the importance of label control during product changeover and of proper product sequencing.

At the federal level in the US, the USDA Food Safety and Inspection Service (FSIS) is responsible for food allergen risk management for meat and dairy products. As for the FDA, the available risk management tools include labeling regulations for ingredients, manufacturing controls to eliminate or reduce cross contact, and monitoring of recall trends. Although FALCPA does not apply to USDA regulated products, USDA encourages the use of allergen statements that are consistent with FALCPA requirements and monitors allergen labeling through a prior approval process. Prevention of cross contact is addressed through the USDA Hazard Analysis and Critical Control Point (HACCP) regulation. Allergens are considered to be chemical food safety hazards that need to be addressed in the hazard analysis and with effective controls and monitoring.

US STATE AND LOCAL ALLERGEN RISK MANAGEMENT

Allergen risk management at the state and local levels in the US is the responsibility of over 3,000 different agencies. These agencies have oversight of more than 1 million food establishments such as restaurants and grocery stores, as well as vending machines, cafeterias, schools, and correctional facilities. The FDA works with these agencies through the Conference for Food Protection and by developing a model Food Code. The model Food Code is a reference document prepared by the agency to assist state and local agencies by providing a scientifically sound technical and legal basis for regulating the retail food industry. The model Food Code helps to promote a uniform system of regulation across the many jurisdictions involved. Individual jurisdictions can adopt all or part of the model code into local laws or regulations, either by reference or by directly incorporating the model's language. As of the end of 2010, 49 of the 50 states and three of the six territories, representing approximately 97% of the US population, had adopted food codes patterned after the model Food Code. The model Food Code was updated after FALCPA to include guidelines for establishments such as restaurants on how to avoid allergen cross contact.

The Conference for Food Protection is a collaborative forum involving producers, regulators at all levels, and academics. The organization provides opportunities to identify emerging problems and to recommend approaches to addressing them. Because these recommendations are reached through a deliberative process of consensus building involving technical and regulatory experts, they carry significant weight in determining appropriate food safety practices. The Conference for Food Protection has an allergen committee that is active in developing suggested wording related to allergen control for inclusion in the model Food Code and guidance and training material for use by local authorities.

CANADIAN NATIONAL ALLERGEN RISK MANAGEMENT

In Canada, responsibility for food allergen risk management is shared at the national level by Health Canada (HC) and the Canadian Food Inspection Agency (CFIA). HC is responsible for establishing policies, regulations, and standards and CFIA is responsible for implementing and enforcing the regulations and standards.

As in the US, the primary risk management tool for food allergens is regulation of what appears on the food label. HC has recently issued a revised regulation updating the list of 'priority allergens' that need to be identified, as well as the appropriate format and terminology for declaring the presence of these allergens. In parallel with the situation in the US, these labeling regulations apply to food ingredients and not to allergens present through cross contact.

The CFIA is part of Agriculture Canada and has the primary responsibility of working with the food industry to implement and enforce labeling and good manufacturing practices. CFIA has developed a Food Safety Enhancement Program (FSEP) to encourage and support the use of HACCP food safety systems. Food allergens are considered to be chemical hazards under the CFIA FSEP/HACCP program. CFIA has also played a leading role in the development and use of allergen testing methods for foods.

The food service sector in Canada is regulated at the provincial and territorial level. The most effective efforts at developing food allergen risk management guidelines and programs for this sector have resulted from collaborations between agencies, government bodies, and patient organizations. In Ontario one such collaboration has focused on the safety of children in schools, while in Quebec another collaboration has developed a reference manual for restaurant and food service managers.

CONCLUSIONS

Food allergen risk management is an evolving process at both the national and local levels. The overall goal remains one of protecting sensitive consumers by ensuring that they have the information that they need to make safe food choices and that this information is complete and accurate. Given the diversity in the number of foods that cause allergies, the widespread and varied use of these foods or their derivatives as ingredients, the number of places in the food production chain where problems can arise, the gaps in our understanding of the biology of food allergy, and the overlapping and complementary legal authorities involved this can seem to be a daunting task. However, the use of transparent risk analysis processes and open communication between all stakeholders has proven to be effective in overcoming the obstacles while protecting public health.

Chapter 12

The Importance of Food Allergy Training for Environmental Health Service Professionals

I.S. Leitch[1], J. McIntosh[2]

[1]Environmental Health Department, Omagh District Council, Omagh, Co. Tyrone, Northern Ireland
[2]Safefood, Eastgate, Little Island, Co. Cork, Ireland

CHAPTER OUTLINE

INTRODUCTION

In July 2006, the United Kingdom House of Lords' Science and Technology Committee appointed a subcommittee to explore the impact of allergy on patients, society, and the economy as a whole. The committee reported in September 2007 that 'allergy in the UK has now reached epidemic proportions' and that new food allergies were regularly being described [1]. During its

Risk Management for Food Allergy. http://dx.doi.org/10.1016/B978-0-12-381988-8.00012-9

deliberations, the committee considered food allergy as well as other allergies and made the key recommendation that:

> *it is imperative that environmental health officers (EHOs) ... and catering workers are adequately and comprehensively trained in practical allergen management.*

The requirement for food allergy training had already been recognized on the island of Ireland, and a comprehensive training program for EHOs and third level catering lecturers was by then underway.

There are no prevalence data for food allergy in the Republic of Ireland or Northern Ireland. Instead, prevalence estimates from Britain are transposed, and this is justified on the basis of similarities in diet, genetics, and geography. Therefore, the estimated prevalence of food allergy throughout the island of Ireland is approximately 1–2% in adults and 5–8% in children. Also, there are no food allergy related mortality data for either jurisdiction. There are at least six confirmed fatal incidents of food related anaphylaxis each year in the rest of the UK in a total population of around 60 million [2]. The majority of deaths from 1996 to 2006 (18 in total) were due to nuts and peanuts, and casualties ranged in age from 5 months to 85 years with a median age of 21 years [3]. In the Republic of Ireland from 1995 to 2004, the Hospital Inpatients Enquiry database recorded that on average 45 people were discharged from hospital with a principal diagnosis due to food related anaphylaxis each year. Apart from cases where the type of food was not specified, discharges with peanut were one of the highest principal causes of hospital discharges due to food-induced anaphylaxis. Tree nuts, eggs, and fish were also significant causes [4].

THE EFFECTIVENESS OF AN AVOIDANCE DIET

A food allergy diagnosis has important consequences for patients and their families, requiring strict avoidance of foods known or thought to contain the offending allergen [5]. However, given the prevalence of allergens in a broad spectrum of foodstuffs, the logistics and anxieties involved in maintaining an avoidance diet are evident. Failure of avoidance is the key event in many severe, or even fatal, allergic reactions [6,7]. In a follow-up study of child peanut allergy sufferers, Bock and Atkins found that half (16/32) of the children had accidentally ingested peanut in the year preceding the review [8]. Only eight out of 32 patients had managed to avoid peanuts completely since the time of their diagnosis. Following an examination of the reasons for peanut allergy sufferers eating or coming into contact with peanuts for their latest reaction, Emmett and Angus noted that 18% of the sufferers

> *didn't think that the type of food which they had eaten could contain peanut* [9].

These findings do not bode well for the food allergy sufferer, and recent research has identified quality of life issues for individuals and families as being a major concern [10]. Peanut allergic children have been shown to have a poorer quality of life than children of similar age with insulin-dependent diabetes [11]. These same children show greater trepidation about eating compared to non-allergic children, especially when away from home where they have less control over their diet. They were also recorded as being more anxious about the potential for experiencing an adverse event: this was shown in 2006 in NI, when, according to Allergy NI, a charity in Northern Ireland for people with severe food allergies, the death of a local teenager from peanut allergy was a source of considerable distress to those with food allergies [12].

LEGISLATIVE BASIS FOR THE RISK MANAGEMENT OF FOOD ALLERGENS

Food allergic consumers can expect legal protection in terms of food labeling and composition, in which a distinction is made between pre-packed and non-pre-packed foods. EU Directive 2003/89/EC amends Directive 2000/13/EC by the addition of Annex IIIa, which requires labeling of 12 major allergenic food groups if they are deliberate ingredients in pre-packed foods [13]. It applies to pre-packed foods delivered to the 'ultimate consumer' and covers the supply of foods to restaurants, hospitals, canteens, and other mass caterers. EU Directive 2006/142/EC added lupin and molluscs to the list [14]. Directive 2003/89/EC has been transposed into national law in the Republic of Ireland (ROI) and Northern Ireland (NI) [15,16].

Currently, there is no requirement to provide information on the allergen content of non-pre-packed food supplied through catering. Protection is however given under the General Food Law Regulation 178/2002/EC, insofar as unsafe food must not be placed on the market (i.e., offered for sale) [17]. The customer must receive sufficient information to make an informed and safe choice about a food product, although in practice this often does not occur. Many anaphylactic reactions have occurred after the in-dividual ingested the allergen unknowingly — at a restaurant, party, in takeout food, etc. — where labeling information about ingredients is not legally required and the liberal use of certain allergens such as peanut is a legitimate feature of many cuisines [18,19,20]. Eating out is a particularly hazardous ac-tivity for those with food allergy.

The protection of the allergic consumer will be further enhanced through the Food Information Regulations 1169 of 2011, which will achieve general applica-tion by the end of 2014 [21]. Key proposals include the highlighting of allergen information on ingredients lists on pre-packed foods through the use of a different typeset and the requirement to identify the allergenic source of each ingredient even where several ingredients originate from a single allergen. A major addition is the requirement that information on the allergenic content of non-pre-packed foods must be provided to the purchaser. Each EU Member State must now decide how best to implement the provisions of the regulation: the information could be on signs, menus, receipts, etc., or just available from a member of staff if and when requested. The implementation of Regulation 1169/2011 will result in an increased emphasis on the importance of food businesses, particularly caterers, deli counters, and bakeries, having a thorough knowledge of the ingredients they use, the maintenance of accurate ingredient records, and the requirement for open and accurate dialog with their customers.

THE ROLE OF THE EHO IN THE RISK MANAGEMENT OF FOOD ALLERGENS

The overarching principle of hygiene/safety legislation is the provision of safe food throughout the food industry, including the catering sector. Food business operators must protect the health and well-being of food allergic customers by ensuring that information about the ingredients they use is managed effectively and available if required. To achieve this they must analyze and control all food safety hazards, including food allergen hazards, and determine the possibility of allergen cross contamination. As of January 2006, Regulation (EC) No. 852/2004 requires that all food safety management systems are based on the principles of Hazard Analysis Critical Control Point or 'HACCP' [22]. The regulation was transposed into national law in ROI and NI [23,24]. The regulatory authorities must ensure compliance with these legal requirements, and they must therefore ensure that HACCP principles are applied to food allergen management.

On the island of Ireland, EHOs are responsible for ensuring that food businesses control physical, chemical, and microbiological hazards. Routine unannounced visits are carried out in retail and catering premises, when advice is given in the context of enforcing compliance with food safety law. Emphasis is on the application of hazard analysis principles to the food production process from ingredient purchases to customer service. To assess in-house allergen risk management procedures and augment in-house allergy awareness, EHOs must understand the legal framework governing this aspect of enforcement while having a sufficient understanding of practical allergen management. This can only be achieved if EHOs have received appropriate training to begin with.

DEFICITS IN TRAINING RESOURCES FOR CATERING STAFF

By law, all catering staff must receive food hygiene training commensurate with their responsibilities, and many food businesses train staff in basic food hygiene using courses provided by the Chartered Institution of Environmental Health in Northern Ireland and by the EHO Association in the Republic. However, prior to this training program the basic grade food safety syllabi of both organizations made no reference to food allergies, despite continuing evidence that the catering industry suffers from a dearth in knowledge and awareness of food allergies and food allergen control [25]. Hence there was a deficit in basic allergen training in both jurisdictions for the catering staff that usually have first contact with an allergic customer. Catering college lecturers also found it difficult to meet their HACCP obligations regarding food allergen control in their own training kitchens, yet alone provide training to their own students on this issue. Professional allergy training for catering lecturers, who would in turn train the caterers of tomorrow, was not available.

DEFICITS IN FOOD ALLERGY AWARENESS AMONG CATERING STAFF

Research carried out in 1999 in Northern Ireland highlighted the lack of knowledge and appropriate training in food allergen control among EHOs, as a result of which it was not incorporated as an aspect of their routine HACCP-related food control work [26]. In 2002, a survey was carried out, also in Northern Ireland, to ascertain if an allergen-free meal could be provided on request in a takeout setting, and also to assess the training and guidance needs of catering staff and EHOs [27]. Using peanut protein as the 'test' allergen, approximately 20% of the takeout premises provided meals that could possibly have triggered a fatal reaction in a peanut allergic customer. Most front-of-house catering staff did not check the allergenic status of the meal with management or the chef, and the majority of EHOs who carried out the actual sampling acknowledged their own need for more training in food allergen control in commercial premises. A similar survey conducted throughout the island of Ireland in 2005 revealed that allergic consumers still faced difficulties in making food purchases [28]. The research showed that some staff in food businesses, including sandwich bars, cafes, supermarkets, and forecourt shops, were unable to give allergy sufferers accurate advice about the food they were ordering. Again using peanut allergen as the test model, one in ten catering staff showed no understanding or awareness of peanut allergy and only a third were confident in the advice they gave (which was frequently incorrect). Across the island of Ireland, 55% of the food samples that tested positive for peanut protein came with the wrong advice. Again the overwhelming majority of participating EHOs expressed a desire to receive training on this issue.

A FOOD ALLERGY TRAINING PROGRAM FOR THE ISLAND OF IRELAND

Recognizing a clear deficit in food allergy training, a joint Health Service Executive (Republic of Ireland)/Local Authority (Northern Ireland) pilot training project sponsored by two cross-border agencies, *safe*food and Cooperation and Working Together, was initiated in 2006. The aim was to deliver training in food allergen control to approximately one hundred EHOs involved in food safety enforcement duties in the border region on the island of Ireland. The scope of the training and the context in which it would be delivered were initially presented at a pre-training conference. This was followed by a series of workshops for EHOs with the provision of online and printed training materials. The objective was to empower EHOs to cascade this newfound knowledge and skill to catering and retail businesses during inspections.

An initial evaluation of the training demonstrated the need to extend it to EHOs in the remaining regions of the island of Ireland. The project was rolled out by *safe*food maintaining the same format as was used previously. However, on this occasion, the training was extended to include catering course lecturers from third level academic institutions and Public Health laboratory scientists. Both projects resulted in almost 600 people involved in food safety enforcement and education being trained by a specialist training contractor in food allergen control [29]. The key areas covered during the training included

- Food sensitivity (allergy, intolerance, celiac) − symptoms, mechanism, prevalence, and the impact on quality of life
- Global nature of the food chain and how this influences the spread of food allergens
- Food allergen alert systems in the EU
- Practical food allergen management in a catering setting
  - Hidden allergens
  - Cross contamination
- Legal aspects of food allergen management and control including labeling requirements for food products
- Communication with the food business operators and customers

Similar training has been provided on an *ad hoc* basis within the Local Councils of England and Scotland. The food allergy training program is currently being evaluated.

REFERENCES

[1] TSO (The Stationery Office). Allergy. House of Lords Science and Technology Committee 6th Report of Session 2006−7; 2007.

[2] Pumphrey RSH. Lessons for the management of anaphylaxis from a study of fatal reactions. J Clin Exp Allergy 2000;30:1144−50.

[3] Pumphrey RSH, Gowland MH. Further fatal allergic reactions to food in the United Kingdom 1996−2006. J Allergy Clin Immunol 2007;119:1018−9.

[4] Hospital In-Patient Enquiry Scheme (HIPE), discharges with a principal diagnosis of ICD-9-CM 995.60 e 995.69. Health Research & Information Division, Economic and Social Research Institute, Whitaker Square, Dublin, Ireland.

[5] Avery NJ, King RM, Knight S, Hourihane JO'B. Assessment of quality of life in children with peanut allergy. Paediatric Allergy Immunol 2003;14:378−82.

[6] Yunginger JW, Sweeney KG, Sturner WQ, Giannandrea LA, Teigland JD, Bray M, Benson PA, York JA, Biedrzycki L, Squillace DL, Helm RM. Fatal food-induced anaphylaxis. JAMA 1988;260:1450−2.

[7] Bock SA, Munoz-Furlong A, Sampson HA. Fatalities due to anaphylactic reactions to foods. J Allergy Clin Immunol 2001;107:191−3.

[8] Bock SA, Atkins FM. The natural history of peanut allergy. J Allergy Clin Immunol 1989;83:900−4.

[9] Emmett S, Angus FJ. Characterisation of Individuals at High Risk of Severe Peanut Anaphylaxis to Produce Targeted Advice and Information; 1996. MAFF / Leatherhead Food Research Assn. Project Report RME/F/08 July.

[10] Marklund B, Ahlstedt S, Nordström G. Food hypersensitivity and quality of life. Curr Opin Allergy Clin Immunol 2007;7:279−87.

[11] Avery NJ, King RM, Knight S, Hourihane JO'B. Assessment of quality of life in children with peanut allergy. Paediatric Allergy Immunol 2003;14:378−82.

[12] Allergy NI (www.allergyni.co.uk/), personal communications. This was a personal verbal communication from the director of AllergyNI, a self help charity advising of the concern of members following the allergy related death of a young person. The web address is for information only.

[13] amending Directive 2000/13/EC as regards indication of the ingredients present in foodstuffs. (OJ L 308/15, 25.11.2003)

[14] Commission Directive 2006/142/EC of 22 December 2006 amending Annex IIIa of Directive 2000/13/EC of the European Parliament and of the Council listing the ingredients which must under all circumstances appear on the labelling of foodstuffs. (OJ L 368/110, 23.12.2006)

[15] S.I. No. 228 of 2005 (see also S.I. No. 483 of 2002 which transposes Directive 2000/13/EC and S.I. 808 of 2007 which transposes Directive 2006/142/EC.)

[16] S.I. No 469 of 2004. Food Labelling (Amendment) (No. 2) Regulations (Northern Ireland) 2004.

[17] Regulation (EC) No 178/2002 of the European Parliament and of the Council of 28 January 2002 laying down the general principles and requirements of food law, establishing the European Food Safety Authority and laying down procedures in matters of food safety (Official Journal L 031, 01/02/2002 P. 0001−0024)

[18] Hourihane J, Roberts SA, Warner JO. Resolution of peanut allergy: case control study. BMJ 1998;316:1271−5.

[19] Hoffman F, Goforth C. Fatal allergy suspected; student reacts after egg roll. Cincinnati Enquirer 1995. Friday March 7th.

[20] Furlong TJ, DeSimone J, Sicherer SH. Peanut and tree nut allergic reactions in restaurants and other food establishments. J Allergy Clin Immunol 2001;108:867−70.

[21] Regulation (EU) No 1169/2011 of the European Parliament and of the Council of 25 October 2011 on the provision of food information to consumers, amending Regulations (EC) No 1924/2006 and (EC) No 1925/2006 of the European Parliament and of the Council, and repealing Commission Directive 87/250/EEC, Council Directive 90/496/EEC, Commission Directive 1999/10/EC, Directive 2000/13/EC of the European Parliament and of the Council, Commission Directives 2002/67/EC and 2008/5/EC and Commission Regulation (EC) No 608/2004. (OJ L304/18; 22.11.2011).

[22] Regulation (EC) No 852/2004 of the European Parliament and of the Council of 29 April 2004 on the hygiene of foodstuffs (Official Journal of the European Union L 226/3, 25.6.2004)

[23] Corrigendum to Regulation (EC) No 852/2004 of the European Parliament and of the Council of 29 April 2004 on the hygiene of foodstuffs. (see also S.I. No. 910 of 2005, S.I. No. 369 of 2006 and S.I. No. 387 of 2006).

[24] S.I. No.3 of 2006. Food Hygiene Regulations (Northern Ireland) 2006.

[25] Bailey S, Albardiaz R, Frew AJ, Smith H. Restaurant staff's knowledge of anaphylaxis and dietary care of people with allergies. Clin Exp Allergy, May 2011;41(5):713−7.

[26] Leitch I, Blair L, McDowell D. Dealing with allergy. Environ Health J October 2000: 335−9.

[27] Leitch IS, Walker MJ, Davey R. Food allergy: gambling your life on a takeaway meal. Int Jl Env Health Research April 2005;15:79−87.

[28] safefood. Evaluating food allergy awareness among catering staff. Published February 2008. Available at: http://www.safefood.eu/SafeFood/media/SafeFoodLibrary/Documents/Food%20Safety/Booklet-Final-April-2008.pdf.

[29] www.hygieneauditsystems.com. The contractor was a partnership of Hygiene Audit Systems Ltd., Albion Mills, 23 Albion Road, St. Albans AL1 5EB, UK and Allergy Action, 23 Charmouth Road, St. Albans AL1 4RS, UK

Chapter | thirteen

Detecting and Measuring Allergens in Food

Joseph L. Baumert
*Department of Food Science & Technology and Food Allergy Research &
Resource Program, University of Nebraska, Lincoln, NE, US*

CHAPTER OUTLINE

INTRODUCTION

Food allergies affect an estimated 2 to 4% of the population around the world [1]. Allergic reactions to foods also account for a high proportion of emergency room visits, some of which result in hospital admissions, thus making food allergies a serious concern for public health around the world [2]. This increased awareness of the public health importance of food allergies has brought about increased regulatory oversight of food allergens. Various countries have highlighted the importance of food allergies by passing labeling legislation that requires a declaration of priority food allergens on the packaged food label [3]. While declaration of ingredients derived from allergenic sources when used as direct ingredients or processing aids has helped to provide allergic consumers with more transparent allergen information, these labeling laws do not address the potential risk involved with undeclared

215

Risk Management for Food Allergy. http://dx.doi.org/10.1016/B978-0-12-381988-8.00013-0

or 'hidden' allergens that may be in the food products due to cross-contact of the product produced on shared equipment, or due to commingling of ingredients at the supply chain level [4]. Cross-contact can occasionally occur despite the food manufacturer's best efforts to remove the allergenic residue. The food industry strives to mitigate this risk through use of allergen control plans and validated cleaning and sanitation procedures. Visual inspection of food contact surfaces is one of the key steps utilized to ensure the effectiveness of the cleaning procedure [5]. Analytical validation to ensure removal of allergenic residue from equipment surfaces or to ensure that the finished product does not contain the allergenic residue of concern is also utilized by the food industry.

Information on minimum eliciting doses of allergic individuals has emerged for various food allergens in recent years [6]. While there is interest by several stakeholder groups (i.e., food industry, regulatory agencies, allergic consumers, and clinicians) to evaluate the efficacy of using clinical threshold information for potential development of regulatory thresholds or action levels, currently many countries have not implemented regulatory thresholds [6]. Japan currently requires source labeling of its defined priority food allergens when the concentration of protein from the allergenic source is > 10 ppm (μg protein/g food) [7]. With the lack of regulatory thresholds, food industry is tasked with complying with essentially a zero threshold level of allergenic residue. This is operationally impossible given the complexity of manufacturing facilities and the numerous routes of allergen contamination and cross-contact that can occur throughout the supply chain.

As mentioned previously, visual inspection and analytical validation can be effectively used monitor the removal of allergenic residue and minimize the risk of hidden allergens in the next product after changeover. There are numerous quantitative and qualitative methods that are available for monitoring residues from allergenic sources [8—9]. It is quite important for food manufacturers to understand the advantages and limitations of the available analytical tools when selecting an appropriate method to ensure that the analytical results provide meaningful data that can be used for risk management purposes. Recently, food industry-led initiatives such as the Australian Allergen Bureau's VITAL (Voluntary Incidental Trace Allergen Labeling) program have been developed in an attempt to curtail widespread use of advisory labeling (www.allergenbureau.net/vital/vital). This voluntary risk management program relies on the accurate assessment of the level of potential allergenic residue that may be present in a packaged food product, along with information about the consumption of the product (i.e., serving size or other estimates of consumption). This is used to assess the need for using advisory statements when the exposure dose is above or below a defined reference dose. These reference doses have been developed based upon the available clinical threshold information for several priority food allergens. Quantitative (probabilistic) methods have also been developed for food allergen risk assessment

[10–13]. These risk assessment models rely on accurate determination of the concentration of allergenic residue. This chapter will discuss the analytical methods that are currently available for the food industry to detect residues from allergenic foods.

IMMUNOCHEMICAL METHODS FOR THE DETECTION OF FOOD ALLERGENS

Immunochemical methods are a broad classification of analytical methods that have been used for either clinical diagnosis of food allergy or for the detection of allergenic food residues. These methods rely on binding of allergen-specific antibodies to the allergenic food protein to be detected. Prior to the mid to late 1990s, rapid analytical methods that could be used by food industry for detection of food allergen residues were not readily available. Methods such as RAST (radio-allergosorbent) or EAST (enzyme-allergosorbent) assays were available primarily for clinical diagnosis of food allergy and for identification of allergenic proteins [9]. RAST and EAST rely upon the serum immunoglobulin-E (IgE) from food allergic individuals for qualitative detection of allergenic proteins. Protein from the allergenic source of interest is coupled to a solid phase. Allergen-specific IgE from allergic individuals is incubated with the allergen bound membrane, followed by detection of any bound IgE with a radioisotope labeled (in the case of RAST, e.g., ^{125}I) or enzyme labeled (in the case of EAST, e.g., horseradish peroxidase or alkaline phosphatase) anti-IgE antibody. Detection of bound IgE is achieved by measuring emitted radiation or color change.

RAST and EAST inhibition allow for quantitative detection of allergenic food proteins based on competitive binding of human IgE [14]. In these assays, protein from the allergenic source of interest is again coupled to a solid phase; however, the sample containing the potential allergenic protein to be quantified is incubated with allergen-specific IgE prior to adding this solution to the solid phase. The specific allergenic protein of interest will bind to the IgE, resulting in a decrease in the IgE that is available to bind to the solid phase. The concentration-dependent inhibition can be compared to a standard curve to allow for quantification of the concentration of food allergen in the sample of interest.

RAST and EAST inhibition do provide quantitative detection of allergenic food proteins; however, one of the main limitations of these immunochemical assays is that they require serum from allergic humans, which is not readily available and which varies from one allergic individual to another. Use of serum from allergic individuals also requires thorough characterization to ensure that IgE is not present that would recognize other allergenic proteins and thus lead to potential false-positive results. Finally, human serum poses a potential biological safety hazard that does not make these assays suitable for use within the food processing facility.

Enzyme-Linked Immunosorbent Assays (ELISAs)

Enzyme-linked immunosorbent assays (ELISAs) are the methods most widely used by the food industry for detecting specific allergenic protein [8,15−16]. ELISAs provide several advantages including:

1) They detect protein(s) from the allergenic source of interest, which make these assays ideal for validation of the removal of specific allergenic proteins,
2) They are sufficiently sensitive to ensure the safety of the allergic consumer (detection limits generally range in the low milligram per kilogram (ppm) range),
3) The reagents used in the assay are suited for use within the food processing facility, and
4) They provide a rapid assessment that can be run in the food processing facility or in a food industry laboratory [8,16].

ELISAs use immunoglobulin-G (IgG) antibody from animal sources such as rabbits, goats, or sheep that are directed against the allergenic protein(s) of interest rather than IgE from human serum. Use of an animal source for IgG antibodies allows the generation of suitable quantities of the antibody and also decreases the variability that is typically observed with human serum IgE.

Quantitative ELISAs can be developed in the sandwich or competitive formats. The sandwich ELISA is the most common format used for detection of food allergens [9]. In this format, an IgG antibody (referred to as a capture antibody) is immobilized onto the surface of a solid phase (typically a polystyrene microtiter plate or strip). The extracted sample is then added to the microwell and allowed to incubate. Any specific allergenic protein of interest will bind to the capture antibody. A second allergen protein specific antibody that is labeled with an enzyme (e.g., horseradish peroxidase or alkaline phosphatase) will bind to any captured allergenic protein. Two IgG binding epitopes must be present on the protein of interest in order to complete the binding to both the capture antibody and the secondary antibody. A substrate is added and allowed to interact with the antibody bound enzyme, which results in the development of a colored product, which can be measured by a spectrophotometer. The intensity of the color is proportional to the concentration of allergen present in the sample. Quantification can be accomplished by comparing the absorbance of each sample to the absorbance of the standard curve [17].

Competitive ELISAs (also referred to as competitive inhibition ELISAs) can also be used to quantitatively determine the presence of allergenic protein. With competitive ELISAs, the antigen (allergenic protein(s) from the source of interest) are coated onto the surface of a microwell plate. The sample extract is pre-incubated with the allergen-specific IgG antibody, which allows the antibody to bind to any specific allergens of interest present in the sample [17]. This solution is then added to the antigen-coated microwells. Any allergenic protein present in the sample will competitively inhibit binding of the IgG to the plate. After several washing steps to remove any unbound antigen, an enzyme-labeled secondary antibody is added, followed by the appropriate substrate, which results in the development of a colored product. Unlike the sandwich ELISA format, the color intensity is inversely proportional to the concentration of the allergen present in the sample. In this format, the more color product produced, the lower the concentration of the allergen in the sample (i.e., less allergen present to compete with the coated antigen for binding to the allergen-specific IgG antibody). One advantage of competitive ELISA is that only one IgG binding epitope is needed on the allergenic protein of interest. This makes this format useful for the detection of fermented or hydrolyzed proteins where the allergenic proteins may be partially digested. Hydrolysis of the proteins can result in disruption of antibody binding epitopes, which can decrease the number of epitopes available for detection with a sandwich ELISA. Currently, commercially competitive ELISAs are available for the detection of gluten peptides that have gone through partial hydrolysis or fermentation.

Lateral Flow Assays

Lateral flow assays (LFAs) are a qualitative immunochromatographic form of an ELISA. These assays are also referred to as lateral flow strips (LFSs) or lateral flow dipsticks (LFDs). LFAs are comprised of five primary components:

1) The sample filter,
2) The conjugate pad,
3) The membrane,
4) The reservoir, and
5) The test and control lines [18].

An extracted food sample, swab, or final rinse water sample is first applied to the sample filter area consisting of a simple paper-like material where any solid food particles are excluded and soluble protein is wicked into the assay. The conjugated pad consists of a fiberglass-type material that is carefully coated within known quantities of allergen-specific IgG antibody coupled to latex or colloidal metals such as gold [8]. The coupled antibody is not bound to the surface of the LFA. When a sample is applied to the LFA, the allergenic proteins of interest will bind to the coupled antibody and continue to wick through the LFA by capillary action. The membrane of an LFA is generally constructed from polyvinylidene difluoride (PVDF), nitrocellulose, or nylon, where allergen-specific IgG is immobilized in the first zone of the LFA called the test zone. The coupled antibody-allergen (if present in the sample) will migrate to the test zone, and the coupled allergen will bind to the IgG present in this zone, forming a visible line that indicates the positive presence of the specific allergen of interest. The intensity of the line can be correlated to the concentration of the allergen present in the sample. Semi-quantitative results can be obtained if a strip reader is utilized. Some of the coupled antibody will not have bound antigen and will continue to migrate towards the second zone (control zone) where species-specific IgG is immobilized. This anti-species IgG antibody is developed to bind to the coupled IgG. For example, if the coupled antibody is peanut-specific IgG developed in rabbits, a goat or sheep IgG antibody developed against rabbit IgG will be used in the control zone to capture any remaining coupled antibody and a visual line will form. The control line allows the user to know that the LFA did run as expected. If a positive result is found, the development of visual lines at both the test and control zones will be observed, whereas a negative result will be indicated by the development of a line in the control zone only. The reservoir mentioned earlier is simply included to absorb any remaining solution that migrates through the entire LFA. In instances where very high levels of the allergen of concern are present (typically greater than 1000−10,000 ppm), the high concentration of allergen can overwhelm the LFA, resulting in the failure of line development in either the test zone or the control zone. If the LFA is not carefully inspected, the user may interpret the result as negative. In these instances, the sample extract should be diluted with the appropriate extraction buffer in order to achieve a concentration of allergen that is suitable for detection with the LFA. A relatively new product being marketed by Neogen Corporation (Neogen Reveal® 3D LFAs) includes an additional line to the LFA, which will allow a visual line to develop even when high concentrations of the specific allergen are present.

Commercial LFAs now exist for a number of the priority allergens and have been widely used by the food industry for validating that allergenic residues have been removed from equipment surfaces [8]. LFAs are relatively inexpensive; rapid (results can be obtained within 5−10 minutes after extraction of the sample); portable; do not require special instrumentation (such as the microplate reader and washer that are needed for quantitative ELISAs); and are extremely simple to perform, so little training is needed to perform the analysis in the food processing plant. Additionally, LFAs are specific to the allergenic protein(s) for the source of interest and have suitable sensitivity, with limits of detection of approximately 5 ppm, so they are well suited for validating cleaning and sanitation procedures.

ELISA methods have become the standard method in the food industry for the qualitative detection and quantitative measurement of specific proteins from allergenic sources. ELISAs can be effectively used as part of a company's overall risk management process; however, it is important for the end users to understand that individual ELISAs do have their inherent differences. Failure to fully understand and carefully consider these differences could lead to incorrect assessment of the allergenic risk associated with a product or cleaning procedure. ELISAs utilize animal IgG antibodies that are directed against either specific allergenic proteins or, in most cases, several proteins from the allergenic source of interest (but not limited to proteins that cause IgE-mediated food allergy in humans). As a result of the different sources of antibody and different target proteins, in addition to different ways in which the results are reported (ppm whole food vs. ppm total protein from the allergenic source vs. ppm of a specific protein from the allergenic source), interpretation of the results of ELISAs can be difficult and can have profound effects on the overall risk assessment outcome. As discussed earlier, clinical threshold doses and reference doses are primarily reported in units of mg of protein from the allergenic source. Analytical data from ELISAs must be converted on occasion to concentration levels reported in ppm total protein from the allergenic source in order to make proper comparisons with the clinical threshold doses or reference doses. It is very important to determine the units in which the ELISA results are reported so that the proper conversions can be made. Detection of milk residue serves as a good example as there are commercial ELISA kits that detect 'total milk' (protein from both the casein and whey protein fractions of milk), caseins, and beta-lactoglobulin (BLG; from the whey fraction of milk). Some of these kits will report the results in ppm (mg/kg) non-fat dry milk (NFDM), ppm casein, or ppm BLG. In typical milk-derived ingredients used in this example, NFDM contains approximately 35% milk protein, so if the ELISA results indicate that 10 ppm NFDM are present in the food product, a level of 3.5 ppm milk protein would be present. Caseins accounts for approximately 80% of milk protein, so 10 ppm casein would correspond to 12.5 ppm milk protein. Milk contains approximately 10% BLG, so a concentration of 10 ppm BLG corresponds to 100 ppm milk protein. Misinterpretation of the analytical results could clearly have significant effects on the overall risk assessment.

Proteins are also known to have varying thermal and proteolytic stability, which can affect the extraction and detection of the allergenic protein residue of interest [19]. It is critical to ensure that the ELISA will detect the allergenic residue of interest reliably, and where appropriate quantitatively, by analyzing a positive control sample (a sample that is known to contain a given amount of the allergenic source of interest).

Surface Plasmon Resonance Immunoassays

Surface plasmon resonance (SPR) biosensor technology has only recently been applied to the detection of allergenic proteins [8,20,21]. These biosensors consist of two primary components: immobilized allergen-specific IgG antibody coupled to a glass chip that is coated with gold film and a transducer that converts the generated signal into a signal that can be measured by an appropriate processing system. Sample extracts are introduced into the system through a microflow cell, which allows in-line use of this assay. On the opposite side of the glass chip is a prism, which is coupled to the sensor. Polarized light from a diode is reflected off the glass chip and detected by a charge-coupled diode array [21]. At specified resonance wavelengths and angles, surface plasmons (free electrons) interact with the photons, resulting in a decrease in the reflected light detected by the diode. When allergenic proteins of interest are bound to the immobilized antibody on the chip, a change in the refractive index and resonance angle of reflected light will be observed. The shift in the refractive index and the resonance angles of the light are proportional to the mass of the bound analyte, thereby allowing quantitative measurement of the allergenic protein of interest. Since reflected light is being measured, this immunoassay does not rely on an enzyme-labeled antibody like traditional ELISAs. SPR biosensor chips can be quickly regenerated hundreds of times, thereby reducing the overall cost of the analysis and allowing this methodology to be used online in real-time. SPR biosensors also have potential for multi-allergen analysis during a single run, which would further decrease the analysis cost and time. SPR is currently a research method that has been applied to detection of allergens such as milk, egg, peanuts, and sesame seed with limits of quantification of 1−10 ppm when spiked into various food matrices [8,21]. One key area of validation that needs to be conducted is the detection of these allergens in food samples after various food processing unit operations have been applied. Processing techniques such as heating have been shown to alter protein conformation and decrease the solubility of allergenic proteins, which can decrease the recovery and detection of the allergen of interest. SPR biosensors rely on the analysis of a soluble extract of the sample and subsequent allergen-antibody binding for detection.

MASS SPECTROMETRY

Mass spectrometry (MS) is an analytical method that has been utilized in the past for the identification and characterization of proteins, but it has only recently been applied to the quantitative analysis of allergenic residues in food. Detection of protein for the allergenic source of interest is one of the major advantages of using MS for food allergen detection. MS also does not rely on the immunochemical antigen-antibody interaction, in which processing can occasionally affect the binding of the antibody to the protein(s) of interest, thereby decreasing detection in analytical methods such as ELISA [22].

Protein detection by MS is achieved using three basic functions: ionization, mass analysis, and detection. With MS analysis of proteins, the protein sample is first digested by proteases such as trypsin or chymotrypsin before being applied to the MS. Protein modification after processing (e.g., Maillard modifications) can modify proteolytic cleavage sites, so selection of an appropriate protease is important and needs to be carefully considered. Digestion of intact proteins from a sample will result in peptides of various sizes that are then separated using various techniques such as liquid chromatography (LC). The separated peptides are ionized by electron ionization, ion bombardment, matrix-assisted-laser-desorption ionization (MALDI), or electrospray ionization [23]. The mass-to-charge ration (m/z) can be measured using a number of different mass analyzers, such as quadruple (Q), ion-trap, time-of-flight (TOF), and Fourier-transform ion cyclotron resonance mass analyzers. These have all been used for amino acid sequence identification in proteins. Commonly, tandem MS/MS techniques such as triple QQQ, Q/TOF, and TOF/TOF have been utilized for protein analysis [24]. Inclusion of a second mass analyzer provides increased sensitivity, resolution, and mass accuracy. Identification of the peptides can be performed using bioinformatics software, such as Mascot, which is linked to public protein sequence databases such as

NCBI. One current limitation to identification of proteins using MS is that sequences for all proteins of interest are not available, making absolute identification of all protein difficult.

Selected Reaction Monitoring (SRM) or Multiple Reaction Monitoring MS approaches can be used for the quantification of food allergens. These methods require the use of internal reference peptides (typically three to four peptides) that must be carefully selected to ensure that they are unique to the protein from the allergenic source of interest so that no false-positive results are obtained. Similarly to all of the other analytical methods mentioned, the effects of processing on the extraction of these proteins/peptides must be evaluated to ensure detection. They must be extracted from the food matrix and included in this soluble extract in order to be detected. Research on quantitative MS methods for the detection of peanut proteins, milk proteins, and gluten protein has been reported with limits of quantification ranging from 1−10 ppm, which is comparable to ELISA-based methods [25]. It is important to note that a limited number of food matrices have been analyzed to date, so additional validation of the MS methods is needed. MS requires expensive equipment and highly trained technicians who can analyze and interpret the large amount of data generated during a single run, which does not make this technique especially appealing for use in the food processing facility. While use of MS is currently in the research phase, it may provide a confirmatory technique that can be used to verify results of rapid methods such as ELISA. MS also has the potential for multi-allergen analysis in a single run, which may also be of benefit for use in regulatory and contract analytical laboratories. As with ELISAs, consideration of the reporting units (i.e., ppm total protein from the allergenic source vs. ppm specific allergen) is important to ensure that the results can be used in conjunction with reference doses or thresholds for sound risk management decisions.

POLYMERASE CHAIN REACTION (PCR)

PCR-based methods detect DNA rather than protein from the allergenic source of interest. PCR consists of three steps:

1) DNA extraction and purification,
2) Amplification of specific DNA sequence(s), and
3) Detection of the amplified DNA [26].

Similarly to the other methods discussed, the initial extraction of the analyte (DNA) of interest is extremely important in order to detect the residue and ensure that a suitable sensitivity is achieved. In PCR methodology, Taq polymerase is used to amplify a specific DNA fragment that is flanked on each end by carefully selected oligonucleotides that serve as primers for the reaction [9]. PCR relies upon thermal cycling − repeated heating and cooling cycles − for DNA denaturation and enzymatic replication with Taq polymerase. A series of denaturation, annealing, and extension cycles (typically 25−45 cycles) takes place, in which the DNA of interest is amplified to produce a detectable level. The amplified product can be qualitatively visualized by staining after agarose gel electrophoresis, which provides information on the size of the amplified product. Southern blotting, in which the amplified product is detected on the basis of hybridization to a labeled version of the target DNA, allows identification. DNA sequencing allows a complete identification of such a PCR product.

Real-time PCR has been the preferred approach for the quantitative analysis of specific DNA in a sample. In this technique, the reaction tube also contains a target-specific oligonucleotide probe together with a fluorescent reporter dye that has a quencher attached to it [9]. The detection of fluorescence is prevented by the proximity of the quencher to the dye. When the probe hybridizes to the amplified target DNA, the 5' exonuclease activity of the polymerase cleaves the probe, thereby separating the quencher from the dye, which is displaced by the newly synthesized DNA strand. The newly synthesized DNA strand becomes soluble and the fluorescence of the free reporter dye can then be measured. An increase in fluorescence is proportional to the amount of target DNA present in the sample.

PCR-based methods do have some key advantages over immunochemical methods. Food processing can affect the conformation and solubility of proteins. Harsh extraction methods cannot typically be used since they could further affect antibody-binding epitopes. With DNA however, harsher extraction buffers can be used without affecting the detection of the target DNA. Amounts of DNA also tend to be more stable than protein levels, which can vary between various species or varieties. An additional advantage is that PCR-based tests are available for detection of DNA from a number of allergenic sources for which ELISA methods may not be available.

It is important to note however that several allergenic foods have very low DNA content compared to their protein content, including eggs and milk. Careful consideration of the food ingredients used in the processing facility is needed in order to select the appropriate detection method. PCR tests do not however detect proteins from the allergenic source, so their utility in food allergy risk assessment is limited.

CONCLUSIONS

Several analytical methods exist for the quantitative and qualitative detection of residues of priority allergenic foods. These include methods such as ELISA, LFAs, and PCR, which are currently commercially available and widely used by the food industry. Methods such as MS and SPR biosensors have only recently been applied to the detection and quantification of allergenic residues. Although they are primarily research tools at this point in time, MS may become suitable as a reference method for detection of allergenic proteins in the near future, while SPR biosensors may one day be applied to in-line analysis of allergenic residues in the processing facility. The analytical methods discussed in this chapter can provide food companies with data on allergen concentrations that are essential to risk assessment and risk management decisions.

ELISA methods are currently favored for the analysis of allergen residues because they specifically detect proteins from the allergenic source of interest, are sufficiently sensitive to protect allergic consumers, and are available in rugged formats such as lateral flow assays that allow quick determination of residue levels within food manufacturing facilities.

REFERENCES

[1] Rona R, Keil T, Summers C, Gislason D, Zuidmeer L, Sodergren E, et al. The prevalence of food allergy: a meta-analysis. J Allergy Clin Immunol 2007;120:638−46.
[2] Worm M, Timmermans F, Moneret-Vautrin A, Muraro A, Malmheden-Yman I, Lovik M, et al. Towards a European registry of severe allergic reactions: current status of national registries and future needs. Allergy 2010;65:671−80.
[3] Gendel SM. Comparison of international food allergen labeling regulations. Regul Toxicol Pharmacol 2012;63:279−85.
[4] Taylor SL, Baumert JL. Cross-contamination of foods and implications for food-allergic patients. Curr Allergy Asthma Rep 2010;10:265−70.
[5] Sheehan T, Baumert JL, Taylor SL. Allergen validation − analytical methods and scientific support for a visually clean standard. Food Saf Mag 2011;17(6):14. 16,18,20,62.
[6] Taylor SL, Moneret-Vautrin DA, Crevel RW, Sheffield D, Morisset M, Dumont P, et al. Threshold dose for peanut: risk characterization based upon diagnostic oral challenge of a series of 286 peanut-allergic individuals. Food Chem Toxicol 2010;48:814−9.
[7] Akiyama H, Imai T, Ebisawa M. Japan food allergen labeling regulation−history and evaluation. Adv Food Nutr Res 2011;62:139−71.
[8] Schubert-Ullrich P, Rudolf J, Ansari P, Galler B, Fuhrer M, Molinelli A, et al. Commercialized rapid immunoanalytical tests for determination of allergenic food proteins: an overview. Anal Bioanal Chem 2009;395:69−81.
[9] Poms RE, Klein CL, Anklam E. Methods for allergen analysis in food: a review. Food Addit Contam 2004;21:1−31.
[10] Rimbaud L, Heraud F, La Vieille S, Leblanc JC, Crepet A. Quantitative risk assessment relating to adventitious presence of allergens in food: a probabilistic model applied to peanut in chocolate. Risk Anal 2010;30:7−19.
[11] Remington B, Baumert JL, Taylor SL. Risk assessment of foods containing peanut advisory labeling. J Allergy Clin Immunol 2010;125:AB218.
[12] Kruizinga AG, Briggs D, Crevel RWR, Knulst AC, van den Bosch LMC, Houben GF. Probabilistic risk assessment model for allergens in food: sensitivity analysis of the minimum eliciting dose and food consumption. Food Chem Toxicol 2008;46:1437−43.
[13] Spanjersberg MQI, Kruizinga AG, Rennen MAJ, Houben GF. Risk assessment and food allergy: the probabilistic model applied to allergens. Food Chem Toxicol 2007; 45:49−54.
[14] Koppelman SJ, Knulst AC, Koers WJ, Penninks AH, Peppelman H, Vlooswuk R, et al. Comparison of different immunochemical methods for detection and quantification of hazelnut proteins in food products. J Immunol Methods 1999;229:107−20.
[15] Wang X, Young OA, Karl DP. Evaluation of cleaning procedures for allergen control in a food industry environment. J Food Sci 2010;75:T149−55.
[16] Jackson LS, Al-Taher FM, Moorman M, DeVries JW, Tippett R, Swanson KM, et al. Cleaning and other control and validation strategies to prevent allergen cross-contact in food-processing operations. J Food Prot 2008;71:445−58.
[17] Yeung J. Enzyme-linked immunosorbent assays (ELISAs) for detecting allergens in foods. In: Koppelman SJ, Hefle SL, editors. Detecting allergens in food, vol. 1. Cambridge England: Woodhead Publishing Limited; 2006. p. 109−24.
[18] Van Herwijnen R, Baumgartner S. The use of lateral flow device to detect food allergens. In: Koppelman SJ, Hefle SL, editors. Detecting allergens in food, Vol. 1. Cambridge England: Woodhead Publishing Limited; 2006. p. 175−81.

[19] Downs M, Taylor SL. Effects of thermal processing on the enzyme-linked immunosorbent assay (ELISA) detection of milk residues in a model food matrix. J Agriculture Food Chem 2010;22:10085—91.

[20] Yman IM, Eriksson A, Johnson MA, Hellenas KE. Food allergen detection with biosensor immunoassay. J AOAC Int 2006;89:856—61.

[21] Jonsson H, Eriksson A, Yman M. Detecting food allergens with surface plasmon resonance immunoassay. In: Koppelman SJ, Hefle SL, editors. Detecting allergens in food, vol. 1. Cambridge England: Woodhead Publishing Limited; 2006. p. 158—74.

[22] Johnson PE, Baumgartner S, Aldick T, Bessant C, Giosafatto V, Heick J, et al. Current perspectives and recommendations for the development of mass spectrometry for the determination of allergens in foods. J AOAC Int 2011;94:1—8.

[23] Smith JJ, Thakur RA. Mass spectrometry. In: Nielsen SS, editor. Food Analysis. 3rd ed. New York: Kluwer Academic; 2003. p. 423—33.

[24] Picariell G, Mamone G, Addeo F, Ferranti P. The frontiers of mass spectrometry-based techniques in food allergenomics. J Chromatogr A 2011;1218:7386—98.

[25] Monaci L, Visconti A. Mass spectrometry-based proteomics methods for analysis of food allergens. Trends Anal Chem 2009;28:581—91.

[26] Holzhauser T, Stephan O, Vieths S. Polymerase chain reaction (PCR) methods for the detection of allergenic food. In: Koppelman SJ, Hefle SL, editors. Detecting allergens in food, vol. 1. Cambridge England: Woodhead Publishing Limited; 2006. p. 125—43.

Chapter | fourteen

Effect of Processing on the Allergenicity of Foods

Clare Mills[1], Phil E. Johnson[1],
Laurian Zuidmeer-Jongejan[2], Ross Critenden[3],
Jean-Michel Wal[4], Ricardo Asero[5]

[1]*Institute of Inflammation and Repair, Manchester Academic Health Science Center,*
Manchester Institute of Biotechnology, University of Manchester, Manchester, UK
[2]*Department of Experimental Immunology, Laboratory of Allergy Research, Academic*
Medical Center, University of Amsterdam, Amsterdam, The Netherlands
[3]*Valio Ltd, Helsinki, Finland*
[4]*INRA, Unité d'Immuno-Allergie Alimentaire, Jouy-en-Josas, France*
[5]*Ambulatorio di Allergologia, Clinica San Carlo, Paderno-Dugnano, Milano, Italy*

CHAPTER OUTLINE

Risk Management for Food Allergy. http://dx.doi.org/10.1016/B978-0-12-381988-8.00014-2

INTRODUCTION

One of the questions in food allergy research that remains to be answered is: What are the attributes of certain foods and food proteins that make them more allergenic than others? Seeking to answer this question is much more difficult than investigating the allergenic potency of inhalant or contact allergens, since the proteins involved in sensitizing or eliciting allergic reactions may have undergone extensive modification during food processing and be present within complex structures within the food. These physicochemical changes will alter the way in which they are broken down during digestion and may modify the form in which they are taken up across the gut mucosal barrier and presented to the immune system. Although in principle such changes can affect both the sensitization and elicitation phases of an allergic condition, the lack of effective animal models for food allergy means our knowledge is largely confined to the latter aspect. Thus, it has long been known that the structure of the food matrix can have a great impact on the elicitation of allergic reactions and that fat-rich matrices may affect the kinetics of allergen release, potentiating the severity of allergic reactions [1]. Three case histories illustrating the effects of processing on the allergenicity of foods are given below.

Case 1: A 36-year-old woman reports a three-year history of slight oral itching immediately after eating several fruits (apple, pear, apricot, cherry, and kiwi) and tree nuts (walnut and hazelnut). Symptoms usually last about 10−15 minutes and subside spontaneously. The disorder occurs mainly during spring and summer, and only when eating fresh foods; in fact, she tolerates commercial fruit juices as well as (in most cases) fruit salads. The woman has been suffering from rhino-conjunctivitis during the early spring for several years. Clinical investigation shows hypersensitivity to birch pollen, various fruits, and nuts.

Case 2: A 29-year-old man presents with a history of inconstant, moderate oral itching immediately following the ingestion of several fruits (apple, peach, apricot, cherry, and plum) and nuts (hazelnut and walnut) for about 3−4 years. Symptoms usually last 15−20 minutes and subside spontaneously. He reports that he has always tolerated peeled fruits well, whereas commercial juices frequently elicit the oral symptoms. About 1 month before the visit, the man experienced generalized urticaria with angioedema, dysphagia, and shortness of breath about 30 minutes after eating a freshly made cake containing apple ('apfelstrudel'). The man was immediately brought to the emergency department of the nearest hospital where symptoms gradually subsided following therapy with intravenous corticosteroids, antihistamines, and inhaled short-acting beta agonist. The man did not have a history of hay fever. Clinical investigation both *in vivo* and *in vitro* showed hypersensitivity to several fruits and nuts but no reactivity to airborne allergens.

Case 3: A 3-year-old boy was rushed to the hospital after the development of a generalized urticaria and angioedema. Twenty minutes prior to this reaction he had eaten scrambled hen's egg. The boy has a history of atopic dermatitis and elevated hen's egg-specific immunoglobulin-E (IgE) antibodies. The parents report that he eats cookies and other baked products containing hen's egg on a regular basis without any symptoms. Therefore, the family had been told to keep hen's egg in the boy's diet. Clinical investigation *in vitro* showed still elevated hen's egg-specific IgE antibodies but not to any other foods tested. On oral food challenge he showed immediate reactions to raw egg but tolerated cooked egg.

How can we explain such observations? Our knowledge of the impact of food processing and the food matrix on the allergenicity of proteins is limited because of the complexity of working with foods that are so variable and heterogeneous and are processed or cooked in a huge variety of ways. Studying the impact of food processing is fraught with difficulties, not least the fact that food processing often renders food proteins insoluble in the simple salt solutions frequently employed in serological or clinical studies. As a consequence, our understanding of the impact of food processing on aller-genicity is limited to the more soluble and extractable residues in foods, and the allergenic potential of insoluble protein complexes is virtually unstudied despite the fact that they represent the vast bulk of food proteins consumed.

How Processing Can Modify the Structure and Composition of Foods

Finding ways of making foods more digestible and palatable, as well as preserv-ing them, came early in the history of mankind. Many of the ways that we use to cook and preserve foods today have their roots in our ancient past and involve treating foods with heat and treatment with chemical agents — smoking, salting, pickling with low pH agents such as vinegar or acidic fruit juices, and treating with lime. Lastly there is fermentation using microorganisms such as yeast and lactic acid bacteria. Such processes can also make foods safe to eat, inactivating toxins such as cyanogens (cassava) through treatment with lime and anti-nutritional factors such as protease inhibitors found in many legumes. In addi-tion, food processing can induce the formation of desirable structures in foods, changing their texture and appearance. Examples of this are the foams formed in foods such as meringue and the emulsions formed in sauces such as hollan-daise or mayonnaise.

Processing of raw ingredients into finished foods encompasses primary processing procedures, in which inedible tissues are removed (such as shelling and skinning) and they are possibly subjected to other treatments such as heating (pasteurization, sterilization) to prevent microbiological spoilage. Other processes are involved in the preparation of ingredients, such as milling wheat grains to prepare flour and preparation of soy and whey isolates by combinations of wet and dry processing, often with thermal and pH treat-ments. Such processing is frequently employed to improve the versatility and functional properties of ingredients. Formulation of finished food products involves the development of recipes comprising mixtures of ingredients, frequently including further thermal processing, to deliver products like baked goods (cakes, breads, and pastries) and ready-prepared meals.

In addition to conventional food processing procedures, novel processes are also being developed that may offer advantages over conventional processing in preserving food textures and flavors, inactivating microbes and extending shelf-life, or even developing new functionalities and properties in food ingredients. Such processes include the application of high pressure, where foods are exposed to hydrostatic pressures in the range of 100−600 MPa (equivalent to 6,000 times atmospheric pressure), and is often accompanied with heating, not least the adiabatic heating associated with compression and decompression cycles. Other types of novel processing include ohmic heating and pulsed electric field processes. One other physical means of processing foods is through γ-irradiation, which, while not used in Europe, is widely used for the preservation of spices to remove insect pests as well as a means to killing microorganisms. Lastly, there is great interest in developing novel functional ingredients by exploiting the technological properties of nanoscale structures (particles < 100 nm in size) in foods, in particular for encapsulating flavors or delivering bioactive molecules and nanoscale emulsions. One topical example is the structuring of salt crystals to maximize flavor while reducing the overall content of salt in foods to improve their nutritional quality.

Food processing has the potential to affect the initial process of initiating an allergic reaction − known as sensitization − but given that the mechanisms through which individuals develop allergies are not fully understood, and that we lack effective animal models of food allergy, data on how processing might affect sensitization to food are currently sparse. However, much more data are available that are beginning to give insights into how food processing may affect the elicitation of allergic reactions, particularly in affecting the binding of IgE, which lies at the heart of triggering an allergic reaction. Since it is largely the protein molecules that cause food allergies, the remainder of this review is focused on how processing affects proteins. However, it should be noted that an α-linked glycan (α-galactose) has been associated with potent reactions to meats following sensitization caused by multiple tick bites, an epitope that is likely to be resistant to many food processing procedures [2].

The Impact of Processing on the Structure of Food Proteins

In addition to their role as a macronutrient, proteins play an important role in forming the structure of processed foods such as foams (for example whipped egg white in meringue) and gel networks (such as the white in boiled egg or protein gels found in cooked meat products) as well as acting as emulsifying agents in sauces such as mayonnaise. In emulsions, the proteins form an interconnected adsorbed layer coating the oil droplets, and in foams, a bubble wall is made up of denatured protein aggregates. In some foods the proteins interact with other food ingredients such as sugar and can form glassy states in low water foods such as biscuits and pasta. The partially denatured and modified conformations they adopt in such processed foods are similar to those found in processed natural food matrices, where fruits, vegetables, nuts, or seeds maybe wet-processed (e.g., boiled) or dry-heated (e.g., roasted or fried). In this case, the interactions are more complex because of the ultrastructure of the natural food matrix. For example, plant seed proteins may be compartmentalized in protein bodies, but such compartments break down during food processing and cooking to an extent that varies in a process-dependent way. Similarly thermal processing alters the natural structure of the casein micelles and fat globules found in milk, notably in pasteurization (i.e., heating milk to 72°C for 15 s followed by rapid cooling), or an ultra-high temperature (UHT) process involving heating to 140−150°C for a few seconds, which reduces the milk fat globule size, while homogenization increases the interface between fat droplets and the aqueous medium and disrupts the casein micelles.

Such complex interactions, combining the chemical modification of proteins with unfolding and aggregation, mean that the same protein can be present in a food in a multiplicity of processing-induced forms. These forms may behave differently to the native protein during digestion, hence affecting the form in which proteins are presented to the immune system and also their ability to both sensitize a naive individual and elicit a reaction in someone who is already sensitized. The extent to which proteins are affected by processing conditions is process dependent, since protein denaturation requires the presence of water and proteins become more thermostable in low water systems [3]. Combinations of time and temperature and the presence of other ingredients such as fats and sugars also affect the patterns and kinetics of food protein denaturation and aggregation. These processes can result in a range of modifications to food proteins including unfolding and aggregation, as well as chemical modifications such as non-enzymatic glycation. Both of these have the potential to affect stability to digestion, and hence the form in which allergens are presented to the immune system with regards to both sensitization and elicitation.

As with all antibody responses, food-specific IgE binding is affected by the conformational state of an allergen molecule. Thus, IgE antibodies developed towards native proteins, as is the case when an individual develops allergies to agents such as pollens, may only recognize homologous allergens in plant-derived foods when they are consumed in their fresh form rather than after cooking. This is because food processing has caused changes in allergen conformation, associated with thermal denaturation and aggregation, which abolishes the conformational epitopes present in the native protein, making the epitopes thermolabile. In other instances, processing may introduce new epitopes through modification of amino acids caused by heating or reaction with other food constituents, like sugars, to form Maillard adducts. In other proteins, which are intrinsically disordered and adopt no fixed conformation even in their native state, cooking does not alter epitope structure because they are heat stable. Such processing-induced changes in food protein structure are complicated further by interactions with other components in a food such as starch, non-starch polysaccharides (fiber), and lipids in addition to the effect of the micro- and macrostructures developed in foods spanning natural cellular structures found in muscle fibers of meat and fish and fresh fruits and vegetables, to those of fabricated foods such as gels, foams, and emulsions, all of which make it difficult to predict the impact of a given thermal processing procedure on the allergenicity of foods.

In addition to thermal processing, foods are often subjected to processes such as hydrolysis and extraction. Thus, it appears that extensive refining of oils, including bleaching and deodorizing processes, result in oils that contain almost no detectable protein. This essentially renders even oils from allergenic sources such as soybean non-allergenic, although this is not true for less refined, crude culinary oils [4]. Similarly, hydrolysis, if sufficiently extensive, appears to reduce the allergenicity of foods, such as lentils [5], although such effects are not well described for hydrolysis of legume-derived ingredients, unlike some other food allergens such as cow's milk proteins. Thus, in general hydrolyzed products have reduced residual allergenicity, although this varies with the extent of hydrolysis [6]. Fermentation is another process that is regularly used to improve or preserve the quality of foods, including dairy products like cheese and yogurt, plant-derived foods such miso, soy sauce, and tempeh, and many others. Such fermentation may reduce allergenic activity as a result of the action of proteases secreted by the fermentation microbes, which has been shown to reduce reactivity of soybean products [7], although the allergenic activity of highly modified foods such as soy sauce does not appear to be completely removed [8].

Differences in processing regimes — for example, combinations of time, temperature, and exposure to low pH — can make comparison of studies and interpretation of results difficult, but in recent years the body of evidence has increased such that some broad conclusions can be drawn. One major difficulty with all studies is the fact that food processing renders the majority of food proteins into an insoluble mass and hence not tractable to many of the techniques used to study allergenicity. Since the three-dimensional structure of allergens is key to determining the IgE reactivity of food proteins, this chapter will seek to summarize our knowledge of how members of the major plant and animal food allergen families respond to food processing procedures [9,10], focusing on examples from those foods that seem to be responsible for triggering the majority of reactions, and sometimes referred to as the 'Big 8'. Initially the effects of processing on the structure and properties of allergen molecules will be summarized based on the major plant food allergen families (the cupin, prolamin, and Bet v 1 superfamilies) followed by the major allergen families involved in allergies to animal foods, namely the tropomyosin, parvalbumins, and caseins together with other notable allergens from milk and egg. This is followed by conclusions regarding how these effects relate to those observed on the 'whole food' structures and aspects relating to post-harvest treatments, especially for fresh fruits and vegetables, which may alter allergen levels in foods.

EFFECTS OF PROCESSING MAJOR ALLERGENIC FOODS OF PLANT ORIGIN

Major Plant Food Allergens

Cupins: Legume allergens include two types of cupin superfamily proteins: the vicilin-like 7S seed storage globulins and the legumin-like 11S seed storage globulins. The former are known as peanut Ara h 1, soybean β-conglycinin (Gly m 5), and lupin conglutin β (Lup-1) [11,12,13]. They are large oligomeric proteins composed of N-glycosylated subunits of relative mobility (Mr) 34−67 kDa, which generally form trimers of Mr 180,000−235 kDa. β-Conglycinin is a heterotrimer made up of three different subunits, α (Mr ∼ 67 kDa), α′ (Mr ∼ 71 kDa), and β (Mr ∼ 50 kDa) [12]. The soybean subunits share a 'core' region with sequence homologies of around 75% between α, α′, and β subunits and around 90% between the α and α′ subunits. The core region is extended at the N-terminus in the α and α′ subunits by 125 and 144 residues, respectively, the extension having 57% sequence identity between the subunits [14]. Conarachin subunits are structurally homologous with the α and α′ subunits of β-conglycinin, a minor Mr 33 kDa component lacking the N-terminal extension being homologous to the β-subunit of β-conglycinin [11]. Lupin 7S seed storage globulin is somewhat different to those from soybean and peanut in that, like pea vicilin, it undergoes postranslational proteolytic processing in the seed, giving rise to several lower molecular weight subunits, with Mr 14−59 kDa, the Mr 20 kDa polypeptide known as the blad protein, having lectin-like activity [15]. The 11S seed storage globulins are known as Ara h3/4 in peanut, Glym 6 in soybean, and Lup-2 in lupin. The proteins are generally hexameric and assembled from polypeptides that are proteolytically processed in the seed to give rise to two subunits, one of acidic and one of basic pI, linked via a disulfide bond. Peanut Ara h3/4 comprises two acidic subunits of Mr 43 kDa and 38 kDa together with an Mr 24 kDa basic subunit [11]. Seed storage protein allergens have been described in a variety of nuts and seeds with both 11S and 7S proteins having been reported as allergens in hazelnut (Cor a 11 [7S globulin] and Cor a 9 [11S globulin] [16−18], cashew nut (Ana c 1 and Ana c 2, [19,20]), and walnut (Jug r 2 and Jug r 4, [21,22]). In addition the 7S globulins of sesame seed (Ses i, [23]) mustard [24], and the 11S globulins from pecan [25] and almond (also known as almond major protein AMP [26]) have also been identified as allergens. Many of the tree nut allergens appear to show IgE cross-reactivity, a property that has been well defined for the 7S globulins of pistachio and cashew, two closely related tree nut species [27].

Effects of processing: In addition to the chemical modification of proteins, thermal treatments cause extensive protein aggregation. In particular the seed storage globulins found in the seeds of docotelydenous plants are prone to forming aggregates, especially after heating, the nature of which depend on protein concentration, pH, and ionic strength. It appears that the 'core' region of the cupin barrel largely determines the thermal properties of soybean β-conglycinin subunits, with the β subunits being the most thermostable [28], forming aggregates on boiling [29], which at high protein concentrations of 3% (w/v) interact to form gelled networks resembling 'strings-of-beads' polymers [30]. These are a generic feature of protein aggregation [31]. Boiling Ara h 1 also results in the formation of aggregates, although these are topographically distinct from those formed by β-conglycinin as they are branched. Boiling appears to reduce the IgE binding capacity of the proteins, although their T cell reactivity is unaltered since glycation has little effect on either protein aggregation or IgE binding capacity [32]. Intriguingly, Arah 1 purified from roasted nuts was highly denatured and was not glycated but retained the IgE binding capacity of the native protein. In order to mimic the conditions found during roasting, purified Ara h 1 was also subjected to dry heating at elevated temperature [33], which resulted in extensive modification of the protein, including hydrolysis. While food processing effects are not as well characterized in other legume allergens, it appears that fragments of the allergenic lentil globulins find their way into cooking water [34], and there are indications that although extensive boiling and retorting can destroy most of the IgE binding activity of legume proteins, some resistant fragments do remain [35], similar effects having been observed for lupin [36].

PROLAMIN SUPERFAMILY
2S albumins

The 2S albumins belong to the prolamin superfamily of allergens, sharing the cysteine skeleton with at least eight conserved cysteine residues and a three-dimensional structure comprising five α-helices arranged in a right-handed super helix characteristic of that family. They are produced as a single chain precursor and are proteolytically processed in peanut seeds into two subunits linked by intramolecular disulfide bonds [37]. They include the potent peanut allergens Ara h 2 [38] and Ara h 6 [37,39], together with a third low abundance 2S albumin [40], Ara h 7, and several important allergens in tree nuts including the walnut allergen Jug r 1 [41,42], almond [43], Ber e 1 from Brazil nut [44], Car i 1 from pecan [45], and Ana o 3 from cashew nut [46]. They have also been identified as allergens in many types of seeds, including oriental and yellow mustard allergens Bra j 1 and Sin a 1 [47,48], sesame Ses i 1 and 2 [49−51], and the 2S albumin from sunflower seeds SFA-8 [52].

Effects of processing: The 2S albumin allergens appear to more thermostable than many other types of allergens and have to be heated to temperatures in excess of 110°C to undergo any type of denaturation, as indicated by many studies focused primarily on the 2S albumin allergens from peanut, Ara h 2 and 6, although these properties are shared by the 2S albumins from sesame and Brazil nut [53,54]. This inherent stability also appears to make the protein structure resistant to novel processes such as high pressure [55]. Once unfolded, the allergens from peanut have reduced IgE reactivity and functionality in eliciting histamine release, although they retain their ability to activate T cells [56]. It may be that boiling also causes a loss in the IgE binding capacity of whole peanut due to leaching of Ara h 2 into the cooking water, a consequence of the protein retaining its monomeric, compact structure and hence its solubility even after boiling [57]. In contrast, after roasting, where heat is applied under conditions when water activity is limited, Ara h 2 and 6 retain their native conformations and have at least the same IgE reactivity as unheated, native Ara h 2/6 [56,57]. Heating native Ara h 2 for several days at 55°C in the presence of different sugars increased its IgE binding capacity compared to protein heated alone, without sugar, which was related to the formation of advanced glycation end (AGE) products [58]. Such observations for peanut are consistent with the observation that heat treatment at 80°C and 120°C for 60 min had no significant effect on the IgE binding capacity of soybean proteins [59], although others have reported that, while heating soybean to 80°C reduced its IgE binding, the reactivity of the 2S albumins of soy actually increased [60].

Lipid Transfer Proteins

Another type of fruit and vegetable allergy that has been described in the Mediterranean area involves a different group of allergens, the lipid transfer proteins (LTPs, [61]). They are structurally homologous to the 2S albumin allergens and share the same conserved cysteine skeleton and α-helical structure common to the prolamin superfamily [10]. LTPs have been characterized as allergens in fruits, notably peach (Pru p 3, [62]) and apple (Mal d 3, [63]), as well as vegetables including asparagus [64], cabbage (Bra o 3, [65]), and tomato (technically a fruit but often consumed as a salad vegetable) [66]. LTPs have also been identified as allergens in cereal foods including maize, spelt, and wheat [67−69], tree nuts and seeds (such as walnut (Jug r 3, [70] and hazelnut [71] (Cor a 8), and legumes such as peanut [72].

Effects of processing: Like the 2S albumins, LTP allergens are highly resistant to food processing as a consequence of their relatively rigid structure. Thus, heat processing at 180°C for 30 min is unable to reduce the allergenicity of apple LTP [73−75]. Similarly, peach LTP (Pru p 3) retains its allergenic activity in commercial juices and following ultrafiltration of peach juice through suitable molecular weight cut-off membranes [76], as does the LTP from maize [67]. However species differences in response to processing are emerging, as cooking wheat modified the IgE binding capacity LTP in some patients [69]. The LTP scaffold is stable enough to resist harsh treatments, such as fermentation, and individuals with LTP allergies react to the protein after vinification or brewing, as shown by reports of adverse reactions following the ingestion of wine [77] and beer [78−80]. The location of LTP allergens in the outer layer of fruits also means peeling is an approach that significantly reduces the allergenicity of that fruit for LTP allergic subjects [81].

Seed Storage Prolamins and α-amylase Inhibitors

Other allergen members of the prolamin superfamily are the seed storage proteins of cereals, known as prolamins because of their high proline and glutamine contents, together with the α-amylase inhibitors (AAI), which have the same conserved disulfide skeleton and α-helical structure as the 2S albumin and LTP families. In the prolamin seed storage proteins this skeleton has been disrupted, and in some instances partially lost, through the insertion of a repetitive domain of varying length. They can form disulfide-linked polymers and are soluble only in aqueous alcohols. Although more commonly associated with celiac disease [82], sensitization to seed storage prolamins is associated with conditions such as atopic dermatitis and exercise-induced anaphylaxis (EIA), when a severe reaction is experienced if intense exercise is undertaken within a couple of hours of consuming a wheat-containing food [83]. EIA has been associated with sensitization to ω-5 gliadins [84−86], while other prolamin storage proteins have been identified as major cereal allergens, including both the polymeric high molecular weight (HMW) and low molecular weight (LMW) subunits of glutenin and the monomeric α, β, and γ gliadins [83,87]. The AAI have also been found to be food allergens in wheat [69,87,89], including a chloroform-soluble protein, CM3 [90], a subunit termed RA 17 that has been described as an allergen in rice [97], and an Mr 16,000 protein that is a major allergen in maize [67].

Effects of processing: Cooking appears to affect the allergenicity of cereal-derived foods, with indications that processes such baking may even be essential for cereal prolamins to become allergens [92]. Like the LTPS, the α-amylase inhibitors (AAIs) can survive extensive food processing procedures, such as brewing [93].

BET V 1 SUPERFAMILY

The third major plant allergen family is the Bet v1 superfamily [9], which includes a diverse range of IgE cross-reactive allergens. Individuals develop allergy to either pollen or a particular food and go on to develop sensitivities to several different types of fruits and vegetables. Reactions are often milder in nature and confined to the oral cavity, with some of the most important Bet v 1 homologues being found in the Rosacea fruits such as apple (Mal d 1, [94] and peach (Pru p 1, [95]) among many others. Homologues have also been identified in fruits such as kiwi, which are emerging as important allergenic foods in Europe [96], and can be found in tropical fruits such as Sharon fruit [97] and jackfruit [98]. Allergenic Bet v 1 homologues are also found in vegetables, of which one of the most notable is celery (Api g 1; [99]), allergy to which is observed in central European countries. Since it can elicit severe reactions and is often a component in spices, it has been included on the list of allergenic foods for which labeling is mandatory. Homologues have also been identified in carrot (Dau c 1, [100] and in legumes including peanut (Ara h 8, [101]) and soybean (Gly m 4 as presented in certain soybean products [102]).

Effects of processing: In general the IgE binding sites on Bet v 1 are conformational in nature [103], and consequently IgE reactivity is lost following processing procedures that result in unfolding of the protein. As a result individuals sensitized to these proteins can generally safely consume cooked fruits and vegetables but not fresh produce. This has been described in foods such as apple [104] and kiwi [105]. The simple preparation of a fresh fruit salad is often sufficient to abolish or dramatically reduce the intensity of the oral itching caused by the same fresh fruits, as can more severe processing, such as syruping [106]. While they have generally been considered thermolabile, studies of Mal d 1, the allergenic Bet v 1 homologue from apple, have shown that the protein requires thermal treatment in excess of 90°C to become denatured [55]. Such data suggest that differences in stability may be the result of matrix effects rather than the inherent stability of the Bet v 1 fold. Thus, the Bet v 1 homologue from soybean, Gly m 4, may elicit severe reactions but only when present in a particular type of processed soybean ingredient, in which its allergenic potency is retained. It may be that peanut processing removes any potential reactivity of the peanut Bet v 1 homologue, Ara h 8, although it may also be that the inherent reactivity of native Ara h 8 is also lower. The allergenicity of vegetables that is associated with pollen sensitization, such as celery root (celeriac, [107,108]), carrot [109], and tomato [110], all decrease upon heating, although oral challenge studies showed that celery spice powder retains its allergenic activity [108].

Effects of Processing on Major Allergenic Foods of Animal Origin

Tropomyosins: The major allergen in most investigated crustacean species is tropomyosin, which belongs to a family of highly conserved structural proteins found in both muscle and non-muscle cells. They are α-helical proteins forming a coiled-coil structure of two parallel helices with two sets of seven actin binding sites. In striated muscle cells they facilitate interactions between the troponin and actin complex, thus regulating muscle contraction. The first recombinant shrimp allergen (Met e 1) was produced from the greasy back shrimp by Leung in the mid 1990s [111], and this confirmed the observation of other groups that the major heat stable allergen in many crustacean species is tropomyosin. While mollusks contain various less well characterized allergens, the allergen repertoire also seems to include tropomyosin [112]. To date, only non-vertebrate tropomyosins that have allergenic activity have been identified, but these are highly cross-reactive because of their close sequence homologies [10].

Effects of processing: Relatively few studies have focused on the effects of processing on seafood allergens. An increase in IgE binding reactivity was observed for tropomyosin from scallops after heating in the presence of both hexose sugars (such as glucose) and pentoses (such as ribose) [113], while a decrease of IgE binding capacity was observed for squid (calamari) tropomyosin after heating in the presence of ribose to form Maillard adducts [114]. There is also some evidence that tropomyosin from cooked shrimps has higher IgE reactivity than the protein isolated from raw shrimp [115].

Parvalbumins: The major allergen that has been identified in the flesh of a range of fish species is the white muscle protein known as parvalbumin. This protein contains a structural motif known as an EF-hand, which can bind calcium. If the calcium is removed, the resulting apo-form of the protein has a sufficiently altered three-dimensional structure as to lose its IgE binding capacity [116]. A number of allergenic fish parvalbumins have been identified from a variety of fish species, and these have high levels of sequence homology (around 70%) [10], explaining why sensitization to parvalbumins can result in allergies to multiple fish species. The levels of parvalbumins expressed in the white, fast-twitch muscles (required for rapid movement) and the dark muscle (more important for continuous swimming) are different, parvalbumin levels being much higher in the white muscle [117]. As a consequence, fish species such as tuna and swordfish have around 20−30 fold lower levels of extractable parvalbumin than the predominantly white flesh muscle flesh of fish such as cod. Since the proteins from fish such as swordfish and cod have very similar IgE binding capacities, it seems the lower levels of allergen in the flesh of fish such as tuna explain the apparently lower allergenicity of these fish [118].

Effects of processing: The effect of cooking on the allergenicity of fish was described in one of the very first published reports of an allergic reaction to a food, when the sensitivity of Kustner towards cooked but not raw fish was reported by Prausnitz [119]. It has subsequently proved difficult to find such individuals, and it seems that in general the allergenic activity of fish is reduced, but not abolished, by thermal processing since, for example, canned fish has a 100−200 fold lower allergenic activity than boiled fish [120]. The thermostable nature of the allergenic activity of fish can be attributed to the stability of the holo-parvalbumin, which as long as calcium is present may either resist denaturation in the first place or refolds [116]. There are however other allergens implicated in fish allergy, but little is known about their responsiveness to cooking procedures.

Caseins: Caseins are mammalian proteins present in milk. In bovine milk the casein fraction (Bos d 8) comprises four proteins, α_{S1}-, α_{S2}-, β-, and κ-caseins [121], all identified as allergens and which have a disordered structure. The four classes of casein have low amino acid sequence homology but display common features, being phosphorylated proteins with a loose tertiary, highly hydrated structure. They are often considered poorly immunogenic because of this flexible, expanded structure. The groups of phosphoserine and phosphothreonine form nano-clusters around amorphous calcium phosphate, allowing milk to contain higher levels of soluble calcium than is possible to maintain in ordinary solution. Numerous IgE epitopes have been identified on cow's milk proteins, both conformational and sequential. The linear epitopes have been shown to be widely distributed all along the protein molecules, including in hydrophobic regions where they are masked and not available for binding.

Effects of processing: Casein is a thermostable molecule, its mobile structure being unchanged by heating. Thus, the linear epitopes particularly involved in allergy are likely to be equally available for IgE binding in native and heated caseins. As a consequence, boiling milk for short periods of time (2, 5, or 10 min) results either in no difference or in a reduction of about 50−66% of the positive reactions as compared to raw milk; similar observations have been reported with raw vs. pasteurized or homogenized and pasteurized milk [122−124]. However, homogenization alone has no effect on allergic responses [125]. Caseins are readily degraded by proteases, which largely explains the observation that extensive hydrolysis reduces the allergenicity of cow's milk proteins. However, allergenic activity is not completely abolished by such processing, possibly as a result of residual intact protein, especially for more digestion-resistant proteins such as whey protein β-lactoglobulin.

Minor Allergen Families

Lipocalins: While these form an important class of inhalant allergens, the only food allergen belonging to this superfamily is β-lactoglobulin (BLG, Bos d 5; [126]). Lipocalins share a conserved three-dimensional structure, although their overall sequence similarity is low. They possess a central calyx into which a range of small molecules such as lipids, steroids, hormones, bilins, and retinoids can bind. The β-barrel structure is stabilized by two disulfide bonds, and depending on the pH, bovine BLG can form dimers or higher order oligomers. It is also able to interact and become covalently linked with other proteins by virtue of its free cysteine residue.

Effects of processing: Heating BLG in solution or in whole milk, at 74°C or 90°C, has little effect on overall IgE binding [127], although complete denaturation of BLG by chemical reduction and S-carboxymethylation of the disulfide bonds did not alter the IgE binding capacity [128]. All those observations are in line with the hypothesis that linear epitopes of BLG, which are heat stable, are most important in cow's milk allergy and confirm that even complete denaturation does not abolish the proteins' allergenicity. On the contrary it may unmask those epitopes that are buried within the tertiary structure of the BLG molecule, which then become available for IgE binding. BLG is thermolabile, although it may be protected from denaturation through interaction with casein, and it retains its IgE-binding capacity after boiling milk for 5 minutes. Reaction with milk sugars causes glycation of BLG. The reaction with galactose occurs at 50°C, and this impairs the digestibility and increases the immunoglobulin-G (IgG) immunoreactivity of BLG (see Figure 14.1) [129].

Interactions with fatty acids can also change the secondary structure of BLG upon heating [130]. BLG is degraded during whole milk fermentation by *Lactobacilli*, although this had little effect on its IgE binding capacity [131]. This is consistent with the observation that BLG retains its allergenic activity following extensive hydrolysis [132], reflecting the resistance of this protein to proteolysis in general. Where partially hydrolyzed formulae are concerned, allergic reactions may be due to the presence of either residual native protein or large fragments derived from them. In the case of extensively hydrolyzed formulae where no protein or large fragments remain, the allergic reaction

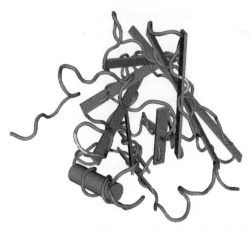

FIGURE 14.1 Molecular structure of β-lactoglobulin A (PDB ID: 1C5J) with positions of two identified sites of lactosylation (Fogliano et al., 1998) (lysines 47 and 100) marked (yellow in black circle). Fogliano V, Monti SM, Visconti A, Randazzo G, Facchiano AM, Colonna G, Ritieni A. Identification of a lactoglobulin lactosylation site. Biochem Biophys Acta 1988;2:295—304.

may be triggered by short peptide fragments comprising IgE binding epitopes that are released during the proteolysis.

Other minor allergen families: Two important milk and egg allergens belong to this minor allergen family, namely the C-type lysozymes and the whey protein α-lactalbumin ([121] ALA, Bos d 4 from cow's milk) and hen's egg lysozyme ([133] Gal d 4). ALA is able to bind calcium and plays a role in lactose synthesis during lactation. However lysozyme is a glycohydrolase found in egg white and has a superimposable three-dimensional structure with ALA. The egg white protein ovomucoid is a Kazal-type protease inhibitor and includes hen's egg ovomucoid, also known as Gal d 1, an extensively glycosylated protein with intramolecular disulfide bonds that may act to stabilize the protein against proteolysis [134]. Another inhibitor family represented in the animal food allergens is the serpin serine protease inhibitor, ovalbumin, Gal d 2 [133]. Lastly, transferrins, sulfur-rich iron-binding glycoproteins, have been identified as minor allergens in milk, lactoferrin [121], and in egg, ovotransferrin, Gal d 3 [133].

Effects of processing: The antigenicity of ovomucoid in particular is relatively thermostable, reflecting the intramolecular disulfide bonds [135,136]. Similarly ovalbumin is relatively stable and, intriguingly, consumption of cooked eggs, as opposed to raw, significantly increases the concentration of ovalbumin in breast milk [137]. However, the allergenic potency and sensitizing capacity of raw and cooked egg proteins transferred via breast milk has not yet been explored, although they may play a role in triggering reactions in infants with egg allergy. The study of both these proteins is made more complex by the presence of N-linked glycans, and, compared to the cow's milk and peanut allergens, the impact of processing on these proteins is poorly understood. In general, both the

C-type glycohydrolases and the transferrin allergens from milk and egg are less thermostable than the other egg and milk allergens, tend to be readily digested, and may contribute less to the thermostability of milk and egg in foods.

WHOLE FOOD EFFECTS OF FOOD PROCESSING
Post-Harvest Treatments

Many fresh fruits and vegetables are subjected to a range of post-harvest treatments and storage during transport to make fresh fruits available throughout the year. Many of the allergens in fresh fruits, including the Bet v 1 homologues and LTPs, are thought to have a role in plant protection, the expression of which is affected by factors such as pathogen attack and abiotic stress, such as temperature and physical damage. As a consequence their expression in fresh fruits and vegetables changes during ripening and storage. Thus, during modified atmosphere storage of apples, the expression of Mal d 1 was increased at both translational and transcriptional levels by 3.5 and 8.5 fold, respectively, over a 5-month period [138]. Under the same conditions the LTP allergen Mal d 3 decreased. This is an allergen whose expression increases during the maturation of apples prior to picking [139]. Similar ripening-related effects have been observed in kiwi fruit [140], and it is likely that similar effects will be observed for these two allergens in a range of fresh fruits and vegetables.

Thermal Treatments

It is becoming evident that more severe thermal treatments employing higher temperatures and longer heating times, especially at high water activity, can extensively modify allergen structure and hence modify the allergenic activity of foods. However, the time-temperature combinations, and the profile of the allergens involved in the allergic patients studied, makes interpreting the results complex. For example, for fruit such as mango, the preparation of purees and nectars did not reduce their allergenicity in one study [141], while it has been reported that canning reduced the IgE reactivity of some proteins but not able to others in lychees [142]. In cow's milk allergy it is becoming evident that some individuals who react to raw, pasteurized, or ultra-heat treated milk may tolerate products containing extensively heated milk (e.g., baked products), probably reflecting the fact that only the latter extensively modifies the proteins compared to raw milk [143]. Thus, milk included in a muffin could be tolerated by children whose allergy was resolving, reflecting both the effect of baking on the allergen structure and the complex interactions with other ingredients, notably the gluten matrix [144]. There are similar whole food effects observed with egg, where cooking generally reduces its allergenic potency, although it does not eliminate all allergenic epitopes [145], presumably because the thermal processes are insufficient to destroy them. For some individuals, the reduction in allergenicity due to heating is sufficient to allow them to tolerate cooked but not necessarily raw egg [146].

Maillard Modifications

Complex interactions with the sugars in the formation of Maillard adducts resulting from browning reactions have also been implicated in affecting the allergenicity of whole foods, notably peanut. Thus, Maillard modification of peanut allergens appears to increase their allergenic activity, rendering them more resistant to gastrointestinal digestion [57,147,148]. Such observations, coupled with the fact that peanut allergy appears less problematic in China despite peanuts being widely consumed, has led some to propose that cooking practices (in particular roasting) are responsible for the apparent allergenic potency of peanuts [149]. In contrast, soybean is rarely consumed whole in the West, generally being used as a processed ingredient and not usually eaten in a roasted form, and it does not seem to be such a problematic legume allergen, further suggesting that processing procedures may modulate the allergenicity of foods. However, Maillard modifications are complex and difficult to characterize because they make the proteins insoluble, and the complex chemical modifications and rearrangements have made it, as yet, difficult to relate structural changes to changes in allergenic activity.

Fermentation and Hydrolysis

In general hydrolysis, like heat treatment, appears to reduce, although not necessarily abolish, the allergenic activity of foods. This has been especially well characterized in dairy products, and especially with the need to provide infants who are allergic to cow's milk with alternative, hypoallergenic formula. These foods have proven especially well adapted to using hydrolysis since it does not adversely affect the texture of beverages in particular. Extensive hydrolysis has also been used to reduce the allergenicity of eggs, although a significant proportion of egg allergic subjects continue to react even when the proteins were hydrolyzed to peptides smaller than 4.5 kDa [150]. However, hydrolysis tends to have only limited application, since it destroys the functional properties of food, which often rely on the presence of intact proteins. Although extensively hydrolyzed formulae have demonstrated a marked decrease of allergenicity, no hypoallergenic milk or milk-derived formulas are available that could provide total safety to the whole population of milk allergic patients, including highly sensitive children. While many lactic fermentations cause hydrolysis of proteins, their impact on allergenicity seems variable, with contradictory reports of efficacy, probably arising from differences in the fermentation conditions and particularly in the microbial strains used.

Novel Processing Methods

Novel preservation methods, including γ-irradiation, ultra-high pressure (UHP), and high voltage impulse, have also been employed to modify the potential allergenic activity of foods. However, these did not result in a decrease in IgE binding activity for celery and tomato (UHP only) [107]. High pressure treatment appeared to reduce that of apple and celeriac, although this may be due to adiabatic heating effects, since the allergen structure is not disrupted by high pressure processing [59,151], suggesting thermal processing is the main means whereby allergenicity of foods is reduced. In contrast, gamma radiation has been applied to crustacean and molluscan shell-fish, resulting in reduced IgE binding capacity of the allergens [152−154]. Similar results were found for the allergenicity of egg allergens following combined gamma-irradiation and heat treatment [155,156]. However, it remains to be seen whether combinations of such processing methods can produce truly hypoallergenic products.

CONCLUSIONS

For many years, there was a lack of knowledge about the effect of food processing on the allergenicity of foods, the ways that cooking can affect the structure and reactivity of molecules, and how this relates to effects observed in whole foods as they are eaten. This review demonstrates how our knowledge has increased greatly over the last 10 years, especially for major allergenic foods such as peanut and cow's milk. Patterns are emerging and it is becoming evident that the food matrix itself can attenuate the effects of food processing on allergen molecules in ways that are poorly understood and that still make it difficult to predict how a given food process may affect the allergenicity of molecules. It also appears that in many instances, food processing reduces but does not abolish allergenicity, and while for some foods the assertion that certain types of processing, such as roasting, appear to enhance allergenicity, precise molecular explanations linked to effective studies of clinical reactivity are still lacking. A close collaboration between food chemists and clinical researchers is essential if such objective evidence is to be obtained and utilized for more effective diagnosis. Further work will also need to be done to translate this into advice for patients to enable them to effectively manage their condition and to assist the food industry in managing allergens in foods more effectively.

REFERENCES

[1] Grimshaw KE, King RM, Nordlee JA, Hefle SL, Warner JO, Hourihane JO. Presentation of allergen in different food preparations affects the nature of the allergic reaction − a case series. Clin Exp Allergy 2003;33:1581−5.
[2] Commins SP, James HR, Kelly LA, Pochan SL, Workman LJ, Perzanowski MS, et al. The relevance of tick bites to the production of IgE antibodies to the mammalian oligosaccharide galactose-α-1,3-galactose. J Allergy Clin Immunol 2011;127: 1286−93.
[3] Gekko K, Timasheff SN. Mechanism of protein stabilization by glycerol: preferential hydration in glycerol-water mixtures. Biochem 1981;20:4667−76.
[4] Crevel RW, Kerkhoff MA, Koning MM. Allergenicity of refined vegetable oils. Food Chem Toxicol 2000;38:385−93.
[5] Cabanillas B, Pedrosa MM, Rodríguez J, González A, Muzquiz M, Cuadrado C, et al. Effects of enzymatic hydrolysis on lentil allergenicity. Mol Nutr Food Res 2010;54: 1266−72.
[6] Tsai E, Yeung J, Gold M, Sussman G, Perelman B, Vadas P. Study of the allergenicity of plant protein hydrolysates. Food Allergy and Intolerance 2003;4:117−26.

[7] Lee JO, Lee SI, Cho SH, Oh CK, Ryu CH. A new technique to produce hypoallergenic soybean proteins using three different fermenting microorganisms. J Allergy Clin Immunol 2004;113:S239.

[8] Hefle SL, Lambrecht DM, Nordlee JA. Soy sauce retains allergenicity through the fermentation production process. J Allergy Clin Immunol 2005;115:S32.

[9] Jenkins JA, Griffiths-Jones S, Shewry PR, Breiteneder H, Mills ENC. Structural relatedness of plant food allergens with specific reference to cross-reactive allergens — an *in silico* analysis. J Allergy Clin Immunol 2005;115:163—70.

[10] Jenkins JA, Breiteneder H, Mills ENC. Evolutionary distance from human homologs reflects allergenicity of animal food proteins. J Allergy Clin Immunol 2007;120:1399—405.

[11] Marsh J, Rigby N, Wellner K, Reese G, Knulst A, Akkerdaas J, et al. Purification and characterisation of a panel of peanut allergens suitable for use in allergy diagnosis. Mol Nutr Food Res 2008;52(S2):S272—85.

[12] Nielsen NC, Dickinson CD, Cho TJ, Thanh VH, Scallon BJ, Fischer RI, et al. Characterization of the glycinin gene family in soybean. Plant Cell 1989;1:313—28.

[13] Guillamon E, Rodriguez J, Burbano C, Muzquiz M, Pedrosa M, Cabanillas B, et al. Characterization of lupin major allergens (*Lupinus albus* l.). Mol Nutr Food Res 2009;54:1668—76.

[14] Maruyama N, Katsube T, Wada Y, Oh MH, Barba De La Rosa AP, Okuda E, et al. The roles of the N-linked glycans and extension regions of soybean beta-conglycinin in folding, assembly and structural features. Eur J Biochem 1998;258:854—62.

[15] Ramos PC, Ferreira RM, Franco E, Teixeira AR. Accumulation of a lectin-like breakdown product of beta-conglutin catabolism in cotyledons of germinating Lupinus albus L. seeds. Planta 1997;203:26—34.

[16] Lauer I, Foetisch K, Kolarich D, F Ballmer-Weber BK, Conti A, Altmann F, et al. Hazelnut (*Corylus avellana*) vicilin Cor a 11: molecular characterization of a glycoprotein and its allergenic activity. Biochem J 2004;382:327—34.

[17] Pastorello EA, Vieths S, Pravettoni V, Farioli L, Trambaioli C, Fortunato D, et al. Identification of hazelnut major allergens in sensitive patients with positive double-blind, placebo-controlled food challenge results. J Allergy Clin Immunol 2002;109:563—70.

[18] Beyer K, Grishina G, Bardina L, Grishin A, Sampson HA. Identification of an 11S globulin as a major hazelnut food allergen in hazelnut-induced systemic reactions. J Allergy Clin Immunol 2002;110:517—23.

[19] Wang F, Robotham JM, Teuber SS, Tawde P, Sathe SK, Roux KH. Ana o 1, a cashew (*Anacardium occidental*) allergen of the vicilin seed storage protein family. J Allergy Clin Immunol 2002;110:160—6.

[20] Robotham JM, Wang F, Seamon V, Teuber SS, Sathe SK, Sampson HA, et al. Ana o 3, an important cashew nut (Anacardium occidentale L.) allergen of the 2S albumin family. J Allergy Clin Immunol 2005;115:1284—90.

[21] Teuber SS, Jarvis KC, Dandekar AM, Peterson WR, Ansari AA. Identification and cloning of a complementary DNA encoding a vicilin-like proprotein, Jug r 2, from English walnut kernel (*Juglans regia*), a major food allergen. J Allergy Clin Immunol 1999;104:1111—20.

[22] Wallowitz M, Peterson WR, Uratsu S, Comstock SS, Dandekar AM, Teuber SS. Jug r 4, a legumin group food allergen from walnut (*Juglans regia* Cv. Chandler). J Agric Food Chem 2006;54:8369—75.

[23] Beyer K, Bardina L, Grishina G, Sampson HA. Identification of sesame seed allergens by 2-dimensional proteomics and Edman sequencing: seed storage proteins as common food allergens. J Allergy Clin Immunol 2002;110:154—9.

[24] Palomares O, Cuesta-Herranz J, Vereda A, Sirvent S, Villalba M, Rodríguez R. Isolation and identification of an 11S globulin as a new major allergen in mustard seeds. Ann Allergy Asthma Immunol 2005;94:586—92.

[25] Sharma GM, Irsigler A, Dhanarajan P, Ayuso R, Bardina L, Sampson HA, et al. Cloning and characterization of an 11S legumin, Car i 4, a major allergen in pecan. J Agric Food Chem 2011;59:9542—52.

[26] Roux KH, Teuber SS, Sathe SK. Tree nut allergens. Int Arch Allergy Appl Immunol 2003;131:234−44.

[27] Willison LN, Tawde P, Robotham JM, Penney RMt, Teuber SS, Sathe SK, et al. Pistachio vicilin, Pis v. 3, is immunoglobulin E-reactive and cross-reacts with the homologous cashew allergen, Ana o 1. Clin Exp Allergy 2008;38:1229−38.

[28] Maruyama N, Sato R, Wada Y, Matsumura Y, Goto H, Okuda E, et al. Structure-physicochemical function relationships of soybean beta-conglycinin constituent subunits. J Agric Food Chem 1999;47:5278−84.

[29] Mills ENC, Huang L, Noel TR, Gunning AP, Morris VJ. Formation of thermally induced aggregates of the soya globulin β-conglycinin. Biochim Biophys Acta 2001;1547:339−50.

[30] Ramlan M, Salleh BM, Maruyama N, Takahashi K, Yagasaki K, Higasa T, et al. Gelling properties of soybean beta-conglycinin having different subunit compositions. Biosci Biotechnol Biochem 2004;68:1091−6.

[31] Krebs MR, Devlin GL, Donald AM. Protein particulates: another generic form of protein aggregation? Biophys J 2007;92:1336−42.

[32] Blanc F, Vissers YM, Adel-Patient K, Rigby NM, Mackie AR, Gunning AP, et al. Boiling peanut Ara h 1 results in the formation of aggregates with reduced allerge-nicity. Mol Nutr Food Res 2011;55:1887−94.

[33] Vissers YM, Iwan M, Adel-Patient K, Stahl Skov P, Rigby NM, Johnson PE, et al. Effect of roasting on the allergenicity of major peanut allergens Ara h 1 and Ara h 2/6: the necessity of degranulation assays. Clin Exp Allergy 2011;4:1631−42.

[34] Sánchez-Monge R, Pascual CY, Díaz-Perales A, Fernández-Crespo J, Martín-Esteban M, Salcedo G. Isolation and characterization of relevant allergens from boiled lentils. J Allergy Clin Immunol 2000;106:955−61.

[35] Cuadrado C, Cabanillas B, Pedrosa MM, Varela A, Guillamón E, Muzquiz M, et al. Influence of thermal processing on IgE reactivity to lentil and chickpea proteins. Mol Nutr Food Res 2009;53:1462−8.

[36] Alvarez-Alvarez J, Guillamón E, Crespo JF, Cuadrado C, Burbano C, Rodríguez J, et al. Effects of extrusion, boiling, autoclaving, and microwave heating on lupine aller-genicity. J Agric Food Chem 2005;53:1294−8.

[37] Bernard H, Mondoulet L, Drumare MF, Paty E, Scheinmann P, Thaï R, et al. Identi-fication of a new natural Ara h 6 isoform and of its proteolytic product as major allergens in peanut. J Agric Food Chem 2007;55:9663−9.

[38] Burks AW, Williams LW, Connaughton C, Cockrell G, O'Brien TJ, Helm RM. Iden-tification and characterization of a second major peanut allergen, Ara h II, with use of the sera of patients with atopic dermatitis and positive peanut challenge. J Allergy Clin Immunol 1992;90:962−9.

[39] Flinterman AE, van Hoffen E, den Hartog Jager CF, Koppelman S, Pasmans SG, Hoekstra MO, et al. Children with peanut allergy recognize predominantly Ara h 2 and Ara h 6, which remains stable over time. Clin Exp Allergy 2007;37:1221−8.

[40] Schmidt H, Krause S, Gelhaus C, Petersen A, Janssen O, Becker WM. Detection and structural characterization of natural Ara h 7, the third peanut allergen of the 2S albumin family. J Proteome Res 2010;9:3701−9.

[41] Teuber SS, Dandekar AM, Peterson WR, Sellers CL. Cloning and sequencing of a gene encoding a 2S albumin seed storage protein precursor from English walnut (*Juglans regia*), a major food allergen. J Allergy Clin Immunol 1998;101:807−14.

[42] Teuber SS, Jarvis KC, Dandekar AM, Peterson WR, Ansari AA. Identification and cloning of a complementary DNA encoding a vicilin-like proprotein, Jug r 2, from English walnut kernel (*Juglans regia*), a major food allergen. J Allergy Clin Immunol 1999;104:1111−20.

[43] Poltronieri P, Cappello MS, Dohmae N, Conti A, Fortunado D, Pastorello EA, et al. Identification and characterisation of the IgE-binding proteins 2S albumin and conglutin gamma in almond (*Prunus dulcis*) seeds. Internat Arch Allergy Appl Immunol 2002;128:97−104.

[44] Pastorello EA, Farioli L, Pravettoni V, Ispano M, Conti A, Ansaloni R, et al. Sensitization to the major allergen of Brazil nut is correlated with the clinical expression of allergy. J Allergy Clin Immunol 1998;102:1021–7.

[45] Sharma GM, Irsigler A, Dhanarajan P, Ayuso R, Bardina L, Sampson HA, et al. Cloning and characterization of 2S albumin, Car i 1, a major allergen in pecan. J Agric Food Chem 2011;59:4130–9.

[46] Robotham JM, Wang F, Seamon V, Teuber SS, Sathe SK, Sampson HA, et al. Ana o 3, an important cashew nut (*Anacardium occidentale* L.) allergen of the 2S albumin family. J Allergy Clin Immunol 2005;115:1284–90.

[47] Menendez-Arias L, Moneo I, Dominguez J, Rodriguez R. Primary structure of the major allergen of yellow mustard (*Sinapis alba* L.) seed, Sin a I. Eur J Biochem 1988;177:159–66.

[48] Monsalve RI, Gonzalez de la Peña MA, Menendez-Arias L, Lopez-Otin C, Villalba M, Rodriguez R. Characterization of a new oriental-mustard (*Brassica juncea*) allergen, Bra j IE: detection of an allergenic epitope. Biochem J 1993;293: 625–32.

[49] Pastorello EA, Varin E, Farioli L, Pravettoni V, Ortolani C, Trambaioli C, et al. The major allergen of sesame seeds (*Sesamum indicum*) is a 2S albumin. J Chromatogr B Biomed Sci Appl 2001;756:85–93.

[50] Wolff N, Yannai S, Karin N, Levy Y, Reifen R, Dalal I, et al. Identification and characterization of linear B-cell epitopes of beta-globulin, a major allergen of sesame seeds. J Allergy Clin Immunol 2004;114:1151–8.

[51] Beyer K, Bardina L, Grishina G, Sampson HA. Identification of sesame seed allergens by two dimensional proteomics and Edman sequencing: seed storage proteins as common food allergens. J Allergy Clin Immunol 2002;110:154–9.

[52] Kelly JD, Hlywka JJ, Hefle SL. Identification of sunflower seed IgE-binding proteins. Internat Arch Allergy Appl Immunol 2000;121:19–24.

[53] Moreno FJ, Maldonado BM, Wellner N, Mills ENC. Thermostability and *in vitro* digestibility of a purified major allergen 2S albumin (Ses i 1) from white sesame seeds (Sesamum indicum L.). Biochim Biophys Acta 2005;1752:142–53.

[54] Moreno FJ, Jenkins JA, Mellon FA, Rigby NM, Robertson JA, Wellner N, et al. Mass spectrometry and structural characterization of 2S albumin isoforms from Brazil nuts (*Bertholletia excelsa*). Biochim Biophys Acta 2004;1698:175–86.

[55] Johnson PE, Van der Plancken I, Balasa A, Husband FA, Grauwet T, Hendrickx M, et al. High pressure, thermal and pulsed electric-field-induced structural changes in selected food allergens. Mol Nutr Food Res 2010;54:1701–10.

[56] Vissers YM, Blanc F, Skov PS, Johnson PE, Rigby NM, Przybylski-Nicaise L, et al. Effect of heating and glycation on the allergenicity of 2S albumins (Ara h 2/6) from peanut. PLoS One 2011;6:e23998.

[57] Mondoulet L, Paty E, Drumare MF, Ah-Leung S, Scheinmann P, Willemot RM, et al. Influence of thermal processing on the allergenicity of peanut proteins. J Agric Food Chem 2005;53:4547–53.

[58] Gruber P, Becker WM, Hofmann T. Influence of the Maillard reaction on the allergenicity of rAra h 2, a recombinant major allergen from peanut (*Arachis hypogaea*), its major epitopes, and peanut agglutinin. J Agric Food Chem 2005;53:2289–96.

[59] Burks AW, Williams LW, Helm RM, Thresher W, Brooks JR, Sampson HA. Identification of soy protein allergens in patients with atopic dermatitis and positive soy challenges; determination of change in allergenicity after heating or enzyme digestion, nutritional and toxicological consequences of food processing. Adv Exp Med Biol 1991;289:295–307.

[60] Shibasaki M, Suzuki S, Tajima S, Nemoto H, Kuruome T. Allergenicity of major component proteins of soybean. Int Arch Appl Immunol 1980;61:441–8.

[61] Fernández-Rivas M, Bolhaar S, González-Mancebo E, Asero R, van Leeuwen A, Bohle B, et al. Apple allergy across Europe: how allergen sensitization profiles determine the clinical expression of allergies to plant foods. J Allergy Clin Immunol 118, 481–488.

[62] Pastorello EA, Farioli L, Pravettoni V, Ortolani C, Ispano M, Monza M, et al. The major allergen of peach (*Prunus persica*) is a lipid transfer protein. J Allergy Clin Immunol 1999;103:520−6.

[63] Sanchez-Monge R, Lombardero M, Garcia-Selles FJ, Barber D, Salcedo G. Lipid-transfer proteins are relevant allergens in fruit allergy. J Allergy Clin Immunol 1999;103:514−9.

[64] Diaz-Perales A, Tabar AI, Sanchez-Monge R, Garća BE, Gómez B, Barber D, et al. Characterization of asparagus allergens: a relevant role of lipid transfer proteins. J Allergy Clin Immunol 2002;110:790−6.

[65] Palaćn A, Cumplido J, Figueroa J, Ahrazem O, Sánchez-Monge R, Carrillo T, et al. Cabbage lipid transfer protein Bra o 3 is a major allergen responsible for cross-reactivity between plant foods and pollens. J Allergy Clin Immunol 2006;117:1423−9.

[66] Pravettoni V, Primavesi L, Farioli L, Brenna OV, Pompei C, Conti A, et al. Tomato allergy: detection of IgE-binding lipid transfer proteins in tomato derivatives and in fresh tomato peel, pulp, and seeds. J Agric Food Chem 2009;57:10749−54.

[67] Pastorello EA, Pompei C, Pravettoni V, Farioli L, Calamari AM, Scibilia J, et al. Lipid-transfer protein is the major maize allergen maintaining IgE-binding activity after cooking at 100°C, as demonstrated in anaphylactic patients and patients with positive double-blind, placebo-controlled food challenge results. J Allergy Clin Immunol 2003;112:775−83.

[68] Pastorello EA, Farioli L, Robino AM, Trambaioli C, Conti A, Pravettoni V. A lipid transfer protein involved in occupational sensitization to spelt. J Allergy Clin Immunol 2001;108:145−6.

[69] Pastorello EA, Farioli L, Conti A, Pravettoni V, Bonomi S, et al. Wheat IgE-mediated food allergy in European patients: α-amylase inhibitors, lipid transfer proteins and low-molecular-weight glutenins. Int Arch Allergy Immunol 2007;144:10−22.

[70] Pastorello EA, Vieths S, Pravettoni V, Farioli L, Trambaioli C, Fortunato D, et al. Identification of hazelnut major allergens in sensitive patients with positive double-blind, placebo-controlled food challenge results. J Allergy Clin Immunol 2002;109: 563−70.

[71] Pastorello EA, Farioli L, Pravettoni V, Robino AM, Scibilia J, Fortunato D, et al. Lipid transfer protein and vicilin are important walnut allergens in patients not allergic to pollen. J Allergy Clin Immunol 2004;114:908−14.

[72] Krause S, Reese G, Randow S, Zennaro D, Quaratino D, Palazzo P, et al. Lipid transfer protein (Ara h 9) as a new peanut allergen relevant for a Mediterranean allergic population. J Allergy Clin Immunol 2009;124:771−8.

[73] Primavesi L, Brenna OV, Pompei C, Pravettoni V, Farioli L, Pastorello EA. Influence of cultivar and processing on cherry (*Prunus avium*) allergenicity. J Agric Food Chem 2006;54:9930−5.

[74] Asero R, Mistrello G, Roncarolo D, Amato S, Falagiani P. Analysis of the heat stability of lipid transfer protein from apple. J Allergy Clin Immunol 2003;112:1009−11.

[75] Sancho AI, Rigby NM, Zuidmeer L, Asero R, Mistrello G, Amato S, et al. The effect of thermal processing on the IgE reactivity of the non-specific lipid transfer protein from apple, Mal d 3. Allergy 2005;60:1262−8.

[76] Brenna O, Pompei C, Ortolani C, Pravettoni V, Farioli L, Pastorello EA. Technological processes to decrease the allergenicity of peach juice and nectar. J Agric Food Chem 2000;48:493−7.

[77] Schad SG, Trcka J, Vieths S, Scheurer S, Conti A, Brocker EB, et al. Wine anaphylaxis in a German patient: IgE-mediated allergy against a lipid transfer protein of grapes. Int Arch Appl Immunol 2005;136:159−64.

[78] Asero R, Mistrello G, Roncarolo D, Amato S, van Ree R. A case of allergy to beer showing cross-reactivity between lipid transfer proteins. Ann Allergy Asthma Immunol 2001;87:65−7.

[79] Curioni A, Santucci B, Cristaudo A, Canistraci C, Pietravalle M, Simonato B, et al. Urticaria from beer: an immediate hypersensitivity reaction due to a 10 kDa protein derived from barley. Clin Exp Allergy 1999;29:407−13.

[80] Garcia-Casado G, Crespo JF, Rodriguez J, Salcedo G. Isolation and characterization of barley lipid transfer protein and protein Z as beer allergens. J Allergy Clin Immunol 2001;108:647—9.

[81] Fernandez-Rivas M, Cuevas M. Peels of Rosaceae fruits have a higher allergenicity than pulps. Clin Exp Allergy 1999;29:1239—47.

[82] Hischenhuber C, Crevel R, Jarry B, Mäki M, Moneret-Vautrin DA, Romano A, et al. Review article: safe amounts of gluten for patients with wheat allergy or coeliac disease. Aliment Pharmacol Ther 2006;23:559—75.

[83] Tatham AS, Shewry PR. Allergens in wheat and related cereals. Clin Exp Allergy 2008;38:1712—26.

[84] Matsuo H, Morita E, Tatham AS, Morimoto K, Horikawa T, Osuna H, et al. Identification of IgE-binding epitope in ω-5 gliadin, a major allergen in wheat-dependent exercise-induced anaphylaxis. J Biol Chem 2004;279:12135—40.

[85] Palosuo K, Alenius H, Varjonen E, Koivuluhta M, Mikkola J, Keskinen H, et al. A novel wheat gliadin as a cause of exercise-induced anaphylaxis. J Allergy Clin Immunol 1999;103. 917-917.

[86] Palosuo K, Varjonen E, Kekki OM, Klemola T, Kalkkinen N, Alenius H, et al. Wheat omega-5 gliadin is a major allergen in children with immediate allergy to ingested wheat. J Allergy Clin Immunol 2001;108:634—8.

[87] Battais F, Pineau F, Popineau Y, Aparicio C, Kanny G, Guerin L, et al. Food allergy to wheat: identification of immunoglobulin E and immunoglobulin G-binding proteins with sequential extracts and purified proteins from wheat flour. Clin Exp Allergy 2003;33:962—70.

[88] Akagawa M, Handoyo T, Ishii T, Kumazawa S, Morita N, Suyama K. Proteomic analysis of wheat flour allergens. J Agric Food Chem 2007;55:6863—70.

[89] James JM, Sixbey JP, Helm RM, Bannon GA, Burks AW. Wheat α-amylase inhibitor: a second route of allergic sensitization. J Allergy Clin Immunol 1997;99:239—44.

[90] Nakayama MK, Iwamoto M, Shibata R, Sato M, Imaizumi K. CM3, one of the wheat α-amylase inhibitor subunits and binding of IgE in sera from Japanese with atopic dermatitis related to wheat. Food Chem Toxicol 2000;38:179—85.

[91] Nakase M, Adachi T, Urisu A, Miyashita T, Alvarez AM, Nagasaka S, et al. Rice (Oryza sativa L.) alpha amylase inhibitors of 14—16 kDa are potential allergens and products of a multi gene family. J Agric Food Chem 1996;44:2624—8.

[92] Simonato B, Pasini G, Giannattasio M, D.B.Peruffo A, De Lazzari F, et al. Food allergy to wheat products: the effect of bread baking and in vitro digestion on wheat allergenic proteins. A study with bread dough, crumb and crust. J Agric Food Chem 2001;49:5668—73.

[93] Curioni A, Santucci B, Cristaudo A, Canistraci C, Pietravalle M, Simonato B, et al. Urticaria from beer: an immediate hypersensitivity reaction due to a 10 kDa protein derived from barley. Clin Exp Allergy 1999;29:407—13.

[94] Vanek-Krebitz M, Hoffmann-Sommergruber K, Laimer da Camara Machado M, Susani M, Ebner C, Kraft D, et al. Cloning and sequencing of Mal d 1, the major allergen from apple (Malus domestica), and its immunological relationship to Bet v 1, the major birch pollen allergen. Biochem Biophys Res Comm 1995;214:538—51.

[95] Gaier S, Marsh J, Oberhuber C, Rigby NM, Lovegrove A, Alessandri S, et al. Purification and structural stability of the peach allergens Pru p 1 and Pru p 3. Mol Nutrition Food Res 2008;52(Suppl 2):S220—9.

[96] Oberhuber O, Bulley SM, Ballmer-Weber BK, Bublin M, Gaier S, DeWitt AM, et al. Characterization of Bet v. 1 related allergens from kiwifruit relevant for patients with combined kiwifruit and birch pollen allergy. Mol Nutrition Food Res 2008;52(Suppl 2): S230—40.

[97] Bolhaar ST, Ree R, Bruijnzeel-Koomen CA, Knulst AC, Zuidmeer L. Allergy to jackfruit: a novel example of Bet v 1-related food allergy. Allergy 2004;59:1187—92.

[98] Bolhaar ST, van Ree R, Ma Y, Vieths S, Hoffmann-Sommergruber K, Knulst AC, et al. Severe allergy to sharon fruit caused by birch pollen. Int Arch Allergy Appl Immunol 2005;136:45—52.

[99] Breiteneder H, Hoffmann-Sommergruber K, O'Riordain G, Susani M, Ahorn H, Ebner C, et al. Molecular characterization of Api g 1, the major allergen of celery (*Apium graveolens*), and its immunological and structural relationships to a group of 17-kDa tree pollen allergens. Eur J Biochem 1995;233:484−9.

[100] Hoffmann-Sommergruber K, O'Riordain G, Ahorn H, Ebner C, Laimer Da Camara Machado M, Pühringer H, et al. Molecular characterization of Dau c 1, the Bet v 1 homologous protein from carrot and its cross-reactivity with Bet v 1 and Api g 1. Clin Exp Allergy 1999;29:840−7.

[101] Riecken S, Lindner B, Petersen A, Jappe U, Becker WM. Purification and characterization of natural Ara h 8, the Bet v 1 homologous allergen from peanut, provides a novel isoform. Biol Chem 2008;389:415−23.

[102] Mittag D, Vieths S, Vogel L, Becker WM, Rihs HP, Helbling A, et al. Soybean allergy in patients allergic to birch pollen: clinical investigation and molecular characterization of allergens. J Allergy Clin Immunol 2004;113:148−54.

[103] Neudecker P, Lehmann K, Nerkamp J, Haase T, Wangorsch A, Fötisch K, et al. Mutational epitope analysis of Pru av 1 and Api g 1, the major allergens of cherry (Prunus avium) and celery (*Apium graveolens*): correlating IgE reactivity with three-dimensional structure. Biochem J 2003;376:97−107.

[104] Bohle B, Zwolfer B, Heratizadeh A, Jahn-Schmid B, Antonia YD, Alter M, et al. Cooking birch pollen-related food: divergent consequences for IgE- and T cell-mediated reactivity in vitro and in vivo. J Allergy Clin Immunol 2006;118:242−9.

[105] Fiocchi A, Restani P, Bernardo L, Martelli A, Ballabio C, D'Auria E, et al. Tolerance of heat-treated kiwi by children with kiwifruit allergy. Pediatr Allergy Immunol 2004; 15:454−8.

[106] Besler M, Steinhart H, Paschke A. Stability of food allergens and allergenicity of processed foods. J Chromatogr B Biomed Sci Appl 2001;756:207−28.

[107] Jankiewicz A, Aulepp H, Baltes W, Bogl KW, Dehne LI, Zuberbier T, et al. Allergic sensitization to native and heated celery root in pollen-sensitive patients investigated by skin test and IgE binding. Int Arch Allergy Immunol 1996;111:268−78.

[108] Ballmer-Weber BK, Hoffmann A, Wuthrich B, Luttkopf D, Pompei C, Wangorsch A, Kästner M, et al. Influence of food processing on the allergenicity of celery: DBPCFC with celery spice and cooked celery in patients with celery allergy. Allergy 2002;57: 228−35.

[109] Gomez M, Curiel G, Mendez J, Rodriguez M, Moneo I. Hypersensitivity to carrot associated with specific IgE to grass and tree pollens. Allergy 1996;51:425−9.

[110] Germini A, Paschke A, Marchelli R. Preliminary studies on the effect of processing on the IgE reactivity tomato products. J Sci Food Agric 2007;87:660−7.

[111] Leung PSC, Chu KH, Chow WK, Ansari A, Bandea CI, Kwan HS, et al. Cloning, expression, and primary structure of metapenaeus-ensis tropomyosin, the major heat-stable shrimp allergen. J Allergy Clin Immunol 1994;94:882−90.

[112] Taylor SL. Molluscan shellfish allergy. Adv Food Nutr Res 2008;54:139−77.

[113] Nakamura A, Watanabe K, Ojima T, Ahn DH, Saeki H. Effect of Maillard reaction on allergenicity of scallop tropomyosin. J Agric Food Chem 2005;53:7559−64.

[114] Nakamura A, Sasaki F, Watanabe K, Ojima T, Ahn DH, Saeki H. Changes in allergenicity and digestibility of squid tropomyosin during the Maillard reaction with ribose. J Agric Food Chem 2006;54:9529−34.

[115] Liu GM, Cheng H, Nesbit JB, Su WJ, Cao MJ, Maleki SJ. Effects of boiling on the IgE-binding properties of tropomyosin of shrimp (*Litopenaeus vannamei*). J Food Sci 2010;75:T1−5.

[116] Bugajska-Schretter A, Grote M, Vangelista L, Valent P, Sperr WR, Rumpold H, et al. Purification, biochemical, and immunological characterisation of a major food allergen: different immunoglobulin E recognition of the apo- and calcium-bound forms of carp parvalbumin. Gut 2000;46:661−9.

[117] Griesmeier U, Vázquez-Cortés S, Bublin M, Radauer C, Ma Y, Briza P, et al. Expression levels of parvalbumins determine allergenicity of fish species. Allergy 2010;65: 191−8.

[119] Prausnitz C, Küstner H. Studien über die Ueberempfindlichkeit. Centbl Bakteriol 1921; 86:160—9.

[120] Bernhisel-Broadbent J, Strause D, Sampson HA. Fish hypersensitivity. II: Clinical relevance of altered fish allergenicity caused by various preparation methods. J Allergy Clin Immunol 1992;90:622—9.

[121] Wal JM. Bovine milk allergenicity. Ann Allergy Asthma Immunol 2004;93:2—11.

[122] Gjesing B, Osterballe O, Schwartz B, Wahn U, Lowenstein H. Allergen-specific IgE antibodies against antigenic components in cow's milk and milk substitutes. Allergy 1986;41:51—6.

[123] Norgaard A, Bernard H, Wal JM, Peltre G, Skov PS, Poulsen LK, et al. Allergenicity of individual cow's milk proteins in DBPCFC-positive milk allergic adults. J Allergy Clin Immunol 1996;97:237.

[124] Werfel T, Ahlers G, Schmidt P, Boeker M, Kapp A, Neumann C. Milk-responsive atopic dermatitis is associated with a casein-specific lymphocyte response in adolescent and adult patients. J Allergy Clin Immunol 1997;99:124—33.

[125] Paajanen L, Tuure T, Poussa T, Korpela R. No difference in symptoms during challenges with homogenized and unhomogenized cow's milk in subjects with subjective hypersensitivity to homogenized milk. J Dairy Res 2003;70:175—9.

[126] Virtanen T, Kinnunen T, Rytkönen-Nissinen M. Mammalian lipocalin allergens — insights into their enigmatic allergenicity. Clin Exp Allergy 2012;42:494—504.

[127] Ehn BM, Ekstrand B, Bengtsson U, Ahlstedt S. Modification of IgE binding during heat processing of the cow's milk allergen beta-lactoglobulin. J Agric Food Chem 2004;52:1398—403.

[128] Selo I, Negroni L, Yvon M, Peltre G, Wal JM. Allergy to bovine-lactoglobulin: specificity of human IgE using CNBr derived peptides. Int Archs Allergy Immunol 1998; 117:20—8.

[129] Corzo-Martínez M, Moreno FJ, Olano A, Villamiel M. Structural characterization of bovine beta-lactoglobulin-galactose/tagatose Maillard complexes by electrophoretic, chromatographic, and spectroscopic methods. J Agric Food Chem 2008;56:4244—52.

[130] Ikeda S, Fogeding E, Hardin C. Phospholipid/fatty acid induced secondary structural changes in beta-lactoglobulin during heat-induced gelation. J Agric Food Chem 2000; 48:605—10.

[131] Ehn BM, Allmere T, Telemo E, Bengtsson U, Ekstrand B. Modification of IgE binding to beta-lactoglobulin by fermentation and proteolysis of cow's milk. J Agric Food Chem 2005;53:3743—8.

[132] Haddad ZH, Kalra V, Verma S. IgE antibodies to peptic and peptic-tryptic digests of β-lactoglobulin: significance in food hypersensitivity. Ann Allergy 1979;42:368—363.

[133] Mine Y, Yang M. Recent advances in the understanding of egg allergens: basic, industrial, and clinical perspectives. J Agric Food Chem 2008;56:4874—900.

[134] Cooke SK, Sampson HA. Allergenic properties of ovomucoid in man. J Immunol 1997;159:2026—32.

[135] Urisu A, Ando H, Morita Y, Wada E, Yasaki T, Yamada K, et al. Allergenic activity of heated and ovomucoid-depleted egg white. J Allergy Clin Immunol 1997;100:171—6.

[136] Faeste CK, Løvberg KE, Lindvik H, Egaas E. Extractability, stability, and allergenicity of egg white proteins in differently heat-processed foods. J AOAC Int 2007;90:427—36.

[137] Palmer DJ, Gold MS, Makrides M. Effect of maternal egg consumption on breast milk ovalbumin concentration. Clin Exp Allergy 2008;38:1186—91.

[138] Sancho AI, Foxall R, Browne T, Dey R, Zuidmeer L, Marzban G, et al. Effect of postharvest storage on the expression of the apple allergen Mal d 1. J Agric Food Chem 2006;54:5917—23.

[139] Sancho AI, Foxall R, Rigby NM, Browne T, Zuidmeer L, van Ree R, Waldron KW, et al. Maturity and storage influence on the apple (*Malus domestica*) allergen Mal d 3, a nonspecific lipid transfer protein. J Agric Food Chem 2006;54:5098—104.

[140] Ciardiello MA, Giangrieco I, Tuppo L, Tamburrini M, Buccheri M, Palazzo P, et al. Influence of the natural ripening stage, cold storage, and ethylene treatment on the

protein and IgE-binding profiles of green and gold kiwi fruit extracts. J Agric Food Chem 2009;57:1565−71.

[141] Dube M, Zunker K, Neidhart S, Carle R, Steinhart H, Paschke A. Effect of technological processing on the allergenicity of mangos (*Mangifera indica* L.). J Agric Food Chem 2004;52:3938−45.

[142] Hoppe S, Neidhart S, Zunker K, Hutasingh P, Carle R, H Steinhart. The influences of cultivar and thermal processing on the allergenic potency of lychees (*Litchi chinensis* SONN.). Food Chem 2006;96:209−19.

[143] Johnson P, Philo M, Watson A, Mills ENC. Rapid fingerprinting of milk thermal processing history by intact protein mass spectrometry with non-denaturing chromatography. J Agric Food Chem 2011;59:12420−7.

[144] Nowak-Wegrzyn A, Bloom KA, Sicherer SH, Shreffler WG, Noone S, Wanich N, et al. Tolerance to extensively heated milk in children with cow's milk allergy. J Allergy Clin Immunol 2008;122:342−7.

[145] Faeste CK, Løvberg KE, Lindvik H, Egaas E. Extractability, stability, and allergenicity of egg white proteins in differently heat-processed foods. J AOAC Int 2007; 90:427−36.

[146] Eigenmann PA. Anaphylactic reactions to raw eggs after negative challenges with cooked eggs. J Allergy Clin Immunol 2000;105:587−8.

[147] Chung SY, Champagne ET. Association of end-product adducts with increased IgE binding of roasted peanuts. J Agric Food Chem 1999;47:5227−31.

[148] Maleki SJ, Chung SY, Champagne ET, Raufman JP. The effects of roasting on the allergenic properties of peanut proteins. J Allergy Clin Immunol 2000;106:763−8.

[149] Beyer K, Morrow E, Li XM, Bardina L, Bannon GA, Burks AW, et al. Effects of cooking methods on peanut allergenicity. J Allergy Clin Immunol 2001;107:1077−81.

[150] Takagi K, Teshima R, Okunuki H, Itoh S, Kawasaki N, Kawanishi T, et al. Kinetic analysis of pepsin digestion of chicken egg white ovomucoid and allergenic potential of pepsin fragments. Int Arch Allergy Immunol 2005;136:23−32.

[151] Husband FA, Aldick T, Van der Plancken I, Grauwet T, Hendrickx M, Skypala I, et al. High-pressure treatment reduces the immunoreactivity of the major allergens in apple and celeriac. Mol Nutr Food Res 2011;55:1087−95.

[152] Li ZX, Lin H, Cao LM, Jamil K. Impact of irradiation and thermal processing on the immunoreactivity of shrimp (*Penaeus vannamei*) proteins. J Sci Food Agric 2007;87: 951−6.

[153] Sinanoglou VJ, Batrinou A, Konteles S, Sflomos K. Microbial population, physicochemical quality, and allergenicity of molluscs and shrimp treated with cobalt-60 gamma radiation. J Food Prot 2007;70:958−66.

[154] Byun MW, Kim JH, Lee JW, Park JW, Hong CS, Kang IJ. Effects of gamma radiation on the conformational and antigenic properties of a heat-stable major allergen in brown shrimp. J Food Prot 2000;63:940−4.

[155] Kim MJ, Lee JW, Yook HS, Lee SY, Kim MC, Byun MW. Changes in the antigenic and immunoglobulin E-binding properties of hen's egg albumin with the combination of heat and gamma irradiation treatment. J Food Prot 2002;65:1192−5.

[156] Lee JW, Lee KY, Yook HS, Lee SY, Kim HY, Jo C, et al. Allergenicity of hen's egg ovomucoid gamma irradiated and heated under different pH conditions. J Food Prot 2002;65:1196−9.

Chapter | fifteen

Communication with Food Allergic Consumers: A Win-Win Experience

George E. Dunaif[1], Susan A. Baranowsky[2]

[1]*Food Safety & Technical Services, Grocery Manufacturers Association, Washington, DC, US*

[2]*Consumer Affairs, Campbell Soup Company, Camden, NJ, US*

CHAPTER OUTLINE

INTRODUCTION

It is estimated that approximately 3—4% of individuals in the United States have an IgE-mediated food allergy that requires strict avoidance of an offending substance [1—4]. This represents between 9,000,000 to 12,000,000 Americans who are afflicted with this medical condition and who must make their dietary choices carefully or suffer consequences of serious illness or even death. Worldwide, many additional millions have food allergies, although the exact numbers are impossible to estimate. In addition to the significant number of Americans directly afflicted by this condition, a large number of persons with various relationships to the food allergic individual, including, but not limited to, caregivers, such as parents, siblings, grandparents, teachers,

Risk Management for Food Allergy. http://dx.doi.org/10.1016/B978-0-12-381988-8.00015-4

and healthcare professionals, as well as friends and acquaintances, must also share in the responsibility for keeping these at-risk individuals safe.

What does that mean? In a practical sense it means that all of these individuals, from the allergy sufferer to the caregivers to the acquaintances, need accurate and reliable food allergen information. Avoidance of specific foods is their only option for living a healthy and safe life. In order to meet this requirement, it is critical that all interested parties have quick and ready access to essential and accurate food ingredient and allergen information. In general, food ingredient statements on the labels of processed foods act as the most common and important source of information for the food sensitive individuals and their caregivers. However, at times the food allergic individuals and/or their caregivers may require a deeper understanding of the extended supply chain in the production of food ingredients and finished food products, including the capability of tracing the production of foodstuffs from the farm gate, through the manufacturing facility, to the retail establishment, and all the way to the consumer's fork.

While on the surface this might seem to be a rather daunting task, it is in fact a tremendous benefit for the food allergic consumers and/or their caregivers and the packaged food industry. This presents a great opportunity for protecting the public health, building relationships, and ultimately enhancing brand loyalty.

Food companies actively seek ways to better connect with their consumers. Nothing is more important to consumers than protecting themselves or a loved one, and nothing is more important to a food company than protecting the health and well-being of their consumers. Not only is this the right thing to do, it also builds an honest and open dialog between the company and its loyal consumers. This relationship serves these consumers by providing them with the information that they require and the peace of mind that they need. For the food company, the rewards of this relationship are many, but not the least of these are satisfied consumers who demonstrate remarkable brand loyalty in their product choices. Clearly, this represents a win-win proposition for both the consumer and the food company.

Up to the last 15−20 years, the food industry paid little attention to food allergies. However, as consumer awareness of food allergies increased, the food industry also began to focus increased attention on this segment of consumers. While food industry awareness continues to grow, many food companies are highly aware and responsive to the needs of food allergic consumers.

Today the average consumer has many options in selecting those products that meet the dietary needs of their families. Those consumers concerned with food allergies must make these decisions carefully, as they can impact the health and well-being of their family. Companies have the opportunity to create a bond with these consumers that is based on trust, and this relationship can be everlasting. This is not a challenge that can be taken on by one individual within a company but must be woven into the culture from the people that craft our recipes, to the associates who operate and clean the equipment on the floor of the food manufacturing facility, to those company representatives that talk to our consumers every day. Those food companies that can consistently demonstrate this level of commitment will win in the marketplace.

START WITH THE BASIC FACT THAT EDUCATION AND SENSITIVITY TRAINING RELATED TO FOOD ALLERGIC CONSUMERS MUST BE HOLISTIC AND INCLUSIVE

Building a trusting relationship with consumers starts with company engagement and sensitivity to an issue, such as food allergens, that then leads to honest and open communication with the concerned consumer. Communication relative to food allergen issues must first be developed on an internal company basis. Then the external communication can be developed once the company is fully sensitized and aligned to the needs of the food allergen community.

The approach to internal food allergen training and education within a food manufacturing company must be holistic and all-inclusive. Food allergen training must be top down and bottom up and must touch, to some degree, all levels of a company over time. As with most types of technical training, one size cannot fit all. For example, the training that is given to senior level executives in Quality and Legal functions will be very different from the training for middle managers, such as the product development, quality and food safety staff, plant management, and line operators.

The first step in designing a successful allergen program is to identify all stakeholders who play a role in keeping consumers safe and developing content that helps each of these to understand the role they play. One possible visual aid would be four concentric rings (see Figure 15.1). The outermost ring consists of senior food safety and quality leaders in the company. Level 1 group needs to be educated about the general principles related to food allergen safety, compliance requirements, and the special needs of the food allergic individuals and their caregivers. The next level, represented by the number 2, comprises the professional staff that support the overall design, development, and execution of new products and the enhancement of existing products. This level 2 group would consist of various functions including, but not limited to, product development, quality and food safety staff, regulatory affairs and labeling, legal, supply chain, and related corporate or centralized functions. This group is typically composed of highly skilled subject matter experts and professionals who can exert a great deal of influence over the design, development, and eventual commercialization of a packaged food product. The level of technical and/or scientific detail that this level requires is much greater and more complex than that needed by the level 1 group. This is where you will find the company's most knowledgeable technical experts, who need continual and consistent access to real-time data, knowledge, and compliance requirements, often on a global basis. They might not know all the answers, but they have a keen ability to be connected to reliable sources of internal and external technical

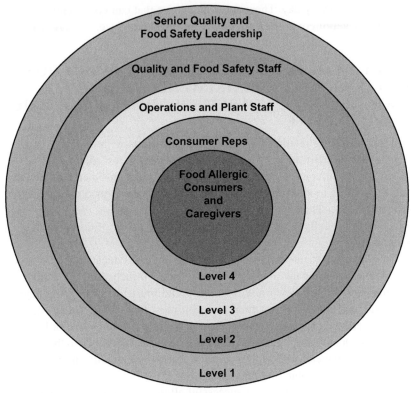

FIGURE 15.1

and regulatory information. Because of the complexity of the food chain, the food company will also be reliant on information from numerous suppliers, and those suppliers must also have considerable technical and scientific knowledge to assure overall success.

The third level consists of the manufacturing facility and production staff. This is a very key group of individuals, since they actually produce the products that are then shipped to the retail establishments for sale to the consumers. The modern food manufacturing facility is an impressive, complex beehive of activity. Ingredients arrive at the plant door in bulk and are then prepared, assembled, processed, and packaged on a daily basis. The knowledge and skills of these manufacturing plant professionals in all phases from ingredient preparation to processing to packaging to sanitation is paramount to the ongoing and consistent production of safe foods. In most cases, numerous products are made in a single facility, and manufacturing equipment is also often shared among various products.

The final level, number 4, encompasses all those responsible for providing the communication conduit between consumers and caregivers dealing with food allergies and the company. This level would include communication teams and also the contact center representatives who are responsible for the daily handling of consumer phone calls, letters, e-mail, and now even social media. Communication with consumers on the subject of food allergies can be particularly challenging considering the growing legislative and regulatory mandated food allergen labeling changes on a global basis and the ambiguities that can be created by conflicting food allergen information as well as the consumers' knowledge about this topic. What is most important is that the consumer has the most accurate information that is required to keep food allergic individuals healthy and safe.

The brand building opportunity is to develop and nurture your relationship with those consumers dealing with food allergies. Services and information targeted toward educating this segment of the population on food allergies as well as your company's position or activities in this area are key components in furthering this relationship. Moreover, by having an ongoing constructive dialog with these key consumers, you enable the company to continually learn and improve its overall offerings in this area. Consumers are very thoughtful and are eager to share their experiences and provide input and guidance on the types of information that they desire and which provides them with the most value when it comes to managing their food allergy challenges, be it for themselves or for others.

The final, and possibly most important, phase of a successful allergen training program focuses on the centermost point of the four concentric rings — services and communication provided by the company to their food allergic consumers and caregivers. On any given day, a consumer response representative can handle hundreds of consumer communications on a myriad of topics, but it will be those consumers contacting your company about food allergies that can have a lasting impact on your brand and the relationship with the consumer concerned with food allergies. All representatives should understand the controls, protocols, and practices put in place by your company to protect the food allergic consumer, as well as food labeling regulations and practices within your geographical area (e.g., a particular country) and your company. This requires extensive training, including regular refresher courses, to ensure that they are comfortable with the information when the food allergic consumer contacts your company. An important aspect of this education is sensitivity training and education. Although many of us know someone who has food allergies, not all people truly understand what it is like to walk in the shoes of an individual, parent, grandparent, or other caregiver who must carefully read labels for each food or beverage that touches his or her lips or the lips of his or her food allergic child.

The real and ongoing challenge is to maintain a heightened awareness and knowledge of information around food allergies. In order to accomplish this, some companies have developed interactive sessions designed to simulate the experiences and struggles of our food allergic consumers. Partnering with the Food Allergy and Anaphylaxis Network (FAAN), Campbell Soup

Company has continually educated its front-line representatives on the basics of food allergies as well as the rules and regulations affecting company products and labels. But even more importantly, Campbell has considered the real life challenges and situations that are faced by many individuals and families dealing with food allergies. In order to accomplish this type of learning, we simulated the experience of our consumers dealing with food allergies by hosting an allergy awareness luncheon that 'assigned' a food allergy to each participant, who was then asked to make food choices that avoided this allergen. Based on the feedback from those attending the session, this was a life-altering experience that taught the representatives the importance of demonstrating compassion and empathy along with overall knowledge of food allergies.

Several additional services designed to build loyalty and trust have also been a major component of the Campbell program. In order to reach a broad number of Campbell consumers, an allergen pledge was developed and posted across over 20 different corporate and Campbell brand websites. This pledge was designed to provide the consumer with company-specific information that empowers them to make informed dietary decisions about Campbell products. For those consumers who are newly diagnosed with food allergies or those needing additional education or support in dealing with food allergies, Campbell in conjunction with FAAN launched the FAAN Membership Program. This provided consumers or their families with a free six-month trial membership to FAAN. This membership includes a newsletter and website containing the latest information on food allergy research, practical tips, dietary advice, recipes, and legal news as well as a newsletter designed exclusively for children and special allergy alerts. To those children who are newly diagnosed with food allergies, Campbell sends a children's book stressing the importance of reading food labels.

Campbell also openly shared its experience with FAAN and the internal programs focused on the challenges of the food allergic consumers with many other leading food companies via our membership in the University of Nebraska Lincoln Food Allergy Resource and Research Program (FARRP) — a leading academic-industry research collaboration dedicated to sharing the best technical and scientific information related to food allergens.

Perhaps one of the most gratifying aspects of this ongoing food allergy commitment to education, communication, and training of our company representatives relates to the highly visible engagement and expressed satisfaction of the staff. Representatives take great personal pride in being able to provide these valued food allergic consumers and/or their caregivers with the information and options that they need to make informed purchasing decisions. The efforts of these company representatives and their training staff are clearly a good example of how to create or enhance the relationship with these important food allergic consumers by empowering all to maintain an open, helpful, and cooperative environment.

PUTTING THE PROGRAM INTO PRACTICE AND SELLING THE BENEFITS BEYOND YOUR COMPANY

The ability to consistently assure that all stakeholders benefit from the diligent efforts to both train and educate internal company stakeholders, as well as provide timely and helpful information to external stakeholders, including the food allergic consumers and/or their caregivers, is the key to success in this allergen communication model.

While all of the stakeholders typically have the best of intentions, it definitely takes a high level of dedication and diligence to maintain consistency and performance over the long term. This will require creative communication strategies in order to keep the message fresh and relevant for all stakeholders.

In fact, good, consistent internal and external communications on issues related to the food allergic consumer and/or their caregivers is a win-win for all involved; it is also the reason that there are benefits to all parties beyond the obvious for individual companies, as well as the food industry as a whole. Said another way, the positive relationship of individual companies with their food allergic consumers is not only an advantage to the individual company, building good will and strong brand loyalty, but also for the food industry as a whole. This is because the development of a strong internal management program combined with a consistent and sincerely empathetic position with the food allergic consumer and/or their caregivers will help to enhance the overall confidence of these important consumers in the safety of the food supply.

The food industry in general has worked diligently to assure the ongoing safety of these key and brand loyal consumers. The benefits to both the individual internal company stakeholders and the external food allergic consumers have been, and continue to be, a win-win model for food safety and building the high value consumer-company relationship.

This is clearly the right approach for all involved and is one that is put into action by many food companies on a daily basis. Each individual company may take a slightly different approach, as there is no right or wrong method. As long as the food allergic consumer and/or their caregivers get the information they require, the outcome will be a benefit for both these individuals and the company.

CONCLUSION AND SUMMARY

Food companies are not simply a corporation, they are at the heart a collection of skilled, caring, and dedicated individuals, people who have lives, families, and interests well beyond the four walls of the buildings in which they work every day. Not surprisingly, many of our associates themselves have had to face many of the challenges of the food allergic consumers, not only for themselves but for others. At Campbell, we live by our mission statement that 'Together we will build the world's most extraordinary food company by nourishing people's lives everywhere, every day.'

Campbell, and many other leading food companies, strongly values the health, happiness, and well-being of all of our consumers. This is expressed in many ways, including the energy and care that we put into developing and producing an array of great tasting, healthful, high quality, safe foods around the world. We recognize the special needs of our consumers and consistently make an effort to address them. Our long culinary heritage and dedication to quality and food safety are clearly part of what makes us engage in a company-wide effort to serve the food allergic community through education and training from the top of the organization to the manufacturing plant staff to the contact center representatives who serve in a long chain of professionals that keeps the needs of the food allergen consumers and their caregivers as a top of mind priority.

The rich and dynamic relationship with the food allergic consumer and/or their caregivers is for Campbell, as well as other like-minded companies within the food industry, a win-win proposition. The goal is to build consumer confidence through honest and consistent communications. Where possible, the food industry must offer viable alternative and sound information in order to assist these individuals in making informed and sound food purchasing decisions. This can result in the enhancement of consumer food choices and the building of brand loyalty, clearly a desirable proposition for all.

The food industry as a whole can be very proud of its long history of serving the needs of food allergic consumers and their caregivers. The industry has worked diligently and cooperatively to provide food allergic consumers with a wide array of safe, high quality, and wholesome food products. We must all continue to strive to improve our service to this important community and to further build and enhance our internal and external relationships and therefore the win-win experience for all stakeholders.

REFERENCES

[1] Liu AH, Jaramillo R, Sicherer SH, Wood RA, Bock SA, Burks AW, et al. National prevalence and risk factors for food allergy and relationship to asthma: results from the National Health and Nutrition Examination Survey 2005–2006. J Allergy Clin Immunol 2010;126:798–806.

[2] Sampson HA. Update on food allergy. J Allergy Clin Immunol 2004;113:805–19.

[3] Sicherer SH, Muñoz-Furlong A, Sampson HA. Prevalence of seafood allergy in the United States determined by a random telephone survey. J Allergy Clin Immunol 2004;14:159–65.

[4] Sicherer SH, Muñoz-Furlong A, Godbold JH, Sampson HA. US Prevalence of self-reported peanut, tree nut, and sesame allergy: 11-year follow-up. J Allergy Clin Immunol 2010;125:1233–6.

Chapter | sixteen

May Contain – To Use or Not?

Robin Sherlock[1], Steven L. Taylor[2], Sylvia Pfaff[3], Kirsten Grinter[4], René W.R. Crevel[5]

[1]*FACTA, Tennyson, Queensland, Australia*
[2]*Food Allergy Research & Resource Program, University of Nebraska, Lincoln, NE, US*
[3]*Food Information Service (FIS) Europe, Bad Bentheim, Germany*
[4]*Nestle Oceania, Rhodes, NSW, Australia*
[5]*Safety and Environmental Assurance Center, Unilever, Sharnbrook, Bedfordshire, UK*

CHAPTER OUTLINE

To be or not to be … Present

INTRODUCTION

Advisory or precautionary allergen labeling is a labeling statement that communicates to the consumer that the food has the potential to contain one or more allergens, which are not intentionally added as ingredients or declared as an ingredient. It is a mechanism to provide advice to the consumer that the product may not be suitable for them due to the possible presence or unintended inclusion of one or more allergens. The use of precautionary

261

or 'may contain' labeling has evolved from a practice that was originally perceived as helpful by allergic consumers to one that they now view with distrust because of the prolific use of the terms.

This chapter examines the reasons for this evolution and how the value of precautionary labeling may be restored. The authors consider that the way forward lies in designing clear, risk-based guidelines based on transparent and consistent science as a basis for the application of precautionary labeling that can restore its value as a valid risk communication strategy and at the same time make it usable as a valid risk management measure for industry. The chapter discusses the evolution of precautionary statements, the issues for industry, and the differing challenges for small to medium enterprises (SMEs) and those that face large companies with complex production sites and international compliance requirements.

HISTORY AND ORIGINS OF PRECAUTIONARY LABELING

The introduction of the requirement for mandatory labeling of allergens raised significant issues for the food industry globally, particularly in the absence of any general agreement on quantitative limits below which an allergen would be deemed to pose effectively no risk to public health and the paucity of data available to establish such limits. While it is relatively simple to accurately define the ingredients used in the production of a food, the situation within the manufacturing facility, and indeed along the whole food supply chain, is somewhat more complex. The unintentional inclusion of allergens from one food product run to the next, and the fact that shared equipment is the reality in most manufacturing facilities, has meant that development of allergen management protocols was necessary to mitigate the risks arising from the unintended presence of allergens.

The decision regarding when to label and when not to label for unintentional introduction of allergens has created great uncertainty for both the industry and the allergic consumer, particularly in the absence of good characterization of the hazard. Significant concern among clinicians and allergic consumers related to the lack of information around allergen thresholds and the initial inability of the food industry to respond definitively to these concerns by giving an assurance that a manufactured food could be considered to be free from risk for the allergic consumer. In response to the uncertainty, and out of the desire to manage the food safety risk and protect the allergic consumer, as well as the product or brand, the industry felt that it needed to communicate as much information as possible about the potential for allergen cross contact.

In an attempt to communicate clearly concerning the risks, two fundamental approaches were applied. The first included the development of a series of statements attempting to differentiate between degrees of risk, around the concept of 'may contain' as part of a manufacturing risk review. The second approach was to apply what is often referred to as 'blanket labeling', where an overarching precautionary label was applied whether a true risk existed or not. It was thought by industry to be an effective way to address cross contact impacts. Both approaches proved problematic. There was, and still is, an enormous range of statements, with precautionary labeling decisions based on different risk assessment protocols or risk judgments and a poorly defined risk hierarchy. In a survey of Australian products performed by the Allergen Bureau in 2009 [1], 53 cross contact statements were identified, which included a range of statements from 'may contain', 'may contain traces', and 'made on a manufacturing line which also processes' (Table 16.1). In some cases the precautionary

Table 16.1 Summary of Results Comparing Data from the Labeling Review Survey 2005 and 2009

Attributes Measured in Survey	2005 Labeling Survey Data		2009 Labeling Survey Data	
Count of different precautionary statement wordings*	31		34	
Count of 'May be present' precautionary statements	0		11	
Most frequently used precautionary statement wordings, in descending order of use (% of products with precautionary statements)	May contain traces of	35%	May contain traces of	38%
	May contain	9%	May contain	6%
	Manufactured on equipment that also processes products containing	8%	Manufactured on equipment that also processes products containing	6%
	Manufactured on equipment that processes products containing added	6%	Manufactured on equipment that processes products containing added	1%
	Manufactured on equipment that processes products containing	5%	Manufactured on equipment that processes products containing	8%
	May be present	0%	May be present	7%
	Manufactured on equipment that processes	5%	Manufactured on equipment that processes	3%
	Made on a production line that also processes products containing	5%	Made on a production line that also processes products containing	3%
	Made on equipment that also processes	3%	Made on equipment that also processes	1%
	Contain traces of	2%	Contain traces of	1%

In 2009, the Allergen Bureau conducted a study to review 340 packaged retail food products to gather information on the different declaration methods used for labeling allergens and allergen-related claims. This information was compared with the results from an Allergen Labeling Survey conducted by industry representatives in 2005.
*NB: The combined data from 2005 and 2009 surveys showed 53 different precautionary statements.
[Australian Food and Grocery Council (AFGC) Allergen Labeling Survey, (2005) (Unpublished).]

statements included references to components that were not allergens, and in one case, after listing all the allergens requiring mandatory declaration as possible cross contact risks, the statement included 'sorry'.

It is possible that a range of suppliers of the same or similar products will have a product with a variety of different precautionary labels saying anything from 'may contain' to 'made in a facility which also handles' to a detailed risk assessment statement. There is some evidence that consumers assign a different degree of risk to differently worded statements [1,2], which adds a layer of complexity to the issue when there is no consistency in application, as revealed in analytical studies.

These types of statements, and the proliferation of precautionary labels, caused significant loss of confidence in the minds of the allergic consumer. The introduction of mandatory allergen labeling had been intended to provide confidence for the consumer, but what actually resulted was a proliferation of foods that the consumer viewed as potentially unsafe due to the wide and often confusing application of precautionary labeling. If there was no precautionary statement present, they were unsure if that meant that the product was inherently safe or that the risk had been assessed and determined to be insignificant. The overuse of precautionary statements thus resulted in significant devaluation of the single most critical tool available to the allergic consumer for their own risk management when determining whether to consume a product. The loss of confidence spread beyond allergic consumers to health care professionals, and indeed was exacerbated by the lack of consistent advice from them in relation to observance of such warnings.

Where a precautionary statement was applied, it was difficult for the allergic consumer to identify its basis, and they did not have a mechanism to differentiate between the application of a statement applied as the result of a risk assessment and one applied by a company that was risk averse but not fully versed in allergen management. The outcome of the proliferation of precautionary statements has been a dilution of their effectiveness and an increasing lack of confidence in validity of the product label.

LEGAL AND REGULATORY STATUS OF PRECAUTIONARY ALLERGEN LABELING

Precautionary labeling is voluntary and has no clear status in law in any country in which it is used. For instance, European food law states that 'Food shall not be placed on the market if it is unsafe' and explains that in determining whether a food might be 'unsafe', 'regard shall be had: (b) to the information provided to the consumer, including information on the label, or other information generally available to the consumer concerning the avoidance of specific adverse health effects from a particular food or category of foods'. However, Recital 16 in the preamble to the Regulation states: 'Measures adopted by the Member States and the Community governing food and feed should generally be based on risk analysis except where this is not appropriate to the circumstances or the nature of the measure', which implies that they should be based on analysis of the risk and therefore a risk assessment. Precautionary labeling used without an appropriate risk assessment might also fall foul of the law's requirement that labeling is not misleading. The Australian and New Zealand Food Standards Code does not carry any guidance regarding precautionary statements, and use of precautionary statements are more likely to be considered under the Competition and Consumer Act and be limited by the requirement to convey accurate information to the consumer.

As a further example, authorities in Germany do not target different statements on packaged foods. However, there is a limitation if a producer uses 'may contain allergens according to the labeling regulation'. This is considered imprecise and will be result in a fine. Rarely, regulatory jurisdictions provide clear instructions against the use of precautionary statements. In Japan, precautionary labeling is contra-indicated, and where the levels of declarable allergens due to cross contact are below 10 ppm, no precautionary statement is allowed.

For further information on the legal and regulatory aspects of labeling, including precautionary labeling, please refer to chapter 17.

Allergic Consumers' Understanding and Perception

The level of awareness of allergic consumers around the use and significance of precautionary labeling varies quite considerably. The perception of risk is often highly individual, and the maturity of the allergic consumer's knowledge base varies from country to country and will often depend on their association or otherwise with key stakeholder groups like allergy associations or allergy clinics [3]. In some countries, this is based around the introduction of regulations and in some cases is based simply on the amount of information that the support groups make available to the allergic consumers and the type of information and interaction that they have with industry organizations. In Germany in 2011, the patient organization DAAB (Deutscher Allergie- und Asthma Bund) launched a roundtable with the food industry to discuss practical allergen management within factories and how best to inform allergic consumers. Both sides continue to benefit from this joint initiative.

A survey in Australia showed that consumers were confused by the range of precautionary labeling possibilities [3], while an informal survey performed through Allergy New Zealand in 2008 indicated that New Zealand allergic consumers were seeking more detailed information in order to make an informed choice. Within many companies there was a perception that statements like 'made on the same line' or 'made in the same facility' conveyed meaning to the consumer, although it is difficult to find empirical evidence to support this assertion. Analytical surveys comparing products with different precautionary statements have found no correlation between the wording of the statement and the actual allergen content [4], although sample sizes have been very limited, making those studies qualitative rather than quantitative. Most allergic consumers do not belong to support groups or consumer organizations, despite in some cases being seriously affected by food allergies. It is therefore extremely important that any precautionary labeling statement reflects a clear and consistent message, which is underpinned by an effective food safety management program. It continues to be important that the food industry works closely with allergy stakeholder groups. Initially, this close association was to share appropriate allergen food safety management practices and help define labeling communication, but ongoing it needs to ensure the effectiveness of communication and to continue to understand the allergic consumer's needs.

Recently published research confirms more anecdotal observations. In a qualitative study, Barnett and colleagues [5] reported considerable confusion about the validity of precautionary labeling, as well as a considerable incidence of non-observance. This picture mirrored that described by Cochrane et al. [6], who found that only 40−50% of allergic consumers, or those buying food for allergic consumers, always respected precautionary labeling statements. While this figure related to the totality of allergic consumers, a high proportion of people who had experienced severe reactions exhibited the same behavior, indicating a significant degree of risk-taking. This may be contributed to by the issue of inhomogeneous contamination in the food production. As an example, the cross contact of milk in chocolate production may be huge (ca. 3000 ppm) in the first lot of the following product. It drops to smaller amounts within the production of the 'milk-free' chocolate. An allergic consumer is not able to identify the batch, and if he ignores the precautionary statement and is fortunate, he will not experience an adverse reaction because of the low allergen amount. It is likely he will continue to buy this chocolate based on his subjective experience and growing confidence in the brand (i.e., his experience will color his risk assessment). However he remains at risk.

The confusion about precautionary labeling and its safety implications extends to the clinical professions involved in providing care and advice to allergic people [4]. As a result, advice about how to take precautionary labeling into account when deciding whether to buy or consume a particular product differs among clinicians and relies more on heuristics than on evidence. Thus some assert that 'may contain' can be safely ignored, while others maintain equally forcefully that products with such labels should be stringently avoided. Of course, in the absence of a common, clear, and well-understood basis for precautionary labeling, such a range of opinion should hardly surprise.

Developing and Implementing Evidence-Based Precautionary Statements

CHALLENGES FOR INDUSTRY

The appropriate use of a 'may contain' or precautionary statement has been poorly circumscribed. In the absence of agreement over the amount of an allergen posing a risk to the allergic consumer, the food industry considered that it was essential to provide as much information as possible in order for the consumer to make an educated decision. In some instances however, this has become of little value as the approach to conservatively apply an overarching blanket label proliferated throughout the industry with the result being very high proportions of certain products being labeled, while precautionary statements sometimes appeared on the most unexpected products (e.g., bottled water). There is a lack of consistency from company to company, and sometimes even from production site to production site within companies. There is also a significant gap between the larger companies, who are aware of the implications and have access to quality control and quality management teams, and small to medium-sized companies, who may have limited access to relevant information and often have insufficient means to manage compliance with regulatory requirements of mandatory declaration for allergens, and potentially less capacity to deal with issues associated with cross contact and precautionary labeling.

The situation is particularly complex in the context of SMEs, where resources are often lacking to deal with complex food safety issues. In many SMEs, the individual who is responsible for decisions associated with labeling will likely perform many roles within the organization, including quality control, production, business, and most likely marketing manager. He or she may not have access to support or information to make informed decisions and may not be sure where to access such information. In addition he or she may be unsure as to what amount of cross contact from one production run to the next is enough to be considered a hazard for the allergic consumer. This is related to the difficulty in accessing data related to levels causing reactions and the level below which one could be confident an allergic consumer would not react.

As SMEs are often suppliers to large companies, their lack of resources to manage allergens optimally, reflected in extensive use of precautionary statements, affects the whole food supply chain. In addition, SMEs often produce specialized and niche market based products with highly specialized ingredients. The availability of these ingredients may be limited, and the success of the allergen management of SMEs is integrally linked to the effectiveness of the allergen management strategies of the ingredient suppliers. Where alternative suppliers are not available, SMEs may make the decision to continue with suppliers, using blanket statements and carrying those statements over to their own product in the absence of clear criteria for the application of advisory statements.

In Germany over 5900 food companies were registered in 2011. Over 90% were SMEs. During the same period around 3000 International Featured Standards (IFS) Food Certificates were issued in Germany. The IFS Food Certificate is mandatory for retail food businesses in Germany. We can therefore assume that almost every second company is following this standard. One section of the standard is dedicated to allergen management. Therefore every certified company is aware of the issue and fulfills the requirements in a certain way. A critical limitation is the knowledge of the auditors. If they are not trained according to a consistent syllabus and standard, then assessment will rely on the individual auditor's understanding of the issues. This issue highlights the difficulty for many professions that provide services to the food industry. Consistency of risk assessment is a critical issue, but monitoring the application of risk assessment and compliance is also significant and needs to be benchmarked against parameters that are aligned with a recognized risk assessment strategy and protocol.

In many cases there is little real understanding of what level might be a sufficient trigger for a consumer to experience an allergic reaction. This issue has in part been contributed to by the clinicians and regulators, who felt that there was insufficient data to support the establishment of management thresholds and a disproportionate representation of the consumer as being acutely or exquisitely allergic. Further studies have shown that the vast majority of allergic consumers would potentially react to levels well above those that can be readily managed in a well-run food business [7—9].

Allergenic foods differ in their allergenic potency [10] (i.e., the doses that are predicted to provoke reactions in a defined proportion of the susceptible population vary), and the implications of this observation are quite significant. It is necessary to understand whether industry should apply the same rules for every allergen or whether management should vary from allergen to allergen. Initially, industry concerned itself with the most critical issue of what amount of allergen was likely to pose a danger to the allergic consumer, in the sense of provoking a severe (or fatal) reaction. Other factors also contributed to lack of progress including, in some cases, a lack of confidence in intervention methodology and a lack of understanding of the degree of control required to produce an appropriate reduction in risk and a satisfactory labeling outcome. In some cases it was clear that there was a lack of willingness to make significant changes to allergen management processes, as the industry was unsure what the best or most appropriate response to the risks should be. The precautionary statement seemed a reasonable and overt mechanism of indicating product allergen risk.

If we consider the aspect of judicious and consistent use as the appropriate way to communicate allergen risk, then we need to reflect on the fact that there is variation, company to company, site to site, line to line, where the allergen risk within a facility can vary enormously. A decision to apply a precautionary label to a particular product or product portfolio becomes quite complex. However a process that ensures that industry applies the same approach and the same standards across the board would be clearly preferable to the existing situation.

APPROPRIATE APPLICATIONS

It is necessary to provide guidance, as a minimum measure of support, for industry to ensure that the application of a precautionary label is a result of a consistent and scientifically based risk assessment, the parameters of which are clearly defined and internationally acceptable. It also becomes essential to provide supporting communication to the consumer, to indicate that the use of the precautionary statement is a clear indication of potential risk arrived at after a thorough risk assessment and not a blanket statement intended to protect the company and not the consumer.

HOW TO DETERMINE THE TOLERABLE LEVEL OF RISK

As industry struggled to deal with the uncertainty of no universally accepted levels, they began to develop their own internal individual thresholds, or risk comfort levels, and made labeling decisions based on this. This had two outcomes: one was associated with the fact that the consumer was uninformed as to the terms of the company's application of risk assessment and therefore had no way to gauge the accuracy or otherwise of that statement. The other issue was that many large companies felt that they were making significant efforts working towards the control mechanisms but that SMEs were more likely perhaps to be risk factors and were not applying process controls or investing in addressing the problem. This aspect has profound implications for SME operations, since in many countries those very same SMEs form an important component of their supply chain, with whom a common understanding of allergen risks is essential to effective management. The complexity of the food industry, where small companies provide raw materials to medium-sized companies, which may be under different regulatory jurisdictions, that then supply to larger international companies contributes to the lack of clarity surrounding precautionary labels. The flow-on effect of blanket type precautionary statements is particularly confusing in this context, where the result is a product, many steps removed from any real risk of allergen cross contact, which carries an unjustified and unverifiable precautionary label.

VITAL AND A THRESHOLD-BASED APPROACH TO RISK ASSESSMENT

The introduction of the requirement for mandatory allergen labeling meant that many countries developed guidelines for allergen risk assessment and management to assist with the complex issue of allergen management and precautionary labeling. Most of these guidelines, while providing significant assistance, did not define the quantitative level of acceptable risk that might be considered when choosing whether or not to apply a precautionary label. One program that differed from the rest in this respect was the VITAL (Voluntary Incidental Trace Allergen Labeling) program developed by the Allergen Bureau, an industry-based organization originally established with the objective of centralizing intelligence concerning allergen management and providing consistent information to all members of the Australian and New Zealand food industry.

The Allergen Bureau established a standardized risk assessment protocol for the determination of potential allergen cross contact and set parameters for acceptable levels of potential cross contact. The VITAL program was developed to provide a systematic risk-based approach for food manufacturers to assess the impact of allergen cross contact and assist in assessing the appropriateness of applying a precautionary labeling statement, and it is currently the most evolved system of its kind. The objective is to provide manufactured food that is safe for the vast majority of food allergic consumers by providing consistent food labels that declare the presence of allergens due to documented, unavoidable, and sporadic cross contact. The labeling outcomes are based on action levels that are linked to the potential for allergic consumers to experience adverse reactions.

The scientific substantiation of the original action level approach was based predominantly on the work completed by the US FDA Threshold Working Group [10] with added uncertainty factors applied. However, new allergen threshold data must always be considered, and with that in mind the Allergen Bureau recognized a need to form a scientific expert panel, known as the VITAL Scientific Expert Panel (VSEP), to review the science underpinning the VITAL Action Level Grid. The scientific review was a critical body of work to ensure that the action levels protect the allergic consumer by enabling industry to make appropriate precautionary labeling decisions. The VSEP is a fundamental part of the Allergen Bureau program, which will ensure that the reference dose levels used to obtain the action level for labeling will remain based on transparent science to determine appropriate threshold data through periodic review and updating.

The importance of VITAL as a risk assessment tool has less perhaps to do with the actual numbers that were used to provide guidelines on labeling and more to do with a unified and open approach to the risk assessment process. By asking companies to review their allergen handling practices and to consider the implications of their production schedules and cleaning practices in order to make quantitative assessment concerning cross contact risk, it contributed to a shift in perspective from blanket statements to science-based determinations of cross contact impact. The decision to make the risk assessment tool, and the scientific arguments supporting it, available to all members of the food industry, local and international, member and non-member, was an intentional choice to provide resources for risk assessment to all members of the food industry irrespective of their size or financial and scientific resources. It was an intentional decision to ensure that poorly resourced companies would have the opportunity to access tools that were consistent with the science available at the time and both comparable and compatible with those being applied by larger facilities.

By providing a quantifiable framework, it gave industry a goal to work towards and an outcome that could be assessed by calculation and verified analytically.

The appropriate application of VITAL should include its incorporation as part of a Hazard Analysis Critical Control Point (HACCP)-based food safety program. The VITAL program consists of a support guideline and process to follow when performing a risk review, a decision tree and a calculator to assist in recording assumptions, and information that should be retained during the review, thereby aiding traceability and documenting the process and also demonstrating due diligence. The program provides a structured approach to examining the risks associated with ingredient and processing impact and the manufacturing process impact to enable both a complete picture of the risk to be obtained and an appropriate precautionary label decision to be made. VITAL uses an action level grid to assist in determining whether the presence of residual protein from allergenic substances through unavoidable cross contact requires a precautionary labeling statement. The VITAL Action Level Grid is a key component of this program and is incorporated into the VITAL Calculator. The action level concentrations are determined using the reference dose information (set by the VSEP) in conjunction with the associated Reference Amount/Serving Size. The VITAL Action Levels Grid (incorporated into the VITAL Calculator) should be used in conjunction with this document.

Further details on the VITAL process and its application are available on the Allergen Bureau website (http://allergenbureau.net, accessed 27 August 2013)

THE FUTURE OF PRECAUTIONARY STATEMENTS

The debasement of the value of the advisory label means that precautionary statements are losing their impact. Industry needs to carefully consider the use of these types of labels and effective communication strategies. The clarity of communication is critical to the success of any advisory statement as a risk assessment tool for the consumer. Clear messages regarding the significance of a precautionary statement and the importance of avoidance of a product that carries an advisory statement are essential.

Meanwhile, many jurisdictions are working towards the establishment of thresholds or action levels in order to provide guidance regarding precautionary statements. While these investigations and discussions take place, industry continues to provide food to the allergic consumer and must make decisions based on the best available science at the time.

REFERENCES

[1] Allergen Bureau. Labeling Review Survey; 2009. www.allergenbureau.net/downloads/vital/Labelling-Review-Survey-2009.pdf. Accessed 28 August 2013.

[2] Noimark L, Gardner J, Warner JO. Parents' attitudes when purchasing products for children with nut allergy: a UK perspective. Pediatr Allergy Immunol 2009;20(5): 500−4.

[3] Food Standards Australia New Zealand. Food labelling issues − Quantitative survey on allergen labelling: benchmark survey; 2003. 2004.

[4] Turner PJ, Kemp AS, Campbell DE. Advisory food labels: consumers with allergies need more than 'traces' of information. BMJ 2011;343:d6180.

[5] Barnett J, Muncer K, Leftwich J, Shepherd R, Raats MM, Gowland MH, et al. Using 'may contain' labelling to inform food choice: a qualitative study of nut allergic consumers. BMC Public Health 2011;11:734.

[6] Cochrane SA, Gowland MH, Sheffield D, Crevel RWR. Clinical and Translational Allergy. Characteristics and purchasing behaviours of food-allergic consumers and those who buy food for them in Great Britain; 2013.

[7] Blom WM, Vlieg-Borstra BJKAGvdHS, Houben GF, Dubois AEJ. Threshold dose distributions for 5 major allergenic foods in children. J Allergy Clin Immunol 2013; 131:172—9.

[8] Taylor SL, Crevel RW, Sheffield D, Kabourek J, Baumert J. Threshold dose for peanut: risk characterization based upon published results from challenges of peanut-allergic individuals. Food Chem Toxicol 2009;47(6):1198—204.

[9] Taylor SL, Moneret-Vautrin DA, Crevel RW, Sheffield D, Morisset M, Dumont P, et al. Threshold dose for peanut: risk characterization based upon diagnostic oral challenge of a series of 286 peanut-allergic individuals. Food Chem Toxicol 2010;48(3):814—9.

[10] Buchanan R, Dennis S, Gendel S, Acheson D, Assimon SA, Beru N, et al. Approaches to establish thresholds for major food allergens and for gluten in food. J Food Prot 2008;71(5):1043—88.

Chapter | seventeen

Regulatory Controls for Food Allergens

Sue Hattersley[1], Rachel Ward[2]

[1]*Food Standards Agency, London, UK*
[2]*r.ward Consultancy Limited, Nottingham, UK*

CHAPTER OUTLINE

INTRODUCTION

Any foodstuff that contains protein has the potential to elicit an allergic reaction in a consumer sensitized to that particular protein. Foods reported in the scientific literature to provoke allergic reactions are very diverse. They range from animal-derived products, for example meat, fish, shellfish, milk, and egg, to non-animal-derived foods, such as tree nuts, seeds, legumes (especially peanuts and soy-beans), cereals (especially wheat and buckwheat), and fruit (e.g., peaches, apples, kiwi, tomatoes) and vegetables, especially those in the Umbelliferae family such as celery and carrot, [1,2,3,4,5]. Several protein-containing food additives have also been documented to provoke IgE-mediated allergic reactions, for example cochineal extracts and carmine [6].

Risk Management for Food Allergy. http://dx.doi.org/10.1016/B978-0-12-381988-8.00017-8

Other non-allergic hypersensitivity reactions can occur to foodstuffs, which are often termed 'food intolerance'. Sulfite intolerance is well-known, especially in people with asthma. Sulfites at levels $> 10\,mg/kg$ are therefore required to be declared on pre-packed foods in the Codex standard on labeling [7]. Lactose intolerance occurs as a result of having insufficient lactase enzyme to digest lactose when consumed. Celiac disease is an allergic non-IgE-mediated reaction to gluten-containing cereals (wheat, rye, and barley), and it is also associated with autoimmune and inflammatory skin diseases such as psoriasis, alopecia areata, and bullous pemphigoid [8].

The vast majority of IgE-mediated food allergies worldwide have been attributed to a small number of allergenic foods: milk, eggs, fish, crustacea (for example shrimp, crab, lobster), peanuts, soybeans, tree nuts (for example almonds, walnuts, pecans, cashews, Brazil nuts, pistachios, hazelnuts, pine nuts, and macadamia nuts), and cereals containing gluten, most specifically wheat [1,9] (see also chapter 2). Consumers with food allergies need to know what is in their food so that they can make safe and informed food choices.

HOW BEST TO PROTECT THE PUBLIC AT RISK FROM ADVERSE REACTIONS TO THE MAJOR FOOD ALLERGENS?

There is still no cure or effective medical intervention available to prevent food allergic or food intolerance reactions. It is not appropriate to ban allergenic foods as they derive from major food groups, providing good sources of nutrients, and are consumed safely without reactions by the majority of the population. The only proven treatment option for sensitive individuals is for them to avoid exposure to the provoking substance. Avoidance requires knowledge of the presence of allergenic foods in any and all foods that an allergic individual might consume. In addition, other non-food sources of the allergen can provide sufficient exposure to provoke serious adverse reactions and must also be avoided (e.g., almond oil in cosmetics, lactose and nut/seed oils used as vehicles for oral medicines, peanut oil used in engineering as a lubricant).

The undeclared presence of an allergenic food of public health importance in a foodstuff would constitute a risk to allergic consumers. Therefore, communication through allergen labeling has become the cornerstone in public health protection from allergic and intolerance reactions. The key principle is to provide risk communication that is equivalent to the relevant allergenic composition of foods.

In pre-packed foodstuffs, allergen presence can be effectively communicated by on-pack labeling. In foods sold loose this information can be available at point of sale, and in catering recipe lists could be available on demand for customers. This has proven effective in risk protection when the information is clear, accurate, and up-to-date, and when consumers have a good understanding of the foods to avoid and take notice of information provided.

In order for food producers to provide accurate and up-to-date provision of information on allergen presence, they have to have an understanding that allergenic foods present a hazard and which of these allergenic foods, including all derivatives from these allergenic foods, require risk management. Food operators should have procedures to manage allergen presence built into their Hazard Analysis and Critical Control Points (HACCP), good manufacturing practices (GMP), and product information systems so that allergen declarations are transferred precisely along the food chain.

Foods manufactured under GMP, with capable integrated allergen risk management and accurate allergen declarations, should be expected to be acceptable for human consumption according to their intended use and be well tolerated by allergic consumers [10].

WHY DO WE NEED SPECIFIC ALLERGEN LABELING RULES?

Avoidance of those foodstuffs to which they are sensitive is the key to allergen risk management for consumers. It is therefore important that any significant presence of allergenic foods is distinctly and consistently communicated to allergic consumers. There are several essential elements necessary to make this allergen risk communication effective. Firstly, consumers need to have a clear understanding of the foods they should avoid through access to expert clinical diagnosis and dietary advice. For example, it has been shown that the majority of consumers allergic to peanuts can have anaphylactic reactions to lupin products. Such consumers and their clinicians are often unaware that they should therefore also avoid lupin as a result of this cross reactivity [11,12].

The second element is the clarity of any allergen declarations accompanying food products with respect to their composition. General food labeling requirements will usually provide information about the presence of major ingredients in any recipe, including the major allergenic foods and other less prevalent allergenic ingredients including foods such as apple, banana, carrot, apricot, kiwi, sunflower seeds, and poppy seeds. Historically the widespread use of a general exemption for the declaration of ingredients below 5% or 25% of the recipe meant there could be a significant amount of allergenic derivatives present but undeclared. Furthermore, processing aids and carryover additives that no longer have a function in the finished product are frequently exempted from food labeling requirements, and some food groups can be declared using generic terms such as vegetable oils or starch. Consumers would not know or expect that certain allergenic derivatives are used in food, for example lupin as an alternative to soy in tofu and as a protein source and texture improver in baking or soy as a replacement in many meat and dairy analogs. General labeling requirements alone will not provide assurance that allergenic derivatives present are clearly declared. Mandatory labeling of intentionally present allergenic derivatives, whether ingredients, additives, or processing aids, has been put in place in many jurisdictions to ensure that food allergens are labeled, even when present at very low levels.

This provides a good level of protection provided the labeling is accurate. However, the use of technical or ambiguous terms, or terms unfamiliar to the local population, can render labeling ineffective in communicating the presence of the allergenic food to at-risk consumers. For example, ingredients in unfamiliar or exotic foods may not be recognized (e.g., hummus, halva, and tahini all contain sesame and couscous is made from wheat) and the use of unfamiliar or technical names for ingredients may confuse (e.g., ground nuts is another name for peanuts, edamame is a type of soybean, sodium caseinates and whey are fractions isolated from milk) [13,14]. To further avoid confusion, consistency between languages is important for multi-language packs to ensure that the same allergen declarations are made for each language.

Allergic consumers will often use the pictures or images and product names on a pack to interpret whether the food is safe to consume. 'Hidden allergens' or unforeseen allergenic ingredients in products can unfortunately be missed, despite clear labeling on a pack and good consumer understanding as to which foods to avoid. For example apple and pear products are often used as a natural sweetener in many foodstuffs, and lupin flour is now frequently used in baking. If consumers would not expect the allergen to be present in this product type, additional emphasis of allergen presence would be helpful. The primary and reliable source of risk communication is the ingredients declaration [15,14]. Allergic consumers have to take personal responsibility for their allergy management and will therefore need to regularly and routinely check for the declared presence of allergens as this can vary due to new ingredient usage, recipe changes, or from brand to brand of similar products.

CODEX

Following the FAO Technical Consultation on Food Allergies in 1995, the Codex Committee on Food Labeling [16] issued a list of eight allergenic foods known to cause allergy that were deemed to be of the greatest public health concern. These were cereals containing gluten (i.e., wheat, rye, barley, oats, spelt, or their hybridized strains and products of these); crustacea and products of these; eggs and egg products; fish and fish products; peanuts; soybeans and products of these; milk and milk products (lactose included); and tree nuts and nut products. Protection of allergic consumers from inadvertent consumption of these major allergenic foods was recommended to be carried out through risk communication. This advice was implemented through the Codex General Standard for the Labeling of Pre-Packaged Foods, which states that they shall always be declared, along with any presence of sulfites in concentrations of 10 mg/kg or more [7].

As with other Codex standards, the General Standard for Labeling provides a robust global standard for food and it has been implemented widely, forming the core of many countries' food labeling legislation. The standard does include some exemptions from general labeling, but these exemptions do not apply to the allergenic foods and their derivatives listed in section 4.2.1.4 of the standard. Other allergenic foods not on this Codex list that have less global prevalence, or locally specific relevance only, would be affected by these exemptions. For example, there is a general exemption for compound ingredients constituting less than 5% of the food, the component ingredients need not be declared; where herbs or spices constitute less than 2% of the food they can be declared simply as 'herbs' or 'spices'; food additives carried over and no longer having a technological function in the food and processing aids need not be declared. Other allergenic foods not on this Codex list would come within these general exemptions as they have less global prevalence. Considerations of local food allergy prevalence would be important to assure mandatory allergen labeling lists are relevant for local public health protection.

EUROPEAN LEGISLATION REGARDING FOOD ALLERGEN RISK COMMUNICATION

In the EU, allergen risk management is provided through a combination of food safety and labeling regulatory requirements. Regulation 178/2002 General Food Law [10] provides the overarching supporting principles for food safety, which implicitly includes the risk management of food allergens. For example, Article 14 specifies that determination of whether a food is injurious to health or unsafe should take into account the 'particular health sensitivities of a specific category of consumers', and whether risk communication 'information (is) provided to the consumer … concerning the avoidance of specific adverse health effects from a particular food or category of foods'. In Regulation 852/2004 [17] on the hygiene of foodstuffs, general rules are defined to control hazards and to ensure that a foodstuff is fit for human consumption while taking into account its intended use. The mandated HACCP approach requires identification of any hazards that must be prevented, eliminated, or reduced to acceptable levels. Regulation 852/2004 does not explicitly name the hazards of concern and will therefore require that food business operators understand that allergens need risk management and know which food allergens are of public health importance. Without this awareness, accurate provision of allergen declarations will not be possible down the food chain.

Historically in the EU Member States, general food labeling Directive 2000/13/EC [18] contained a number of exemptions, which meant that some ingredients were not required to be labeled. For example, components of compound ingredients that made up less than 25% of the final food did not need to be specifically declared. Additionally the source of some ingredients, such as flour, did not need to be stated.

It was recognized that there were situations under the general ingredients labeling legislative requirements in Directive 2000/13/EC where the allergic consumer would not necessarily receive sufficient information. This deficit was addressed by developing specific requirements that would require the clear declaration of the use of allergenic ingredients in pre-packed foods in all circumstances. This requirement was provided by Directive 2003/89/EC [19], which came into effect in November 2005. This legislation amended the parent Directive 2000/13/EC governing general food labeling and therefore covers only the deliberate use of the ingredient and relates only to foods sold pre-packed.

The list of allergenic foods that had to be clearly declared that was developed by the EU was based on the Codex list and scientific evidence that identified that a further three allergenic foods (sesame seeds, mustard, and celery) were a public health concern in at least some EU Member States [19,4]. Subsequently a further two allergenic foods (molluscs and lupin) were added to the EU list by Directive 2006/142/EC [20] on the grounds that there was evidence that these were a public health concern in EU Member States [21,22,23]. The current EU labeling list of specified allergenic foods is listed in Table 17.1. Further allergenic foods can be added to the EU list in the future if there is sufficient scientific justification.

Exemptions for Certain Processed Ingredients Derived from the Specified Allergenic Foods

It is recognized that some ingredients derived from the specified allergenic foods would in practice not present an allergenic risk due to the significant processing they undergo. It is unhelpful for allergic consumers if such ingredients are subject to allergen labeling requirements, as this would unnecessarily restrict their food choices. In addition, it might mislead allergic consumers who inadvertently eat such products into believing that their allergy was resolving. The EU therefore agreed that where scientific dossiers supporting exemption from allergen labeling were provided and assessed favorably by the European Food Safety Authority, certain processed ingredients should be exempt from allergen labeling requirements. Directive 2007/68 sets out the list of all allergenic ingredients that must be declared on labels and exemptions to those declarations — see Table 17.1 [24].

More recently, a range of food labeling requirements in the EU have been consolidated into the Food Information for Consumers Regulation (FIR) 1169/2011, which was published on 22 November 2011 [25]. The FIR places greater emphasis on the protection of allergic consumers and has improved the accessibility of information about allergenic ingredients used in food products. The existing provisions relating to the labeling of allergenic ingredients used in pre-packaged foods have been carried over into the FIR, but there have been some changes or new requirements added regarding the presentation of the allergen information. Most importantly, there is a new requirement to highlight or emphasize the allergenic ingredients, which now have to be declared in the ingredients list, where one is required. The FIR also introduces a new requirement to provide allergy information for food sold non-pre-packaged (foods sold loose, pre-packaged for direct sale and catering), although it is the responsibility of Member States to set out national measures on how such information should be provided. This is in recognition that the majority of

Table 17.1 EU List of All Allergenic Ingredients That Must Be Declared on Labels and Exemptions to Those Declarations

1. Cereals containing gluten, namely wheat, rye, barley, oats, spelt, kamut, or their hybridized strains and products thereof, except:
   (a) wheat-based glucose syrups including dextrose;
   (b) wheat-based maltodextrins;
   (c) glucose syrups based on barley;
   (d) cereals used for making alcoholic distillates including ethyl alcohol of agricultural origin;

2. Crustaceans and products thereof;

3. Eggs and products thereof;

4. Fish and products thereof, except:
   (a) fish gelatin used as carrier for vitamin or carotenoid preparations;
   (b) fish gelatin or isinglass used as fining agent in beer and wine;

5. Peanuts and products thereof;

6. Soybeans and products thereof, except:
   (a) fully refined soybean oil and fat;
   (b) natural mixed tocopherols (E306), natural D-alpha tocopherol, natural D-alpha tocopherol acetate, and natural D-alpha tocopherol succinate from soybean sources;
   (c) vegetable oils derived phytosterols and phytosterol esters from soybean sources;
   (d) plant stanol ester produced from vegetable oil sterols from soybean sources;

7. Milk and products thereof (including lactose), except:
   (a) whey used for making alcoholic distillates including ethyl alcohol of agricultural origin;
   (b) lactitol;

8. Nuts, namely almonds (*Amygdalus communis* L.), hazelnuts (*Corylus avellana*), walnuts (*Juglans regia*), cashews (*Anacardium occidentale*), pecan nuts (*Carya illinoinensis* (Wangenh.) K. Koch), Brazil nuts (*Bertholletia excelsa*), pistachio nuts (*Pistacia vera*), macadamia or Queensland nuts (*Macadamia ternifolia*), and products thereof, except for nuts used for making alcoholic distillates including ethyl alcohol of agricultural origin;

9. Celery and products thereof;

10. Mustard and products thereof;

11. Sesame seeds and products thereof;

12. Sulfur dioxide and sulfites at concentrations of more than 10 mg/kg or 10 mg/liter in terms of the total SO_2, which are to be calculated for products as proposed ready for consumption or as reconstituted according to the instructions of the manufacturers;

13. Lupin and products thereof;

14. Molluscs and products thereof.

allergic reactions in people who are well diagnosed and are actively trying to avoid the foods to which they are sensitive occur after eating food that is sold non-pre-packaged [26].

The new provisions in the FIR relating to the provision of information about allergenic ingredients and about improved legibility and clarity of labeling information come into effect on 13 December 2014.

'GLUTEN-FREE' AND OTHER 'FREE FROM' CLAIMS

The market for 'free from' foods has grown considerably in the last twenty years, in particular across North America and Western Europe, mainly for lactose-, dairy-, and gluten-sensitive consumers. These 'free from' products are targeted at those consumers with a diagnosed food allergy or intolerance. However, consumers in general in North America and Europe have been found to perceive consuming 'free from' foods as contributing to a healthy balanced diet and improving digestive health. This suggests that the market for such foods will broaden and expand in the future [27].

Allergen risk management efforts have generally been focused upon the provision of accessible allergen advice for at-risk allergic consumers. Therefore, legislation is usually focused upon governing the labeling of allergenic ingredients in foods where they are deliberately incorporated into the food product. Currently, claims that foods are suitable for consumption by persons allergic to or intolerant to certain ingredients are generally not regulated. The exception to date has been for foods suitable for consumers intolerant to gluten. Making a 'free from' claim has more impact on public health than avoiding misleading consumers. It constitutes a positive statement that the product is designed to be able to be consumed without any adverse effects by consumers with specific allergies.

The Codex Alimentarius established a standard for foods safe for consumption by persons intolerant to gluten [28]. This is intended to be used in conjunction with the Codex General Standard for Labeling of Pre-Packaged Foods [7]. A 'gluten-free' claim was originally permitted only if foods did not contain more than 200 mg/kg gluten. This was subsequently revised when new scientific evidence became available [29,30,31,32] suggesting that the 200 ppm limit for gluten did not provide sufficient protection for all celiac patients. A revised Codex standard was published in 2008, which permits dietary foods to be considered 'gluten-free' when total gluten does not exceed 20 mg/kg [28]. See chapter 7 for more information on celiac disease.

The revised Codex standard was implemented into legislation in the EU in January 2012 via Regulation EC/41/2009 [33], which sets compositional standards and labeling requirements for foods for people with gluten intolerance. Three categories of food are described in the Regulation: foods that are specially prepared and/or processed to meet the special dietary needs of people intolerant to gluten can make the claim 'gluten-free' as long as they do not contain more than 20 ppm gluten in the food as sold to the final consumer; foods for people with gluten intolerance that consist of, or contain, ingredients made from gluten-containing cereals (such as wheat, barley, or rye) that have been especially processed to reduce gluten, can be described as 'very low gluten' provided that the level of gluten in the food as sold to the final consumer does not exceed 100 ppm; and foods for normal consumption can be described as 'gluten-free' provided that the gluten content does not exceed 20 ppm in the food as sold to the final consumer.

There is however a possible confusion for celiac consumers in that some products that can legitimately make a 'gluten-free' claim because the level of gluten in the final food as sold does not exceed 20 ppm could contain a very small amount of an ingredient derived from a gluten-containing cereal (such as barley malt extract), which would have to be declared according to the allergen labeling rules. While some specific exemptions from allergen labeling rules have been agreed on for certain ingredients derived from gluten-containing cereals (such as glucose syrups), there has been no consideration as yet of a generic exemption for ingredients derived from gluten-containing cereals where the final food is able to make a 'gluten-free' claim.

There is no other legislation that sets out requirements for foods that make claims that they are free from other allergenic foods, other than the general food law provisions in Regulation No. 178/2002 [10] that, *inter alia*, prohibits unsafe food being placed on the market and that requires that food labeling should not be misleading. If food businesses want to make 'free from' claims, they need to put in place procedures and checks to ensure that they are justified. 'Free from' claims cannot be reasonably made based only on recipe absence but also need to depend upon the product being manufactured under specialized robust conditions, using confirmed 'free from' raw materials, in order to guarantee compliance to 'free from' standards for every batch of product.

REGULATORY LISTS AROUND THE WORLD

Although the FAO Technical Consultation on Food Allergies confirmed eight major allergenic foods as being of greatest concern, there are many other foods that have been reported to produce allergic reactions to varying degrees across the world. It is not surprising to find that different countries show variable prevalence of allergenic foods given the different dietary patterns. In many of the major developing markets around the world there are currently no specific allergen labeling legislative requirements, and this can impact on the ability of companies importing food commodities or ingredients from such markets to provide full allergen information on the foods that they make using such ingredients or commodities. The legislative situation [34] in a number of countries is set out in Table 17.2, although it should be recognized that legislative rules can change over time and the current situation should be checked.

WHAT IS NOT COVERED BY LEGISLATION?

There are additional forms of allergen labeling used by some food businesses that are outside current legislative controls. Many food manufacturers, especially in Europe, also voluntarily provide additional information to help food allergic consumers access the information they need to make safe and informed food choices. This can take the form of, for example, additional statements immediately following the formal ingredients declaration text, or 'allergy advice' boxes, with phrases such as 'contains egg, soy and peanuts'. Such

Table 17.2 Regulatory Allergen Labeling Lists Around the World

ALLERGEN	CODEX[a]	European Union	USA	Canada	Japan	South Korea	Australia New Zealand
Peanut	✓	✓	✓	✓	✓	✓	✓
Tree nuts	✓	✓[e]	✓[e]	✓	✓[d]	✓	✓
Egg	✓	✓	✓	✓	✓	✓	✓
Fish	✓	✓	✓	✓	(Salmon and salmon roe; Mackerel[d])	Mackerel	✓
Milk	✓	✓	✓	✓	✓	✓	
Crustacea/Shellfish	Crustacea	Crustacea Molluscs	Crustacea	Crustacea Molluscs	Crab; Shrimp; Prawn (Abalone; Squid[d])	Crab Shrimp	Crustacea
Sesame seeds	✓	✓		✓			
Soy	✓	✓	✓	✓	(✓[d])	✓	✓
Sulfite[b]	✓	✓	✓	✓			✓

Cereals containing gluten[c]	✓	Wheat	✓	Wheat	Wheat	✓
Mustard	✓		✓			
Other	Celery/Celeriac Lupin			Buckwheat: (Orange, Peach, Apple, Kiwi fruit, Matsutake mushroom, Yam, Gelatin, Beef, Chicken, Pork[d])	Buckwheat: Pork, Peach, Tomato	Bee pollen, Propolis, Royal jelly
Comments	Label all derivatives No quantitative limit for labeling Specific exemptions from labeling approved Accept thresholds in principle	Label all derivatives Exemption for highly refined oils and derivatives No lower limit	Label all derivatives No lower limit	Quantitative limit for labeling	Label all derivatives No quantitative limit for labeling	Label all derivatives No labeling exemptions yet given

[a] Codex Alimentarius Commission of FAO/WHO
[b] Where the concentration is 10 mg/kg or higher
[c] Wheat, rye, barley, oats, spelt, or their hybridized strains (the Codex, Canadian, and EU lists include gluten-sensitive enteropathy (celiac disease) as a food allergy)
[d] Allergen labeling recommended but not mandatory
[e] Defined list of tree nuts prescribed

additional advisory statements are not controlled by legislation, and although they can provide short-cuts for consumers, the concern with providing information in this way is that consumers may come to rely upon this alone and not read the ingredients list. Manufacturers choosing to use such statements should therefore take great care that all the allergenic ingredients declared in the ingredients list are included.

While many consumers find such additional information helpful in identifying products suitable for their needs, they should be aware that such statements are not legal requirements in the EU and that the absence of such a 'contains' statement should not be assumed to mean that none of the major allergenic foods are used in the product.

Cross Contamination Risks and Advisory Labeling

At present, allergen labeling legislation covers only the deliberate use of an ingredient in a pre-packed product. However for the allergic consumer, there may be a health risk if a food product contains a significant level of an allergen as a result of accidental cross contamination at some point in the food chain. While food manufacturers put in place a number of checks and processes to try to control the risk of the accidental presence of an allergenic food ingredient in a product, it is not always possible to completely avoid such a risk, particularly in premises that make a wide range of products with multiple ingredients [35,36].

In such situations, many food manufacturers opt to use some form of advisory labeling to alert allergic consumers to such risks, using phrases such as *'may contain nuts'*, *'made in a factory that also uses nut ingredients'*, or *'not suitable for someone with a nut allergy'*. While the intention of such warnings is to help allergic consumers to make safe food choices, the use of such warnings has become widespread, and it can be difficult to find products without such warnings [37]. Currently there are no internationally agreed upon action levels for cross contamination with allergenic foods below which advisory labeling is not appropriate. Therefore manufacturers often feel obliged to provide additional warning labels upon determination of any qualitative risk of allergen cross contamination following risk assessment of their operations, however low or remote. In addition, improvements in allergen analytical detection methodologies have also meant that the presence of lower and lower levels of allergen can now be detected, which may also be a factor in the increasing use of allergen advisory labeling. Qualitative guidance on allergen management and advisory labeling was published by the UK Food Standards Agency in 2006 [38] to help businesses identify where allergen cross contamination risks could arise and to advise on how such risks could be controlled and minimized as far as possible.

There is evidence from consumer research that many food allergic consumers consider such allergen advisory warnings to be overused, and therefore they are often ignored [39]. A further challenge to the value of precautionary labeling for consumer risk protection is the inconsistency in the range of phrases used to identify products as potentially containing food allergens from unintentional cross contamination. Recent research [40,41,42] has confirmed that this diversity and inconsistency of precautionary warning statements has left consumers confused and anxious. The absence of a 'may contain' label has even been mistaken as synonymous with the product being 'free from' allergens. However, it is also recognized that advisory labels are helpful if they can provide reliable information on the allergen content, steering sensitive allergic consumers away from foods unsuitable for them. A consistent approach to clear discrimination between foods and their suitability for allergic consumers is needed.

A recent proposal by the Food and Drink Federations Allergen Steering Group [36] suggest three classifications for foodstuffs to achieve this goal: 'free from' foods that would carry a 'free from' label and be produced specifically to assure absence of allergens, which the vast majority of highly sensitive allergic consumers would be able to consume without even mild adverse reactions; foods for normal consumption produced such that allergen cross-contact control is managed to below an action level (quantitative dose below which the vast majority of sensitive consumers would be protected from even mild objective adverse reactions) and thus not bearing precautionary labeling as expected to be well-tolerated by the vast majority of allergic consumers; and 'not suitable for' foods, which would be foods carrying a precautionary warning statement and produced where unavoidable traces/very small amounts of allergenic materials are present above the action level in a significant proportion of product units, despite capable allergen risk management efforts.

Such a model would harmonize manufacturing standards and assure consistency of food allergen labeling, based on quantitative action levels derived from scientific knowledge of the profile of reactivity of the allergic population. The major barrier to this approach remains the lack of agreed upon tolerable action levels, although significant progress is now being made towards determining them for specific allergenic foods [43,44,45].

In the past, food allergic consumers have been advised to totally avoid the foods to which they are sensitive — essentially taking a zero tolerance approach assuming that any level of presence could trigger an adverse reaction. There is, however, now clear evidence to suggest that threshold levels do exist, below which reactions are not provoked in allergic individuals [46,47]. In addition, debate among international multi-disciplinary experts over recent years has formed a consensus that zero risk is not a realistic option for the public health management of food allergy and that it is essential to address the current lack of agreed upon action levels for cross contamination with allergens if food allergen management practice is to be improved [44,45]; see also chapter 5.

Significant work is underway to establish what tolerable risk management/regulatory values should be based upon, for example by the Swiss Authorities, the Australian Food and Grocery Council Voluntary Incidental Trace Allergen Labeling (VITAL) system [48], and jointly led UK FSA/EuroPrevall activities [43,45]. Such a risk-based approach to allergen management will be of enormous value to all stakeholders across the supply chain. It would allow industry to consistently manage cross-contact and provide reliable, consistent consumer information for safe food choices.

CONCLUSIONS

In conclusion, many countries have reacted to the risks to public health posed by the undeclared presence of allergenic food ingredients by introducing specific labeling legislation. While there is a large range of different allergenic foods, some present a greater risk to public health by virtue of their potency and prevalence in the population of at-risk consumers. The allergens of public health importance differ in different parts of the world, and this is dependent on a number of genetic, environmental, and dietary factors. Models now exist to support the determination of allergenic foods that are of public health importance, which can be tailored using local prevalence data to inform specific regulatory control measures.

The food allergic consumer can best be protected by:
* Clear, legible, and available scientifically based risk communication matching relevant allergen composition of foods,
* Sound GMP, with allergens integrated into HACCP risk management,
* Clear and precise allergen declarations down the food chain,
* Well-educated food operators and consumers,
* Regulation to set the food safety standards expected and to ensure consistency of allergen risk communication.

It is important to recognize that 'free from' foods are separate from mainstream foods as they are produced specifically to assure absence of allergenic foods, and because they are aimed specifically at the 'at-risk' population. The controls needed to justify such claims are necessarily very stringent. However the food allergic consumer cannot consume only foods specifically prepared to meet their particular allergy requirements. It is therefore critical that clear information about the use of allergenic food ingredients and the possible risks of allergen cross contamination for mainstream foods are clearly communicated to the allergic consumer in a way that allows them to make informed choices about the foods they can safely consume.

However it is not sufficient just for the food business to provide allergy labeling and information; the allergic consumer has to take responsibility for their own health by looking for the allergen information provided and using that information in their food consumption decisions.

REFERENCES

[1] FAO. Report of the FAO Technical Consultation on Food Allergies. Rome, Italy: Food and Agricultural Organization of the United Nations; 1995, 13−14 November.
[2] Hefle S, Nordlee J, Taylor S. Allergenic foods. Crit Rev Food Sci Nutr 1996;36S: S69−89.
[3] Taylor S. Emerging problems with food allergens. Food Nutr Agric 2000;26:14−23.

[4] European Food Safety Authority. Opinion of the Scientific Panel on Dietetic Products, Nutrition and Allergies on a request from the Commission relating to the evaluation of allergenic foods for labelling purposes. EFSA J 2004;32:1–197.

[5] Burks A, Tang M, Sicherer S, Muraro A, Eigenmann P, Ebisawa M, Fiocchi A, Chiang W, Beyer K, Wood R, Hourihane J, Jones S, Lack G, Sampson H. ICON: food allergy. J Allergy Clin Immunol 2012;129(4):906–20.

[6] JECFA. Evaluation of certain food additives and contaminants: fifty-fifth report of the Joint FAO/WHO Expert Committee on Food Additives: cochineal extract and carmines. Geneva: WHO Library; 2000.

[7] Codex Alimentarius. Codex General Standard for the Labelling of Pre-Packaged Foods CODEX STAN 1–1985. FAO/WHO 2010.

[8] Humbert P, Pelletier F, Dreno B, Puzenat E, a Aubin F. Gluten intolerance and skin diseases. Eur J Dermatol 2006;16:4–11.

[9] Bousquet J, Bjorksten B, Bruijnzeel-Koomen C, Huggett A, Ortolani C, Warner J, Smith M. Scientific criteria and selection of allergenic foods for labelling. Allergy 1998;53(Suppl. 47):3–21.

[10] European Commission (EC). Regulation (EC) No. 178/2002 of the European Parliament and of the Council of 28 January 2002 laying down the general principles and requirements of food law, establishing the European Food Safety Authority and laying down procedures of food safety. Offical J Eur Communities 2002;L31:1–24.

[11] Jappe U, Vieths S. Lupine, a source of new as well as hidden food allergens. Mol Nutr Food Res 2010;54(1):113–26.

[12] Campbell C, Yates D. Lupin allergy: a hidden killler at home, a menace at work; occupational disease due to lupin allergy. Clin Exp Allergy 2010;40(10):1467–72.

[13] Cornelisse-Vermaat J, Voordouw J, Yiakoumaki V, Theodoridis G, Frewer L. Food allergic consumers' labelling preference: a cross-cultural comparison. Eur J Public Health 2008;18(2):115–20.

[14] Sakellariou A, Sinaniotis A, Damianidou L, Papdopoulos N, Vassilopoulou E. Food allergen labelling and consumer confusion. Allergy 2009;65:531–6.

[15] Barnett J, Leftwich J, Muncer K, Grimshaw K, Shepherd R, Raats M, Gowland M, Lucas J. How do peanut and nut-allergic consumers use information on the packaging to avoid allergens? Allergy 2011;66(7):969–78.

[16] FAO. Report of the Twenty-sixth Session of the Codex Committee on Food Labelling (unpublished FAO document ALINORM 99/22). Ottawa: Food and Agricultural Organization of the United Nations; 1998.

[17] European Commission (EC). Regulation (EC) No 852/2004 of the European Parliament and of the Council of 29 April 2004 on the hygiene of foodstuffs. Official J Eur Union 2004. L139/1.

[18] European Commission (EC). Directive 2000/13/EC of the European Parliament and of the Council of 20 March 2000 on the approximation of the laws of the Member States relating to the labelling, presentation and advertising of foodstuffs. Official J Eur Union 2000;L109:29–42.

[19] European Commission (EC). European Commission Directive 2003/89/EC of the European Parliament and of the Council of 10 November 2003 amending Directive 2000/13/EC as regards indication of the ingredients present in foodstuffs. Official J Eur Union 2003;L308:15–8.

[20] European Commission (EC). Commission Directive 2006/142/EC of 22 December 2006 amending Annex IIIa of Directive 2000/13/EC of the European Parliament and of the Council listing the ingredients which must under all circumstances appear on the labeling of foodstuffs. Official J Eur Union 2006;L368:110–1.

[21] Radcliffe M, Scadding G, Brown HM. Lupin flour anaphylaxis. Lancet 2005;365:1360.

[22] European Food Safety Authority. Opinion of the Scientific Panel on Dietetic Products, Nutrition and Allergies on a request from the Commission related to the evaluation of lupin for labelling purposes (Request No. EFSA-Q-2005–086) (adopted 6 November 2005). EFSA J 2005;302:1–11.

[23] European Food Safety Authority. Opinion of the Scientific Panel on Dietetic Products, Nutrition, and Allergies on a request from the Commission related to the evaluation of molluscs for labelling purposes (Request No. EFSA-Q-2005−84) (adopted 15 February 2006). EFSA J 2006;327:1−25.

[24] European Commission (EC). Commission Directive 2007/68/EC of 27 November 2007 amending Annex IIIa to Directive 2000/13/EC of the European Parliament and of the Council as regards certain food ingredients. Official J Eur Union 2007;L310:11−4.

[25] European Union (EU). Regulation (EU) No 1169/2011 of the European Parliament and of the Council of 25 October 2011 on the provision of food information to consumers, amending Regulations (EC) No 1924/2006 and (EC) No 1925 of the European Parliament and of the Council, and repealing Commission Directive 87/250/EEC, Council Directive 90/496/EEC, Commission Directive 1990/10/EC, Directive 2000/13/EC of the European Parliament and of the Council, Commission Directives 2002/67/EC and 2008/5/EC and Commission Regulation (EC) No 608/2004. Official J Eur Union 2011;L304:18−63.

[26] Pumphrey R, Gowland M. Further fatal reactions to foods in the United Kingdom 1999−2006. J Allergy Clin Immunol 2007;119(4):1018−9.

[27] Leatherhead Food Research. Food allergies and intolerances: consumer perceptions and market opportunities for free from foods. Leatherhead Food Res 2011.

[28] Codex Alimentarius. Standard for foods for special dietary use for persons intolerant to gluten 118−1979. FAO/WHO 2008.

[29] Collin P, Thorell L, Kaukinen K, Mäki M. The safe threshold for gluten contamination in gluten-free products. Can trace amounts be accepted in the treatment of coeliac disease? Aliment Pharmacol Ther 2004;19:1277−83.

[30] Gibert A, Espadaler M, Angel Canela M, Sánchez A, Vaqué C, Rafecas M. Consumption of gluten-free products: should the threshold value for trace amounts of gluten be at 20, 100 or 200 ppm? Euro J Gastroenterol Hepatol 2006;18(11):1187−95.

[31] Catassi C, Fabiani E, Iacono G, D'Agate C, Francavilla R, Biagi F, Volta U, Accomando S, Picarelli A, De Vitis I, Pianelli G, Gesuita R, Carle F, Mandolesi A, Bearzi I, Fasano A. A prospective, double-blind, placebo-controlled trial to establish a safe gluten threshold for patients with celiac disease. Am J Clin Nutr 2007;85:160−6.

[32] Collin P, Mäki M, Kaukinen K. Safe gluten threshold for patients with celiac disease: some patients are more tolerant than others. Am J Clin Nutr 2007;86(1):260.

[33] European Commission (EC). Commission Regulation (EC) No. 41/2009 of 20 January 2009 concerning the composition and labelling of foodstuffs suitable for people intolerant to gluten. Official J Eur Union 2009;L16:3−5.

[34] Gendel S. Comparison of international food allergen labelling regulations. Reg Tox Pharm 2012;63:279−85.

[35] Alvarez P, Boye J. Food production and processing considerations of allergenic food ingredients: a review. J Allergy 2012.

[36] Ward R, Crevel R, Bell I, Khandke N, Ramsay C, Paine S. A vision for allergen management best practice in the food industry. Trends Food Sci Technol 2010:1−7.

[37] Food Standards Agency. 'May contain' labelling − the consumer's perspective. HMSO Crown Copyright 2002.

[38] Food Standards Agency. Guidance on allergen management and consumer information. HMSO Crown Copyright 2006.

[39] Food Standards Agency. Qualitative research into the information needs of teenagers with food allergy and intolerance. HMSO Crown Copyright 2005.

[40] Turner P, Kemp A, Campbell D. Advisory food labels: consumers with allergies need more than 'traces' of information. BMJ 2011;343:d6180.

[41] Sampson M, Munoz-Furlong A, Sicherer S. Risk-taking and coping strategies of adolescents and young adults with food allergy. J Allergy Clin Immunol 2006;117(6):1440−5.

[42] Zurzolo G, Mathai M, Koplin J, Allen K. Hidden allergens in foods and implications for labelling and clinical care of food allergic patients. Curr Allergy Asthma Rep 2012.

[43] Madsen C, Hattersley S, Buck J, Gendel S, Houben G, Hourihane J, Mackie A, Mills E, Norhede P, Taylor S, Crevel R. Approaches to risk assessment in food allergy: report from a workshop 'developing a framework for assessing the risk from allergenic foods'. Food Chem Toxicol 2009;47(2):480−9.

[44] Madsen C, Crevel R, Chan C, Dubois A, DunnGalvin A, Flokstra-de Blok B, Gowland M, Hattersley S, Hourihans J, Norhede P, Pfaff S, Rowe G, Schnadt S, Vlieg-Boerstra B. Food allergy: stakeholder perspectives on acceptable risk. Regul Toxicol Pharmacol 2010;57:256−65.

[45] Madsen C, Hattersley S, Allen K, Beyer K, Chan C, Godefroy S, Hodgson R, Mills E, Munoz-Furlong A, Schnodt S, Ward R, Wickman M, Crevel R. Can we define a tolerable level of risk in food allergy? Report from a EuroPrevall / UK Food Standards agency workshop. Clin Exp Allergy 2012;42(1):30−7.

[46] Moneret-Vautrin D, Kanny G. Update on threshold doses of food allergens: implications for patients and the food industry. Curr Opin Allergy Clin Immunol 2004;4(3): 215−9.

[47] Taylor S, Gendel S, Houben G, Julien E. The key events dose response framework: a foundation for examining variability in elicitation thresholds for food allergens. Crit Rev Food Sci Nutr 2009;49:729−39.

[48] The Allergen Bureau. VITAL. Retrieved 2012, from http://www.allergenbureau.net/vital; 2009. Accessed 2 August 2012.

Chapter | eighteen

Keeping Updated

David Reading[1], Erna Botjes[2], Pia Nørhede[3], Marjan van Ravenhorst[4]

[1]*Food Allergy Support Ltd, Aldershot, UK*
[2]*Stichting Voedselallergie (Dutch Food Allergy), Nijkerk, The Netherlands*
[3]*Division of Toxicology and Risk Assessment, National Food Institute, Technical University of Denmark, Lyngby, Denmark*
[4]*Allergenen Consultancy, Scherpenzeel, The Netherlands*

CHAPTER OUTLINE

Risk Management for Food Allergy. http://dx.doi.org/10.1016/B978-0-12-381988-8.00018-X

INTRODUCTION

The previous chapters of this book are based on up-to-date knowledge at the time of publication. However, research and knowledge are continuously developing, for example in the area of thresholds. With more information about thresholds for individual allergens it will be possible to use more advanced risk assessment methods as a basis for risk management.

Many countries have labeling legislation for allergenic foods based on recommendations from Codex Alimentarius. However, the Codex recommendations do not cover allergens that enter foods inadvertently through, for example, the use of common processing equipment or adjacent production lines. Several countries have written guidelines on how to manage and label products that may contain allergen residues because of cross-contamination. With new knowledge, best practice may change and national guidelines are likely to be updated. The same goes for food allergy regulation.

In this chapter we have made a collection and description of the most important web links published in English so that readers can stay updated. The collection includes websites of regulatory bodies, databases containing information about food allergens, and sites providing guidance on allergen risk management. The websites are in alphabetical order.

The authors of this chapter are:

Erna Botjes, chair of Dutch Food Allergy (Stichting Voedselallergie); Resource Access Committee, Europrevall; EFA Working Group on Food Allergy.

Pia Norhede, toxicologist at the National Food Institute, Technical University of Denmark; creator of the website Food Allergy Information in collaboration with a group of leading allergy experts.

David Reading, director of Food Allergy Support Ltd; co-founder of the Anaphylaxis Campaign in the UK; research director of the European Anaphylaxis Taskforce.

Marjan van Ravenhorst, director of Allergenen Consultancy, The Netherlands.

THE ALLERGEN BUREAU (AUSTRALIA)

www.allergenbureau.net

The Allergen Bureau was established in 2005 as an initiative of the Australian Food and Grocery Council Allergen Forum and operates on a membership basis.

Although the Bureau's main objective is to help the food industries in Australia and New Zealand, its website offers information that is relevant to a worldwide audience. Almost 20% of visitors to the website come from Canada and the US and over 10% from Europe, with the majority from the UK.

Food Allergy News and Information

The home page provides a summary of topical allergy-related news from around the world. It is possible to subscribe in order to receive updates by email.

There is a general information section on food allergens, including clinical information and a Q&A page tackling questions such as: What is the safe limit for food allergens in products? How do we check that our cleaning procedure stops cross-contamination?

VITAL

The Voluntary Incidental Trace Allergen Labeling system (VITAL) is a major part of the Allergen Bureau's work. This is a standardized allergen risk assessment tool for food producers, enabling them to assess the impact of allergen cross-contact and to provide appropriate and consistent precautionary labeling. VITAL allows the assessment of likely sources of allergen cross-contact plus an evaluation of the amount present and a review of the ability to reduce it.

Downloads available include an auditor guide, the VITAL decision tree, the VITAL calculator (which helps food companies to make decisions about 'may be present' labels), and a protein reference table showing total protein per 100 g of individual foods. Also available for downloading are VITAL case stories and supporting material showing the application of VITAL.

THE AMERICAN ACADEMY OF ALLERGY, ASTHMA, AND IMMUNOLOGY

www.aaaai.org

The American Academy of Allergy, Asthma, and Immunology (AAAAI) is the largest professional medical organization devoted to allergy/immunology in the United States.

Although its primary aim is to represent and inform professionals working in allergy and immunology, the AAAAI website contains much of interest to the food industry and the public generally. Under the heading Patients and Consumers, visitors to the site will, for example, find the latest e-headlines, while the site's Media page contains press releases that outline latest research developments and other news.

CANADIAN FOOD INSPECTION AGENCY

www.inspection.gc.ca/english/fssa/labeti/allerg/allerge.shtml

The Canadian Food Inspection Agency (CFIA) is the government body that enforces Canada's labeling laws and works with associations, distributors, food manufacturers, and importers to ensure complete and appropriate labeling of all foods.

As well as including information about Canada's food labeling regulations, the CFIA website provides additional help for industry on the specific subject of allergens. The site has a question-and-answer section and advice on precautionary labeling.

To assist the industry further there is a tool for managing allergen risk in food products, which is an allergen checklist that food suppliers and manufacturers can use.

CATERING FOR ALLERGY (UK)

www.anaphylaxis.org.uk/catering-for-allergy

The Catering for Allergy website was set up by the Anaphylaxis Campaign, the UK charity that supports people with life-threatening food allergies. The site aims to help catering businesses develop strategies to assess and manage the risks associated with sourcing, storing, preparing, and serving food to people with food allergy.

Strategies

Training and communication strategies are an important part of the site. Guidance is provided on key areas in catering, including food preparation, serving food, and describing the food on sale. Case studies and checklists are published as a way of offering practical help.

Assessing Risk

Key areas of focus include risks in delivery, risks in storage, cross-contamination, risks from serving food, and what to do when someone has an allergic reaction.

THE EUROPEAN ACADEMY OF ALLERGY AND CLINICAL IMMUNOLOGY

www.eaaci.net

The European Academy of Allergy and Clinical Immunology (EAACI) is an association of clinicians, researchers, and allied health professionals dedicated to improving the health of people affected by allergy. Food technologists and scientists wishing to look at allergy and immunology in some depth, including recent research, can find theallergynews.com accessible from the home page.

THE EUROPEAN FOOD SAFETY AUTHORITY

The European Food Safety Authority (EFSA) is the keystone of European Union risk assessment regarding food safety. In collaboration with national authorities and in consultation with stakeholders, EFSA provides independent scientific advice and communication on existing and emerging risks.

EFSA documents pertaining to food allergens include an important evaluation of allergenic foods for labeling purposes. This report formed the basis for the formation of the list of food allergens that became subject to mandatory labeling. The report includes comprehensive information on individual allergenic foods including clinical features, the proteins that are responsible for triggering reactions, cross-reacting foods, and the effects of processing on the allergenicity of the foods.

The full document is accessed via the web page:

www.efsa.europa.eu/en/efsajournal/pub/32.htm

Separate reports focusing on lupin and mollusks, which joined the list of allergens subject to mandatory labeling, are also published.

Lupin: www.efsa.europa.eu/en/efsajournal/pub/302.htm

Mollusks: www.efsa.europa.eu/en/efsajournal/pub/327.htm

The site also includes a register of current and previous applications to EFSA where answers have been sought on matters pertaining to allergens (for example, on whether specific derivatives of allergens were considered to be allergenic). The web address is:

http://registerofquestions.efsa.europa.eu/roqFrontend/questionsListLoader? panel=NDA. (In the box 'Food Sector Area' enter 'Food Allergy.')

FOOD ALLERGY INFORMATION (EU)

www.foodallergens.info

Food Allergy Information is a website that offers information geared to the needs of the food industry and national authorities as well as to general readers such as people with food allergies.

The content was developed by allergy experts from the food industry, patient organizations, clinical centers, and research institutions in Europe as part of the European research project EuroPrevall.

Food Allergy Facts

Basic information is provided on what a food allergy is, the extent of the problem, possible causes, prevention, symptoms, diagnosis, and treatment.

Foods Causing Allergy

This section contains links to pages in the InformAll Database, a searchable database with comprehensive information on many allergenic foods, hosted by the Institute of Food Research, Norwich, UK.

EU Legal Requirements

Information is provided for manufacturers operating in the EU or wishing to export to the EU about the legal requirements in Europe regarding allergenic foods. Sub-sections cover the EU Labeling Directive on food allergens (including a list of foods subject to mandatory labeling and derivatives that are exempt from mandatory labeling), the Codex recommendations, EU General Food Law on the general obligations to provide safe food, and contact details for national authorities worldwide that are responsible for food allergy.

Food Manufacturing

A section on food manufacturing offers guidance for the food industry on how to deal with allergens in food production including how to avoid cross-contamination. There are sub-sections on GMP and HACCP, risk assessment, risk management, risk communication, and national guidelines.

Catering

There are links to relevant websites with information on what caterers can do to provide safe foods for allergic customers.

Food Allergy Portal

This is a collection of critically assessed websites about food allergy in various European languages.

THE FOOD ALLERGY RESEARCH AND RESOURCE PROGRAM (US)

http://farrp.unl.edu

The Food Allergy Research and Resource Program (FARRP) is a co-operative venture between the University of Nebraska, US, and its food industry members. It aims to provide industry with information, expert opinion, tools, and services relating to allergenic foods and novel foods. Services include analytical testing for food companies worldwide, workshops, on-site consultation, and training.

The general heading 'FARRP resources' contains sub-sections that offer specific information on allergen-related subjects, including the following:

AllergenOnline Database

AllergenOnline provides access to a peer-reviewed allergen list and sequence searchable database intended for the identification of proteins that may present a risk of cross-reactivity. This database was designed to help in assessing the safety of proteins that may be introduced into foods through genetic engineering or through food processing methods. The objective is to identify proteins that may require additional tests, such as serum IgE binding, basophil histamine release, or *in vivo* challenge to evaluate potential cross-reactivity.

Thresholds for Allergenic Foods

FARRP is working with organizations worldwide to establish scientific and clinical evidence and approaches needed to determine threshold levels that are safe for the vast majority of food-allergic consumers. A section on thresholds on the FARRP website provides background information on the progress so far. While the site does not contain clinical threshold data at the time of publication of this book, it does outline the approaches that are being used to determine thresholds. Information on the site is likely to be updated as further progress on thresholds is made.

Regulatory Situation

FARRP's international regulatory chart names the allergens that are subject to mandatory labeling in countries across the world.

Allergen Control in the Food Industry

A framework for developing an effective allergen control plan is provided in some detail in a 15-page document that can be downloaded as a pdf. Areas covered include research and development, segregation of allergenic foods, storage and handling, prevention of cross-contact, product labeling, validation of cleaning, and staff training.

FOOD ALLERGY SUPPORT (UK)

www.foodallergy-support.com

Food Allergy Support Ltd is a UK consultancy that works with industry to find practical solutions to allergy-related problems, with the ultimate objective of ensuring that food is safe and well labeled for the allergic population. Food Allergy Support offers food companies access to allergy-related expertise through its wide range of contacts.

Anyone wishing to stay updated can find the latest allergy news and research developments on its website.

FOOD AND DRUG ADMINISTRATION (US)

www.fda.gov/Food/LabelingNutrition/FoodAllergensLabeling/default.htm

The Food and Drug Administration (FDA) is the US agency responsible for protecting public health by assuring the safety and efficacy of food, drugs, cosmetics, biological products, and medical devices.

The FDA website includes sections covering a wide range of food allergy issues in order to guide and inform industry. The key information for industry is covered in a section headed 'Food allergens labeling.'

This section contains industry/regulator information with particular focus on the all-important Food Allergen Labeling and Consumer Protection Act of 2004 (FALCPA). The legislation is published in full and there is a supplementary question-and-answer section that is intended to make the legislation understandable to industry staff at all levels. The answers published cover a wide range of topics including individual allergens, advisory labeling, and 'contains' statements.

In a separate sub-section, guidance is provided on the labeling of certain uses of lecithin derived from soy.

A sub-section entitled 'More guidance, compliance and regulatory information' covers subjects including thresholds and gluten-free labeling.

THE FOOD STANDARDS AGENCY (UK)

www.food.gov.uk/safereating/allergyintol

The Food Standards Agency (FSA) is the government body responsible for food safety and food hygiene across the UK. The agency's website covers a wide range of food allergy issues in order to guide and inform industry and the public. Subjects covered include the following:

Food Allergen Labeling

There is an outline of the European labeling regulations for allergens including a list of those subject to mandatory labeling. Links are provided to the relevant EU Directives. The rules covering ingredients derived from allergens that have gained exemption are included among these links.

There is a sub-section that helps food businesses to understand the requirements for the labeling of gluten-free foods.

Research

A section is devoted to the FSA's research work that investigates food allergy and intolerance. There are sub-sections on consumer surveys, current objectives, and past projects.

Guidance for Food Businesses

This important section includes a link to an interactive online training tool highlighting steps to be followed to ensure good practice in food production. It offers practical advice to local authority enforcement officers, managers, and staff in manufacturing and catering.

Advice for caterers is provided in a separate sub-section in question-and-answer form.

A sub-section on allergen management and labeling of pre-packed food includes links to FSA guidance documents including 'Guidance on Allergen Management and Consumer Information.' This document pays special attention to 'may contain' labeling, providing voluntary best practice advice on how to assess the risks of cross-contamination and determine whether advisory labeling is appropriate.

Guidance notes on the food allergen labeling legislation provide informal advice on labeling for pre-packed foods.

A sub-section covering non-pre-packed foods includes a link to the voluntary best practice guidance document 'The Provision of Allergen Information for Non Pre-packed Foods.'

THE INFORMALL DATABASE (UK)

http://foodallergens.ifr.ac.uk

The InformAll Database is a searchable database with comprehensive information on many individual allergenic foods. It was developed with funding from the European Union and is hosted by the Institute of Food Research, Norwich, UK.

Allergenic foods are listed alphabetically, from abalone to wheat, and summaries about each one are written for a wide audience (lay people as well as scientists). Information is provided about the forms the allergy can take. For example, some allergens such as fruits are associated with pollen allergy, and where relevant this is explained.

Sub-sections on each food show clinical and biochemical data, and information provided is supported by references to papers in the scientific literature, with links to the abstracts of those papers.

Information in the database is subject to a review process, so there is a time-lag between a scientific publication appearing and a database entry being compiled.

Food Science and Technology International Series

Amerine, M.A., Pangborn, R.M., and Roessler, E.B., 1965. Principles of Sensory Evaluation of Food.

Glicksman, M., 1970. Gum Technology in the Food Industry.

Joslyn, M.A., 1970. Methods in Food Analysis, Second Ed.

Stumbo, C. R., 1973. Thermobacteriology in Food Processing, Second Ed.

Altschul, A.M. (Ed.), New Protein Foods: Volume 1, Technology, Part A—1974. Volume 2, Technology, Part B—1976. Volume 3, Animal Protein Supplies, Part A—1978.

Volume 4, Animal Protein Sup-plies, Part B—1981. Volume 5, Seed Storage Proteins—1985.

Goldblith, S.A., Rey, L., and Rothmayr, W.W., 1975. Freeze Drying and Advanced Food Technology.

Bender, A.E., 1975. Food Processing and Nutrition.

Troller, J.A., and Christian, J.H.B., 1978. Water Activity and Food.

Osborne, D.R., and Voogt, P., 1978. The Analysis of Nutrients in Foods.

Loncin, M., and Merson, R.L., 1979. Food Engineering: Principles and Selected Applications.

Vaughan, J. G. (Ed.), 1979. Food Microscopy.

Pollock, J. R. A. (Ed.), Brewing Science, Volume 1—1979. Volume 2—1980. Volume 3—1987.

Christopher Bauernfeind, J. (Ed.), 1981. Carotenoids as Colorants and Vitamin A

Precursors: Technological and Nutritional Applications.

Markakis, P. (Ed.), 1982. Anthocyanins as Food Colors.

Stewart, G.G., and Amerine, M.A. (Eds.), 1982. Introduction to Food Science and Technology, Second Ed.

Iglesias, H.A., and Chirife, J., 1982. Handbook of Food Isotherms: Water Sorption

Parameters for Food and Food Components.

Dennis, C. (Ed.), 1983. Post-Harvest Pathology of Fruits and Vegetables.

Barnes, P.J. (Ed.), 1983. Lipids in Cereal Technology.

Pimentel, D., and Hall, C.W. (Eds.), 1984. Food and Energy Resources.

Regenstein, J.M., and Regenstein, C.E., 1984. Food Protein Chemistry: An Introduction for Food Scientists.

Gacula Jr. M.C., and Singh, J., 1984. Statistical Methods in Food and Consumer Research.

Clydesdale, F.M., and Wiemer, K.L. (Eds.), 1985. Iron Fortification of Foods.

Decareau, R.V., 1985. Microwaves in the Food Processing Industry.

Herschdoerfer, S.M. (Ed.), Quality Control in the Food Industry, second edition. Volume 1—1985. Volume 2—1985. Volume 3—1986. Volume 4—1987.

Urbain, W.M., 1986. Food Irradiation.

Bechtel, P.J., 1986. Muscle as Food.

Chan, H.W.-S., 1986. Autoxidation of Unsaturated Lipids.

Cunningham, F.E., and Cox, N.A. (Eds.), 1987. Microbiology of Poultry Meat Products.

McCorkle Jr. C.O., 1987. Economics of Food Processing in the United States.

Japtiani, J., Chan Jr., H.T., and Sakai, W.S., 1987. Tropical Fruit Processing.

Solms, J., Booth, D.A., Dangborn, R.M., and Raunhardt, O., 1987. Food Acceptance and Nutrition.

Macrae, R., 1988. HPLC in Food Analysis, Second Ed.

Pearson, A.M., and Young, R.B., 1989. Muscle and Meat Biochemistry.

Penfield, M.P., and Campbell, A.M., 1990. Experimental Food Science, Third Ed.

Blankenship, L.C., 1991. Colonization Control of Human Bacterial Entero-pathogens in Poultry.

Pomeranz, Y., 1991. Functional Properties of Food Components, Second Ed.

Walter, R.H., 1991. The Chemistry and Technology of Pectin.

Stone, H., and Sidel, J.L., 1993. Sensory Evaluation Practices, Second Ed.

Shewfelt, R.L., and Prussia, S.E., 1993. Postharvest Handling: A Systems Approach.

Nagodawithana, T., and Reed, G., 1993. Enzymes in Food Processing, Third Ed.

Hoover, D.G., and Steenson, L.R., 1993. Bacteriocins.

Shibamoto, T., and Bjeldanes, L., 1993. Introduction to Food Toxicology.

Troller, J.A., 1993. Sanitation in Food Processing, Second Ed.

Hafs, D., and Zimbelman, R.G., 1994. Low-fat Meats.

Phillips, L.G., Whitehead, D.M., and Kinsella, J., 1994. Structure-Function Properties of Food Proteins.

Jensen, R.G., 1995. Handbook of Milk Composition.

Roos, Y.H., 1995. Phase Transitions in Foods.

Walter, R.H., 1997. Polysaccharide Dispersions.

Barbosa-Canovas, G.V., Marcela Góngora-Nieto, M., Pothakamury, U.R., and Swanson,

B.G., 1999. Preservation of Foods with Pulsed Electric Fields.

Jackson, R.S., 2002. Wine Tasting: A Professional Handbook.

Bourne, M.C., 2002. Food Texture and Viscosity: Concept and Measurement, second ed.

Caballero, B., and Popkin, B.M. (Eds.), 2002. The Nutrition Transition: Diet and Disease in the Developing World.

Cliver, D.O., and Riemann, H.P. (Eds.), 2002. Foodborne Diseases, Second Ed.

Kohlmeier, M., 2003. Nutrient Metabolism.

Stone, H., and Sidel, J.L., 2004. Sensory Evaluation Practices, Third Ed.

Han, J.H., 2005. Innovations in Food Packaging.

Sun, D.-W. (Ed.), 2005. Emerging Technologies for Food Processing.

Riemann, H.P., and Cliver, D.O. (Eds.), 2006. Foodborne Infections and Intoxications, Third Ed.

Arvanitoyannis, I.S., 2008. Waste Management for the Food Industries.

Jackson, R.S., 2008. Wine Science: Principles and Applications, Third Ed.

Sun, D.-W. (Ed.), 2008. Computer Vision Technology for Food Quality Evaluation.

David, K., and Thompson, P., (Eds.), 2008. What Can Nanotechnology Learn From Biotechnology?

Arendt, E.K., and Bello, F.D. (Eds.), 2008. Gluten-Free Cereal Products and Beverages.

Bagchi, D. (Ed.), 2008. Nutraceutical and Functional Food Regulations in the United States and Around the World.

Singh, R.P., and Heldman, D.R., 2008. Introduction to Food Engineering, Fourth Ed.

Berk, Z., 2009. Food Process Engineering and Technology.

Thompson, A., Boland, M., and Singh, H. (Eds.), 2009. Milk Proteins: From Expression to Food.

Florkowski, W.J., Prussia, S.E., Shewfelt, R.L. and Brueckner, B. (Eds.), 2009. Postharvest Handling, Second Ed.

Gacula Jr., M., Singh, J., Bi, J., and Altan, S., 2009. Statistical Methods in Food and Consumer Research, Second Ed.

Shibamoto, T., and Bjeldanes, L., 2009. Introduction to Food Toxicology, Second Ed.

BeMiller, J. and Whistler, R. (Eds.), 2009. Starch: Chemistry and Technology, Third Ed.

Jackson, R.S., 2009. Wine Tasting: A Professional Handbook, Second Ed.

Sapers, G.M., Solomon, E.B., and Matthews, K.R. (Eds.), 2009. The Produce Contamination Problem: Causes and Solutions.

Heldman, D.R., 2011. Food Preservation Process Design.

Tiwari, B.K., Gowen, A. and McKenna, B. (Eds.), 2011. Pulse Foods: Processing, Quality and Nutraceutical Applications.

Cullen, P.J., Tiwari, B.K., and Valdramidis, V.P. (Eds.), 2012. Novel Thermal and Non-Thermal Technologies for Fluid Foods.

Stone, H., Bleibaum, R., and Thomas, H., 2012. Sensory Evaluation Practices, Fourth Ed.

Kosseva, M.R. and Webb, C. (Eds.), 2013. Food Industry Wastes: Assessment and Recuperation of Commodities.

Morris, J.G. and Potter, M.E. (Eds.), 2013. Foodborne Infections and Intoxications, Fourth Ed.

Berk, Z., 2013. Food Processing Engineering and Technology, Second Ed.

Singh, R.P., and Heldman, D.R., 2014. Introduction to Food Engineering, Fifth Ed.

Han, J.H. (Ed.), 2014. Innovations in Food Packaging, Second Ed.

Index

Note: Page numbers followed by "f" denote figures; "t" tables.

Printed and bound by CPI Group (UK) Ltd, Croydon, CR0 4YY

08/05/2025

01864827-0001